Astronomers' Universe

Series Editor
Martin Beech

More information about this series at http://www.springer.com/series/6960

Edward van den Heuvel

The Amazing Unity of the Universe

And Its Origin in the Big Bang

Second Edition

Edward van den Heuvel
Astronomical Institute
University of Amsterdam
Amsterdam, The Netherlands

ISSN 1614-659X ISSN 2197-6651 (electronic)
Astronomers' Universe
ISBN 978-3-319-23542-4 ISBN 978-3-319-23543-1 (eBook)
DOI 10.1007/978-3-319-23543-1

Library of Congress Control Number: 2016934059

Original Dutch edition published by Veen Media, Amsterdam, 2012
Original Dutch title: Oerknal: Oorsprong van de eenheid van het heelal © 2012 by Edward van den Heuvel & Veen Media, Amsterdam
Cover Image credit: Cover illustration: Star cluster Pismis 24 and the emission Nebula NGC 6357, (c) NASA Hubble Space Telescope

Printed on acid-free paper

This Springer imprint is published by Springer Nature
The registered company is Springer International Publishing AG Switzerland

Preface

This book resulted from the amazing fact that human beings, with only some three pounds of brain cells, on a small planet in an insignificant corner of the universe, have been able to unravel in considerable detail how the universe is built and has evolved with time, from a fraction of a second after its beginnings until the present.

This is possible thanks to the fact that the laws of physics that during the past centuries were discovered here on Earth have been found to be valid throughout the observable universe. Already in the eighteenth century, it was discovered that stars in double-star systems orbit each other according to the laws of gravity and motion discovered by Isaac Newton a century earlier for our solar system. It thus was found that these laws are valid not only on Earth and for the motions of the planets but equally well for stars at distances of hundreds of light-years. Subsequently, in the past one-and-a-half century, spectroscopic analysis of the light of distant stars and galaxies has shown that all stars—which in fact are 'other suns'—and galaxies consist of exactly the same chemical elements that we know here on Earth, from the lightest, hydrogen, to the heaviest, uranium. Not more and not less. This discovery means that matter and radiation behave exactly the same throughout the universe and that the laws of quantum mechanics and relativity discovered by Einstein and his successors are valid throughout the universe. This great unity of the universe, and the universal validity of the physical laws that were discovered by laboratory experiments on Earth, is the basis of the miraculous fact that we, as little human beings, are able to study and understand the physical processes taking place throughout the universe. This has enabled us to understand in large part how our cosmos and the structures in it, from planets to stars and galaxies, are built and have developed in the course of time.

The present thinking is—on good physical grounds—that the laws of physics that govern everything that happens in the universe came into being during the first fraction of a second after the beginning of the Big Bang. The Big Bang, and the many solid observational facts that show that this event has taken place, therefore plays a central role in this book.

In the course of this story, we will encounter many facts that we already understand, but also a number of important things that we do not understand, for example, why does our universe have a finite age and is not infinitely old? Although we do not know the answers to such fundamental questions, scientists are presently speculating about the answers, and we will tell the story of such speculations and discuss the likelihood of them being true or untrue.

The basis of this entire book is the fact that we humans with brains that in structure are not very different from those of our apelike ancestors of a few million years ago have been able to understand in large part the physical processes that have shaped the universe as we nowadays observe it—an absolute miracle by any account.

Acknowledgements I am particularly grateful to Professor Jaap Goudsblom and Dr. Fred Spier of the University of Amsterdam who in the 1990s of the last century took the initiative to organize the university-wide lecture course 'Big History' that from its start every year has been followed by many hundreds of students. They invited me to present in the first 4 h of this course 'the structure, origin and evolution of the universe', which I have done for a number of years, both in Amsterdam and later also at Eindhoven's Technical University. These lectures were my inspiration for writing this book.

Further I would like to thank for inspiring conversations Vincent Icke, Michiel van der Klis, my much too early deceased colleague Jan van Paradijs, Huib Henrichs, John Heise, Rien van de Weygaert, Ralph Wijers, Tim de Zeeuw, Simon Portegies Zwart, Sander Bais, Karel Gaemers, Robert Dijkgraaf, Johan Bleeker, Martin Rees, Rashid Sunyaev, Jerry Ostriker, John Bahcall, Geoffrey Burbidge, David Gross, Marty Einhorn, Jayant Narlikar, Naresh Dadhich, Ajit Kembhavi, Ganeshan Srinivasan, Venkataraman Radhakrishnan, Dipankar Bhattacharya and last but not least my teachers at Utrecht University Marcel Minnaert, Kees de Jager and Nico van Kampen.

For their help in the production of the Dutch-language version of the book, I thank very much Tom Kortbeek, Doortje Gorissen and Eddy Echternach. For their help in producing the English-language version, I am most grateful to Esther de Wit of Veen Publishers and Jennifer Satten, Nora Rawn and Gnanasekhar Harish of Springer.

The original version of this book was written in large part when I was guest of the Kavli Institute for Theoretical Physics at the University of California, Santa Barbara, and I am most grateful to the director and staff of the institute for their hospitality.

I am grateful to Johan Bleeker, Rik Gheysens, Lidewijde Stolte and Rob van de Water for suggestions for improvements for the several Dutch editions of the book, which have been implemented here.

Amsterdam, The Netherlands
April 2016

Edward van den Heuvel

Contents

Distant galaxies in the Hubble Ultra Deep Field

Chapter 1
Our Strange Universe

The important thing in science is not so much to obtain new facts as to discover new ways of thinking about them.

Sir William Bragg, British physicist (Nobel laureate)

Astronomical observations of the past century have shown that we live in an expanding universe that originated a long but finite time ago in an incredibly dense and hot initial state called the Big Bang. Apart from matter, also space is an essential ingredient of our universe, and the observed expansion of the universe implies that the amount of space increases in the course of time. In the past there was less space and in the future there will be more. This appears strange and opposite to our daily experience, from which we know space to be a fixed quantity, such as the volume of our room. Physics tells us, however, that space—even if it is pure vacuum (completely empty space that contains no atoms or molecules)—is an essential ingredient of the universe, that contains hidden particles and energy, and can expand or contract. A strange discovery made in 1998, thanks to the measurements of the brightness of very distant exploding stars, is that the empty space of the universe contains the bulk (about 70 %) of all energy of the universe. This energy manifests itself by a mysterious

E. van den Heuvel, *The Amazing Unity of the Universe*,
Astronomers' Universe, DOI 10.1007/978-3-319-23543-1_1

force, still not understood, that causes the expansion of the universe to *accelerate*. The remaining about 30 % of the energy of the universe (according to Einstein, mass and energy are equivalent) manifests itself as the mass of "real" matter, which exerts gravitational attraction. Of this real matter, only about one sixth is ordinary matter, consisting of atoms and molecules, and five sixth is mysterious Dark Matter, which does exert gravitational attraction, but whose nature is still completely unknown.

Already in 1916 Albert Einstein predicted the existence of "anti-gravity". He was then not aware that this would produce a force that would accelerate the expansion of the universe. This was discovered in 1917 by Dutch astronomer Willem de Sitter. Later, Einstein called the idea of anti-gravity his "greatest blunder". It was only in 1998 that it was discovered that this, after all, was not a blunder, but a wonderful and great insight.

Also the fact that time had a beginning appears very strange to us. We are accustomed that a street has a beginning and an end, but in our feeling time is streaming forward endlessly, from a past infinitely long ago, towards an infinitely distant future. Nevertheless, a great variety of astronomical observations have shown us that time has had a beginning, about 13.8 billion years ago. Before that there was no time.

This all seems strange to us and opposite to our intuition and feelings. However, if one comes to think of it, many physical phenomena and laws appear to conflict with our day to day experience and intuition. Only when we have gotten accustomed to them, we no longer consider these phenomena and laws "strange". For example: when we look out of the window, Earth appears flat and at rest. But we know that it is a sphere, and that seen from the perspective of the Northern hemisphere, people in New Zealand are walking upside down, and we know that Earth spins around its axis in about 24 h. At the equator, the speed of rotation is about 0.5 km/s, some 1800 km/h, twice the speed of a commercial jet plane. At the geographical latitude of New York the speed of rotation is still some 1300 km/h. We also know that the Earth moves in its orbit around the sun with a speed of about 30 km/s, about 108,000 km/h. In about three and a half hours all of us move over about the distance from here to the Moon. But we quietly sit in our chair and notice nothing of these large speeds with which we are moving. It seems to us that we are completely at rest. The strange fact that we do not notice that we are moving with a fixed speed was first realized by Italian physicist and astronomer Galilei Galileo (1564–1642). He saw men in Venice on a ship passing by, who were throwing sacks of cargo to each other on the ship, and he noticed that the sacks moved between them as though the men were at rest and throwing the sacks to each other. He noticed that as long as we are *moving with a fixed speed*, all things that we do (and all laws of physics) proceed as though we are at rest. If we are in a vehicle without windows, there is no way for us to find out whether the vehicle is standing still or is moving with a fixed speed. A person that is moving with a fixed speed experiences all physical phenomena in exactly the same way as when he/she is at rest (in so far as it is possible to define "being at rest", which in fact is not possible in an absolute way). Everyone who has travelled in a plane knows this: the plane flies high in the sky with a fixed speed of some 900 km/h, but we walk through the aisle, drink our coffee, and do all the same things that we normally do at home on the ground. Galilei discovered that the only thing one can notice is a *change* in speed, that means: an acceleration or a deceleration (a negative acceleration). When a train or plane accelerates, we are pressed against the back of our seats, and when the vehicle suddenly brakes, we shoot forward. This is the reason we have to wear seatbelts. Being pushed against the back of our seat or shooting forward are consequences of the *law of inertia*: an object wishes to keep moving with the same speed and it "opposes" a change in speed.

Also the physical laws that govern the behaviour of elementary particles and of light are extremely strange. Here we enter the domain of quantum mechanics, where the energy of a particle or of a light wave can no longer take any arbitrary value, as we would according to our intuition would have expected energy to behave. The energy of particles appears here to be distributed in discrete packages, the "quantums", while energy values between these discrete values are not allowed for the particle (for example for an electron in an atom). And also: particles appear to behave as waves, and light waves as particles, the so-called *photons*. This behaviour all seems extremely strange, but has nevertheless been found to be true in thousands of laboratory experiments. The laptop or tablet on our desk or the cell phone in our pocket would not be able to function if these strange laws of quantum mechanics would not be true. These devices work thanks to the quantum behaviour of the atoms of silicon and other elements in the processors and memory chips.

So, if accurate astronomical observations tell us that space is expanding and that time had a beginning, we will have to accept these properties of nature, even though from the point of view of humans this all seems very strange. After all, it is already very surprising that we, with our 3 lb of brain cells and the intuition inherited from our ape-like ancestors, have succeeded in the past five centuries to discover so many of the "strange" and counter-intuitive laws of nature. The knowledge of these laws has led to the development of our high-tech society with its computers, airplanes and space vehicles. Similarly, strange physical laws determine the structure and evolution of the universe.

The foundation of our present ideas about the nature and development of space and time was laid down in 1915 by Albert Einstein's *General Theory of Relativity*, a new theory for accelerated motions and for gravity. Einstein discovered in 1916 that his theory predicts that the universe should shrink or expand. Apart from this, in 1905 Einstein had laid the foundation of quantum mechanics—the other great revolution of physics—which earned him the 1921 physics Nobel prize. Just as General Relativity, quantum mechanics is indispensable for understanding the nature and history of the universe. The discovery that the universe is indeed expanding was made in 1929 by American physicist and lawyer Edwin Hubble, by studying the velocities of galaxies. (In recent years it was found that already 2 years earlier Belgian priest-astronomer George Lemaitre had independently made the same discovery, which went largely unnoticed; he never disputed the priority of Hubble's discovery). The discovery of Hubble and Lemaitre forms the basis for our present-day insights about how the universe originated, has developed and will develop in the future.

Einstein's General theory of Relativity also has answered the question what causes the forces that we experience when we are accelerated, the *inertial forces*. As had already been suggested in 1872 by Austrian physicist-philosopher Ernst Mach, these forces appear to be due to the fact that all stars and galaxies in the universe pull at us with their gravity. So, every time our car or bus rounds a corner, or our train or plane changes speed, we feel that the entire universe is pulling at us! Who would like to maintain that the universe has nothing to do with us, or that we have nothing to do with the universe?

This book describes our present picture of the universe, with the place of Earth and Man in space and time, and how this picture has been obtained, thanks to the works and discoveries of many generations of astronomers and physicists. It describes how the traces of the Big Bang in which the universe originated can still be observed all around us, in the form of radio waves and particles, how the chemical elements originated, and how astronomical observations of the past decades have shown that the universe for over 95 % consists of two still completely un understood ingredients: the mysterious *dark matter* that with its gravity has enabled stars and galaxies to form, and the *dark energy*, that causes the expansion of the universe to accelerate.

Before being able to explain how these mysterious ingredients of the universe have been discovered, we explore in the first chapters what the observations have taught us about the structure of the world around us, from the solar system that is the home of Earth, to the nearby starry world of our Milky Way system, and the distant world of other galaxies and clusters of galaxies. We will see that the measured velocities away from us led Belgian priest-astronomer Lemaitre to the idea that the universe started with a Big Bang. In the subsequent chapters we pay attention to how these observations of the universe and the starry world can be understood in terms of the ground-breaking ideas about gravity, space and time of leading scientists such as Einstein, de Sitter, Friedmann, Eddington, Gamow, Hoyle and Guth. With this all we will at every point also indicate what we still do not understand well, and where great gaps are still present in our knowledge of the cosmos, for example concerning the nature of dark matter and dark energy, and where there is room for interesting speculations, such as about the existence of a multiverse and about the possibility of intelligent life elsewhere in the universe.

The eight planets of our solar system

Chapter 2
The Sun's Backyard: Our Solar System

There are more things in heaven and earth, Horatio,
than are dreamt of in your scholarly philosophy.

Shakespeare—Hamlet (1.5.167-8)

There are eight planets in our solar system. The originally ninth planet Pluto, since 2006 is no longer counted as a planet, but was moved by the International Astronomical Union to the category of *dwarf planets*, of which several more have been discovered at the outer edge of the solar system.

Going outwards from the sun, Earth is the third planet. Closer to the sun move Mercury and Venus, and beyond the Earth's orbit one finds going outwards: Mars, Jupiter, Saturn, Uranus and Neptune (Fig. 2.1). In addition we find in the solar system many hundred thousands of asteroids—also called minor planets—and about a hundred billion comets. Asteroids are rocks with sizes ranging from a few tens of meters to almost 1000 km (Fig. 2.2). They mostly describe orbits between those of Mars and Jupiter. Several hundreds of them are known to have orbits reaching the inner solar system, crossing Earth's orbit. Comets in general are not larger than a few tens of kilometres and consist of ices of mostly water, carbon-dioxide and

E. van den Heuvel, *The Amazing Unity of the Universe*,
Astronomers' Universe, DOI 10.1007/978-3-319-23543-1_2

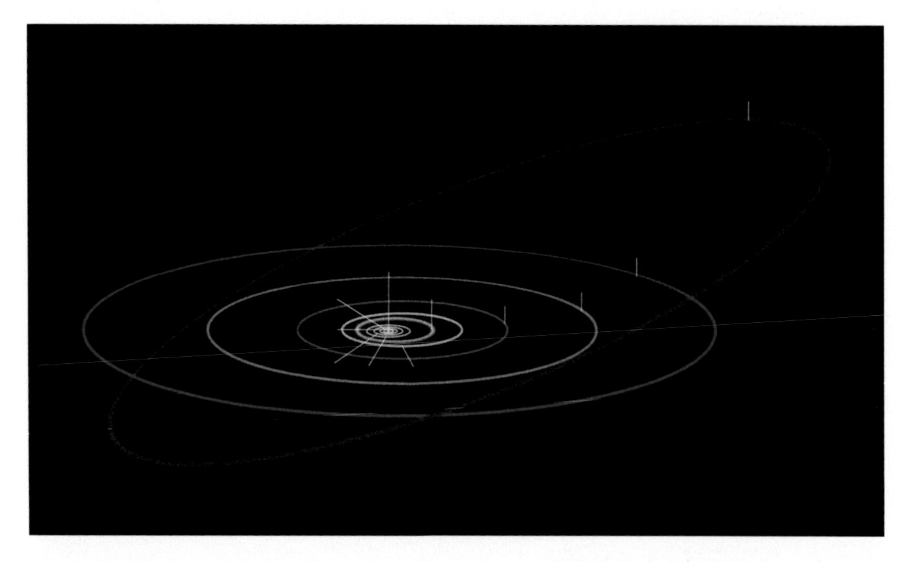

Fig. 2.1 Orbits of the planets of our solar system, to scale. The asteroid belt and the orbit of the ice dwarf Pluto are also indicated

methane, mixed with stones, dust and rocks. They in fact are "dirty snowballs". They are living in a region of the outer solar system stretching outwards from beyond Neptune's orbit. When their orbits are disturbed, they may fall towards the inner solar system, where the heat of the sunlight causes their ices to evaporate, such that a cloud of gas and dust forms around them. The pressure of the sunlight, in combination with the solar wind—a flow of electrically charged particles (ions and electrons) that blows outwards from the solar atmosphere with a speed of 300–500 km/s—causes the evaporated material to be blown outwards, away from the sun. In this way the tail of a comet is formed (Fig. 2.4), which emits light by fluorescence of sunlight by atoms and reflection against dust particles. The billions of comets form a colossal more or less spherical cloud outside Neptune's orbit. The existence of this cloud was discovered around 1950 by Dutch astronomer Jan Hendrik Oort (1900–1992), through a study of a collection of comet orbits gathered by his collaborator Adriaan J.J. van Woerkom (1915–1991). The Oort cloud extends outwards to a distance of 1–2 light-years (Fig. 2.3), that is: halfway to the nearest stars. Oort calculated that when a star passes at a distance of a few light-years from the sun, it will slightly disturb the orbits of a number of comets in the cloud. As a result, some of them will start moving towards the inner solar system, where they will appear many thousands of years later as "new comets", while another number of comets will be ejected from the cloud and will disappear forever into interstellar space. The inner part of the comet cloud, just beyond Neptune's orbit, is flattened and is called the "Kuiper Belt" after Dutch-American astronomer Gerard P. Kuiper (1905–1973). In 1951 Kuiper suggested the existence of this belt of comet-like objects, which he deduced from the fact that the *periodic* comets, such as the famous

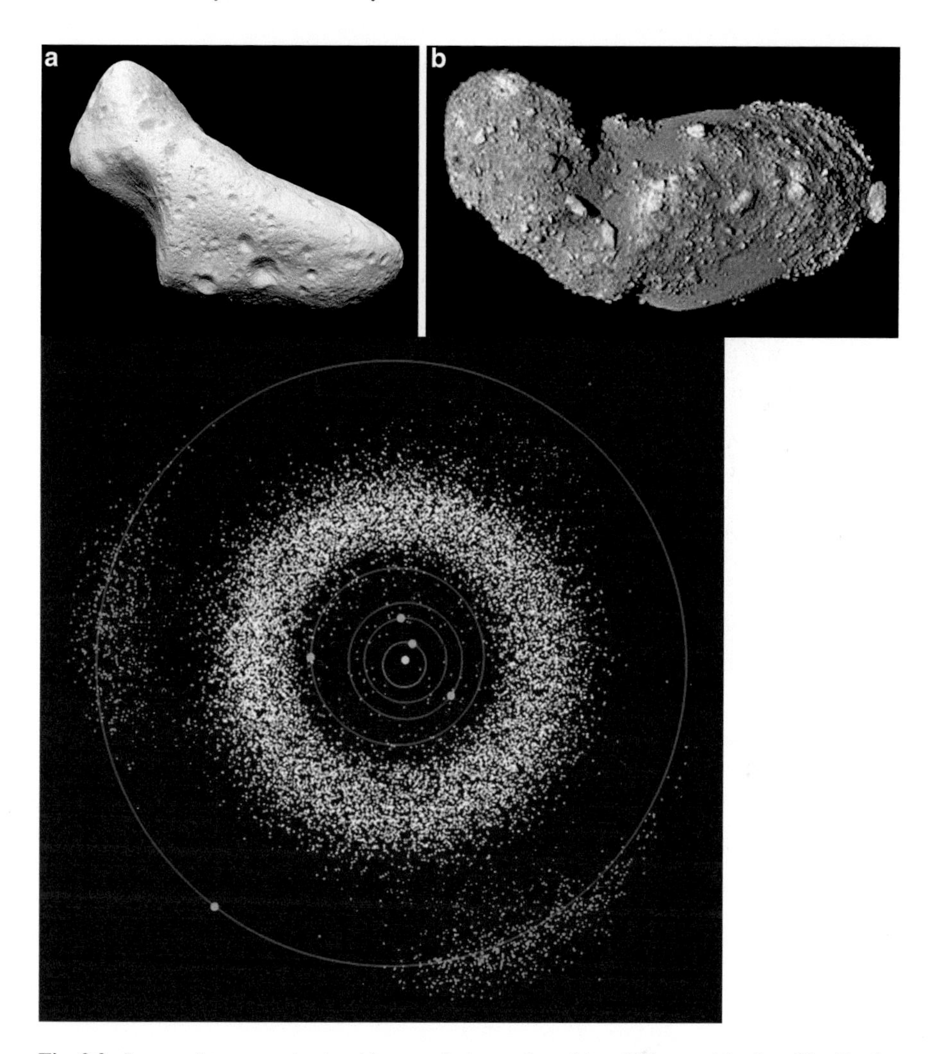

Fig. 2.2 *Lower picture*: most asteroids move between the orbits of Mars and Jupiter. The Greeks and the Trojans are two groups of asteroids that move along with Jupiter in its orbit. The orbits of a number of asteroids cross the orbit of Earth and could, in principle, at some time collide with Earth. (**a**) Asteroid Eros, photographed in 2000 by NASA's space mission NEAR, from a distance of 330 km. The dimensions of Eros are 33 × 13 × 13 km. (**b**) Asteroid Itokawa has a length of only half a kilometre and is in fact a pile of rubble kept together by gravity. Japanese spacecraft Hayabusa photographed it from a distance of 7 km, briefly touched it and took a sample of the dust stirred up by this touch, and brought it back to Earth in 2010

Halley's comet, tend to move around the sun in the same direction and form a somewhat flattened system. This is in contrast to "new comets" which are observed to orbit the sun in all possible directions and have orbital planes that make all possible angles with Earth's orbital plane. With modern large telescopes, equipped with

Fig. 2.3 The solar system is surrounded by the spherical *Oort Cloud* of comets, which contains some 100 billion comets, with orbits stretching to about 1 light year. The flattened inner part of this cloud is the *Kuiper Belt*, which also contains much larger objects, the so-called "ice dwarfs", to which Pluto belongs

Fig. 2.4 *Middle and lower figures:* comets Hale-Bopp (1997) and West (1975), respectively. The picture of Hale-Bopp clearly shows the two different types of tails that comets can have: a *plama tail* of ionized gas carried along by the solar wind and directed straight away from the sun, and a more curved *dust tail*, consisting of dust particles released from the evaporating ices of the comet, which stay somewhat behind with respect to the orbital motion of the comet. This orbital motion is sketched in the drawing in the top figure, which also depicts the two types of tails, always directed away from the sun. Due to evaporation of the ices of the comet, these tails begin to form when the comet has arrived in the region of the asteroid belt

Comet
Hale-Bopp
Orbit of the
comet
Tails point away from
the sun, also when the
comet moves away
from the sun
Sun
Tails get longer when
the comet approaches
the sun

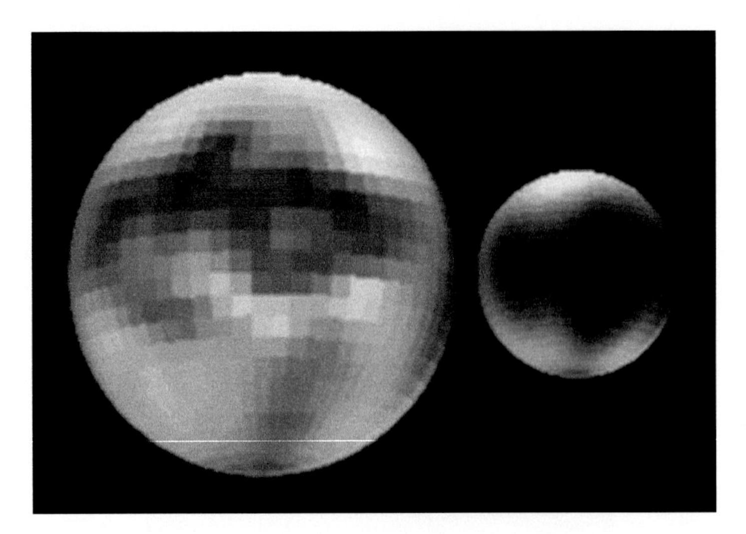

Fig. 2.5 Pictures of Pluto and its largest moon Charon, taken by the Hubble Space Telescope

highly sensitive digital cameras, astronomers have in the past decades discovered over 1000 icy objects outside Neptune's orbit. Kuiper had thought that these would be comet-like, with sizes of a few tens of kilometres. However, many of the discovered trans-Neptunian objects tend to be much larger: sometimes hundreds or even thousands of kilometres in size. A comet of a few tens of kilometres is very hard to detect at a distance beyond Neptune's orbit. Some of the "ice dwarfs" discovered beyond Neptune's orbit are even larger than Pluto, for example the "tenth planet" Eris, discovered in 2005 by American astronomers Mike Brown, Chad Trujillo and David Rabinowitz. Eris has a diameter of 2400 km and even has its own moon. The discovery of these large "ice dwarfs" has led to the insight that Pluto is, in fact, not a planet, but the innermost ice-dwarf of the Kuiper Belt. This was in 2006 the reason for the International Astronomical Union to decide to no longer count Pluto as a planet. Pluto is round (that a celestial object is round is related to the strength of its gravity, which in turn is determined by its mass and density, as will be explained in Chap. 7). Since 2006, round objects in the solar system that are not planets are called "dwarf planets". Also Pluto has a large moon (Charon, Fig. 2.5) plus four smaller ones. The last ones were all discovered with the Hubble Space Telescope, the first two in 2005, and the other two in 2011.

All comets and ice dwarfs together have a mass (amount of matter) of about one-hundredth of the mass of Earth. This is about the mass of the Moon. One sees from the above that the nature of the Kuiper belt is very different from that of the Oort cloud. The present idea is that the Kuiper belt is—just as the asteroid belt between Mars and Jupiter—the remnant of the disk of small objects, so-called *planetesimals*, that were the original building blocks from which some 4.6 billion years ago the larger planets originated by a snow-ball type of growth under the influence of gravity. The much more extended an non-flattened Oort cloud is thought to have originated when Uranus and Neptune formed and with their strong gravity kicked enormous numbers of planetesimals outwards to very large distances. Many of these icy objects must have been ejected out of the solar system, but those which fell just short to receive a sufficient speed to escape, kept hanging in a very extended region at the edge of the solar system and form the Oort cloud.

Sizes and Masses of the Planets

Figure 2.6 depicts the planets and the sun to the same scale. One notices how enormous the sun is in comparison to the planets. The diameter of the sun is 109 times that of Earth, and its mass is some 330,000 times that of Earth. One therefore could make out of the sun some 330,000 little balls like Earth! If we picture the sun as big as a grapefruit, Earth is only the size of a pinhead. Jupiter, the largest of all eight planets, has a diameter of only one tenth of that of the sun and a mass of 318 times that of Earth, which is about one thousandth that of the sun. The sun has 700 times the mass of all eight planets combined (all other objects together have a mass of only a few per cent of that of Earth and can be neglected here). Some 99.86 % of the mass of the solar system is in the sun, and only 0.14 % in the planets and other objects. The conclusion from this is that the sun is, in fact, the only really important object in the solar system, literally the centre around which “everything is moving”.

Fig. 2.6 Sun and planets depicted on the same scale. The mass of the sun is some 330,000 times that of Earth, and its diameter is 109 times that of Earth. The sun has 700 times the mass of all eight planets combined. Even Jupiter, the largest planet, has a mass of only one-thousandth of that of the sun. Some 98 % of the mass of the sun consists of the two lightest gases known: hydrogen (73.9 %) and helium (24.9 %). Also the gas giant Jupiter has this composition

Fig. 2.7 Earth and Jupiter depicted on the same scale. The picture of Jupiter was taken by the Cassini space probe of NASA and ESA. Earth is slightly smaller than the "Big Red Spot", a hurricane that has been blowing in Jupiter's atmosphere for over 300 years. Also the white spots are hurricanes, of shorter duration (decades)

The planets are just some left-over rubble from when the sun formed. And we live on such a grain of rubble, smaller than a tropical cyclone on Jupiter (Fig. 2.7). Appendix A lists some of the main characteristics of the planets, such as size, mass, density and distance to the sun.

Distances in the Solar System

Earth's orbit is an ellipse that closely resembles a circle, and has the sun in one of its focal points. Our mean distance to the sun is about 150 million kilometres. The velocity of light is about 300,000 km/s (to be exact: 299,792,458 km/s), which means that sunlight takes about 8.3 min (500 s) to reach us. If the sun would switch off now, we would notice this 8.3 min later. Measured in human scales, the distance to the sun is enormous: a commercial jet plane, that flies continuously 24 h a day at a speed of 900 km/h would take about 16 years to cover the distance to the sun. A rocket that brings satellites in orbit around Earth has a speed of about 8 km/s (28,500 km/h) and will take about half a year to fly from here to the sun. Sunlight takes about 50 min to reach Jupiter and about 4 h to reach the most distant planet Neptune. With the speed of a commercial jet plane one would need about 480 years to reach Neptune, and with a rocket that brings satellites in orbit, about 20 years. Radio waves move with the same speed as light—they are "light waves" with a much longer wavelength than ordinary light. The wavelength of light is typically about half a micron (half of one thousandth of a millimetre), while radio waves have wavelengths between a few millimetres and several kilometres. If we send a massage to a spacecraft near Neptune and instruct it to take a picture of the planet and send it to us, it will take the message about 4 h to reach the spacecraft. The spacecraft then takes the picture and sends it off, and it will take another 4 h before the picture reaches Earth. So, only after 8 h we will know whether the spacecraft has correctly carried out our instruction.

Although these distances in the solar system seem enormous, they are only tiny on an astronomical scale. The real world of the stars—"other suns"—starts much further away, at distances that cannot be counted in light-minutes or light-hours, like in the solar system, but in light-years and thousands of light-years: distances that are several hundred thousand times larger than the distances of the planets in the solar system, as we will see in the next chapter. The solar system therefore literally is "the sun's backyard".

Chemical Composition of the Sun and Planets

The chemical composition of the sun is very different from that of Earth. The sun consists for 98.8 % of the two lightest gases known in nature: 73.9 % hydrogen (symbol: H) and 24.9 % helium (symbol: He). On Earth we use these gases to fill balloons for our kids. All heavier elements, which are the most important ingredients of Earth, such as Silicon (Si), Magnesium (Mg), Aluminum (Al), Iron (Fe), Calcium (Ca), Carbon (C), nitrogen (N) and Oxygen (O), together make up only 1.2 % of the matter (mass) of the sun.[1]

The sun is a glowing ball of gas. The temperature at its surface is some 5500 K (about 5200 °C, where K denotes the so-called *absolute temperature* in Kelvins: 0 K = −273 °C). Its internal temperature is much higher still: in its centre some 14 million K. At such temperatures all materials, even metals, rocks, etc., have completely evaporated. Every material in the sun is therefore only present in the form of vapour.

Perhaps the most amazing property of the sun is its gigantic emission of energy: 4×10^{26} J/s. This means that the sun is a "lamp" with a power of 4×10^{26} W: a number

[1] M. Asplund, N. Grevesse, A.J. Sauval, *ASP Conf. Series*, vol. 336 (2005), p. 25ff.

4 followed by 26 zeros! In order to have some idea of how large the energy output of the sun is, one may compare this number with the total energy that the seven billion people on Earth consume for heating their houses, powering their factories and airplanes, driving their cars, trains and buses, etc. What the sun emits *every second* is sufficient, at the present rate of energy consumption of the world population, to supply the energy needs of this population for a period of *about 100,000 (one hundred thousand) years*. The sun has been pouring out this energy into space already for some 4.6 billion years (4600 million years), and it will keep doing this for another five billion years. A tiny part of the energy emitted by the sun—one part in two billion—falls on Earth and warms us, drives air and sea currents, makes plants and trees grow and makes all life on Earth possible. The amount of solar energy falling on Earth is still over 10,000 times the energy consumption of the world population. We would, therefore only need to capture one hundredth of a per cent of this solar energy input to supply all human energy needs.

The planets and their moons do not themselves emit light—they just only reflect the light of the sun. Since planets move in orbits around the sun we see in the course of a year that they move with respect to the stars that twinkle in the night sky. These stars are glowing balls of gas just like our sun. As we will see in the next chapter, their distances are so large that, seen from Earth, they just are only faint specks of light on the dark night sky. The sun is "our star". Since 1995 it has been discovered that, like the sun, many stars have planets. We return to this in Chap. 17.

The stars and also the interstellar gas between them have about the same composition as the sun. Most stars in our Milky Way system consist for some 98–99 % of the same two light gases as the sun: hydrogen and helium, and only for at most a few per cent of the well-known heavier elements. This is called a "cosmic composition".

Why Earth Does Not Have a Cosmic Composition

The material from which the solar system formed must originally also have had a cosmic composition, dominated by hydrogen and helium. Indeed, the largest two planets, Jupiter and Saturn, have almost the same composition as the sun: they consist mainly of hydrogen and helium. This is due to the fact that, thanks to their large gravity and relatively low temperature (they are far from the sun) the atoms of their hydrogen and helium had too low velocities to escape. The gravitational attraction of Earth is, however, too low, and its temperature too high, to be able to retain hydrogen and helium gas in our atmosphere. If we release the molecules, respectively atoms, of these gases (H_2, and He) in the atmosphere, they diffuse upwards and end up floating on top of the layers of the heavier atmospheric gases (nitrogen, oxygen, argon). Arriving at the tenuous top of the atmosphere, at a height of some 300 km, the molecules and atoms of H and He move fast enough to escape from Earth's gravity and disappear into space. Only in the form of water molecules, in which two hydrogen atoms are bound to one atom of oxygen, Earth has been able to retain a sizeable amount of hydrogen. No less than 71 % of Earth's surface is covered by oceans, with an average depth of several kilometres. Earth is therefore rightfully called the "water planet". The oceans contain about half a per cent of the mass of Earth. With this, Earth has vastly more water than air. If one would compress and cool the atmosphere such that it becomes liquid (it should then be cooled to below −180 °C), one would retain a layer of liquid with a thickness of only eleven-and-a-half meters (38 ft), not higher than the top of the roof of a two-storey building. This is some 300 times less than the thickness of the layer of water covering Earth. We thus see that the atmosphere is only a very thin and fragile layer. Nowadays, everyone dumps dirty exhaust gases into this tiny thin layer: cars, factories, planes, home chimneys, etc. Knowing that the atmosphere is such an impossibly thin layer, it is not astonishing that the disturbances which the activities of seven billion people produce on Earth, are to be noticed first in our atmosphere. For example: the strongly increased carbon-dioxide content of the atmosphere produced in the past century, causing an extra greenhouse effect (global warming), and the gases from spray-cans, refrigerators and air-conditioners, which have produced the ozone-holes surrounding the North and South poles of our planet.

Also the other three small planets Mercury, Venus and Mars, have too little gravity and too high a temperature to retain hydrogen and helium. Just like Earth they consist of the heavier material that was left behind after the hydrogen and helium gases from their primordial cosmic material had escaped. The materials of which the four inner planets, and also many of the asteroids, consist are often called "rocky" materials. The specific density of this material is high, like that of rocks and metals: roughly some five times the density of water. The four much larger outer planets have much lower densities: about the density of water (one kilogram per cubic decimeter), Saturn even lower than that. The material of Jupiter and Saturn is called "gaseous", and these large planets are called "gas giants". Uranus and Neptune consist of material that originally was composed of various types of *ices*: frozen water, carbon dioxide, methane, etc.: the same material that we find in comets and Kuiper-belt objects. Their composition is therefore called "ice-like".

Mercury and the Moon (Fig. 2.8) are so small and have so little gravity that they have not even been able to retain an atmosphere of heavier gases such as nitrogen, oxygen and water. The surfaces of these air-less worlds are pock-marked by countless craters, the results of the impacts of thousands of smaller and larger asteroids, which hit their surfaces, mostly during the first 600 million years after the formation of the solar system, in the epoch between 4.6 and 4.0 billion years ago. Due to the absence of an atmosphere, these craters have not been eroded away by the works of water and wind, like on Earth which, in addition to plate tectonics (absent on the Moon and Mercury) act to erase impact craters. Also Earth, Venus and Mars underwent this large asteroid bombardment during the first 600 million years, but on these planets with atmospheres, many craters have disappeared in the course of time. On Earth one finds a few hundred, often highly eroded, impact craters, which are recognizable only on photographs taken from high-flying planes or from satellites (Fig. 2.9). On Mars, which has a much thinner atmosphere than Earth, and no plate tectonics, erasing proceeds much more

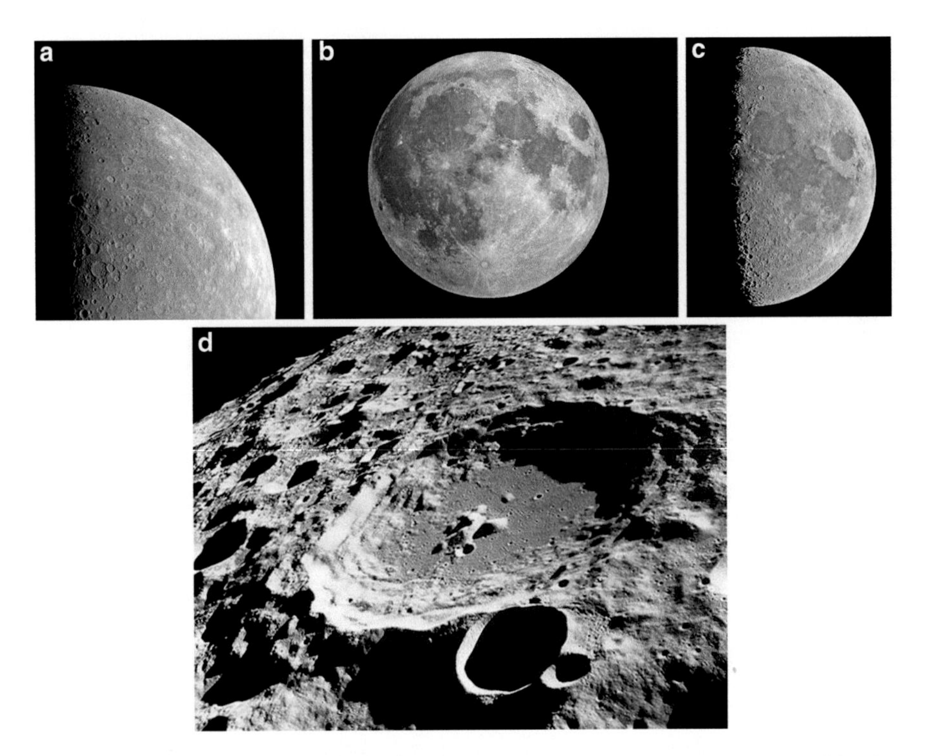

Fig. 2.8 (**a**) The surface of the planet Mercury is covered by impact craters and resembles the lunar surface. Mercury and the Moon have no atmosphere and thus no erosion by wind and water, nor do they have plate tectonics. As a result craters formed by impacts of asteroids billions of years ago remain visible almost forever. (**b**) At full moon the sunlight shines perpendicularly to the lunar surface, such that there are no shadows and lunar craters are hardly visible. On the other hand the large dark 'maria' (lunar seas) are clearly visible. They originated almost four billion years ago due to lava flows, following several very large impacts. (**c**) On the moon at first quarter, many craters on the lunar highlands near the boundary between light and dark are clearly visible, thanks to the long shadows produced by the sunlight shining under a small angle with the lunar surface. (**d**) Close-up of the highly cratered lunar highlands

slowly, and still large numbers of impact craters can be observed (Fig. 2.10). Also on Venus, which has an about hundred times denser atmosphere than Earth, and hardly any winds, erosion proceeds much more slowly than on Earth, and a considerably larger number of impact craters is observable on radar images. The atmosphere of Venus consists mainly of carbon-dioxide, which causes an enormous greenhouse effect. As a result the surface temperature of Venus is between 400 and 500 °C—high enough to make tin and lead melt. With, in addition, a hundred times higher atmospheric pressure, the circumstances in the Venus atmosphere have rightfully been compared with those in hell.

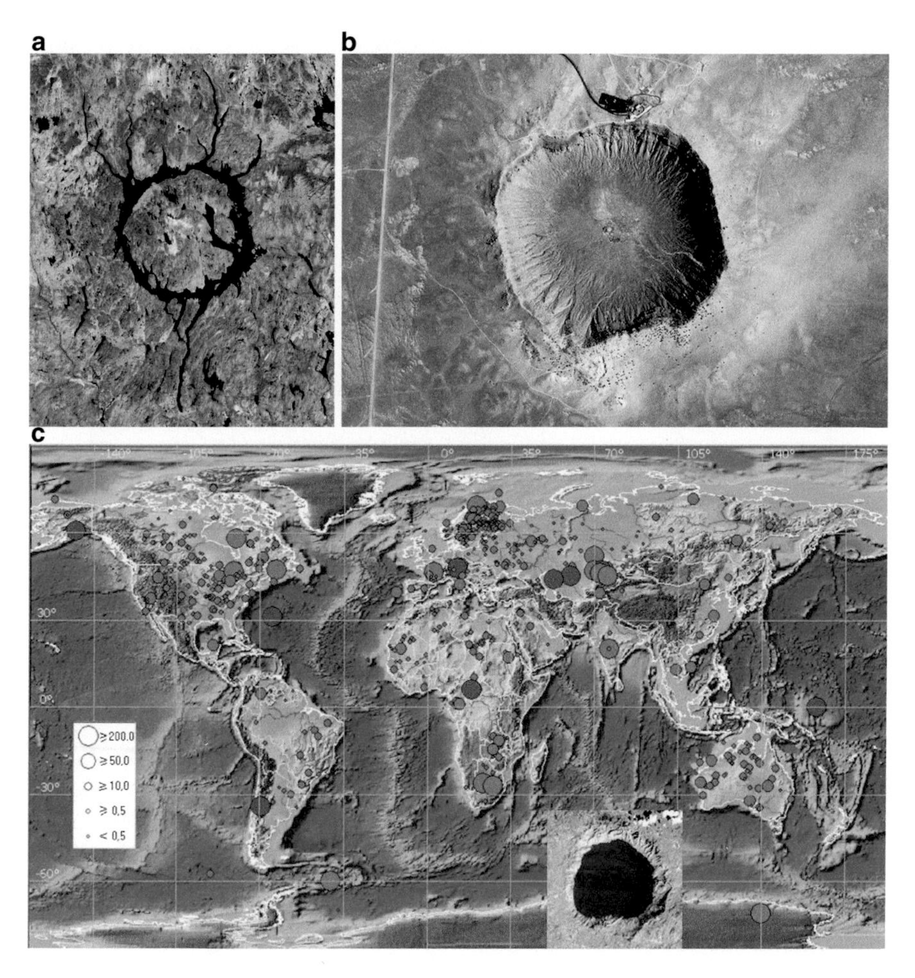

Fig. 2.9 (**a**) The 60 km diameter Manicouagan-crater in Canada was formed by the impact of an about 5 km-size asteroid 214 million years ago. (**b**) The Meteor Crater near Winslow, Arizona has a diameter of 1.5 km and formed by the impact of an about 50 m-size nickel-iron meteorite some 30,000 years ago. (**c**) Locations of the several hundreds of known impact craters on Earth. The scale gives the diameters of the craters in kilometres. Inset: The 8.5 km-size Bosumtwi crater in Ghana was formed by an impact 1.07 million years ago

Moons

Each of the giant planets has a large number of moons. In contrast, among the earth-like planets, only Earth and Mars have moons. Mars has only two tiny moons: Phobos and Deimos, with diameters of only a few tens of kilometres, which most

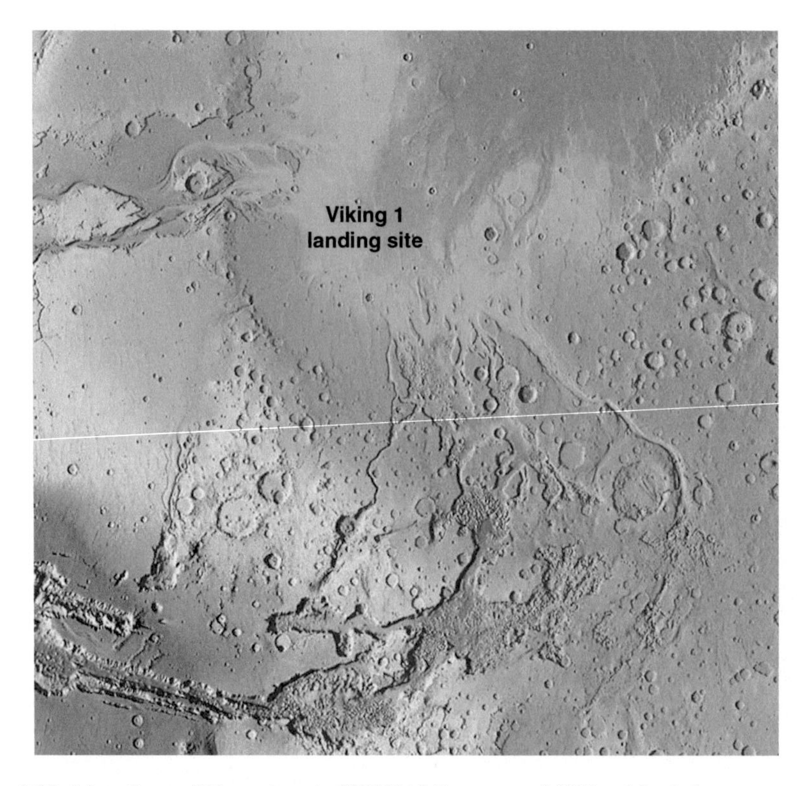

Fig. 2.10 Map of part of Mars where in 1977 NASA's spacecraft Viking 1 landed

probably are captured asteroids. On the other hand, Earth has an extraordinarily large moon, in comparison to the mass of the planet, particularly if one compares it with the largest moons of Jupiter and Saturn, which are hardly larger than our moon (see Fig. 2.11). The four large moons of Jupiter: Io, Europa, Ganymede and Calisto, were discovered in 1609 by Galilei, by using the telescope which in the previous year had been invented by Dutch spectacle-maker Lipperhey of Middelburg. The only large moon of Saturn, Titan, as well as the beautiful rings of this planet, was discovered in 1656 by Christiaan Huygens in the Netherlands, with a telescope which he and his brother Constantijn had constructed. In 1944 Gerard Kuiper discovered that Titan has an atmosphere with a pressure similar to that of Earth. Later, the combined European-American space mission Cassini-Huygens discovered that Titan's atmosphere consists largely of nitrogen, just like Earth's atmosphere. Also Neptune has a large moon, Triton, which has a special property: it is the only large moon in the solar system that orbits its planet in the "wrong" direction, that is: in a direction opposite to the one in which all other large moons in the solar system orbit their planets, which is the same as the direction in which almost all planets spin around their axes and in which they orbit the sun.

Apart from these large moons, Jupiter, Saturn, Uranus and Neptune have many smaller moons. Most of these were discovered in the past 40 years, thanks to

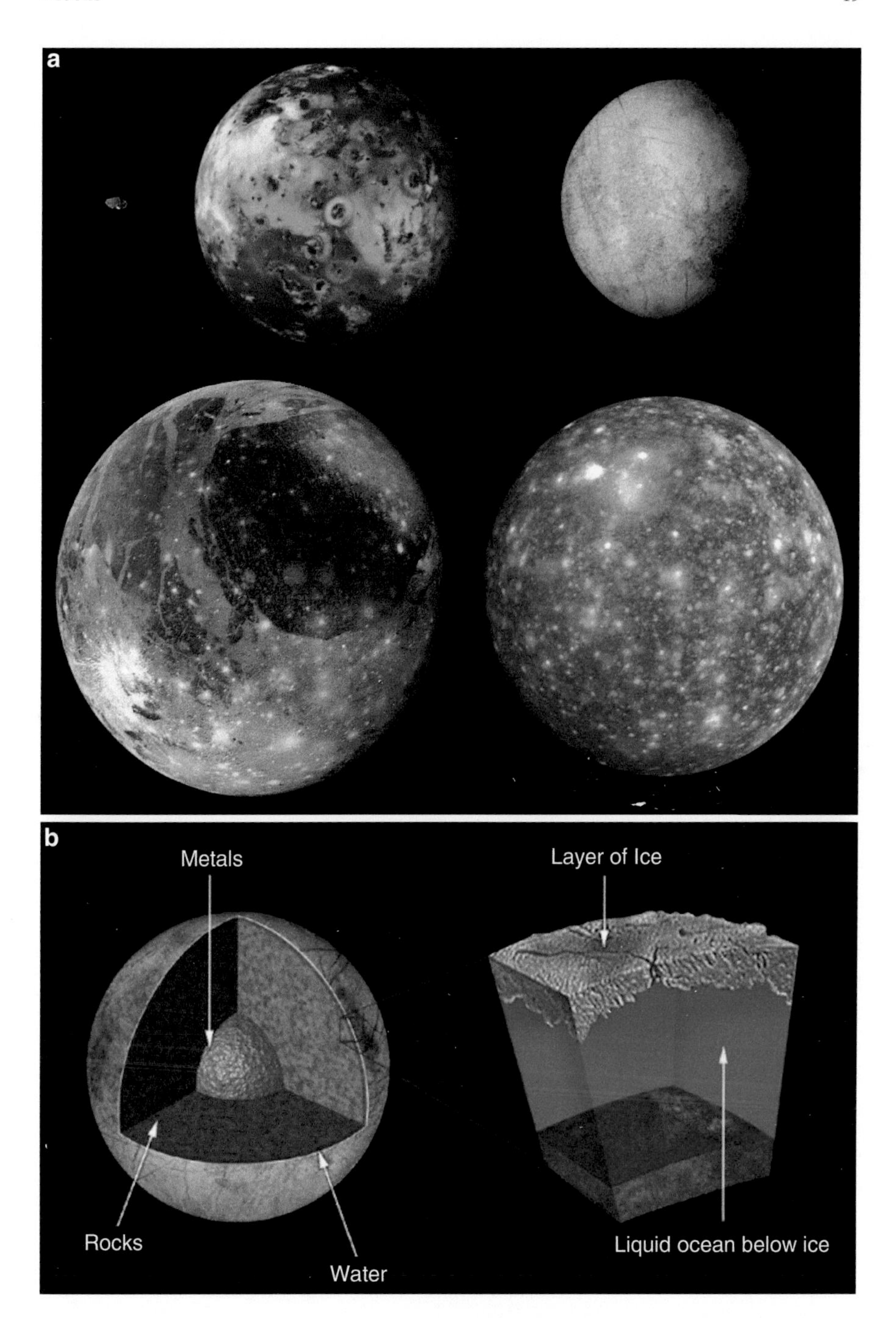

Fig. 2.11 (**a**) Jupiter's largest four satellites (*upper picture* from *left* to *right*: Io and Europa, *lower picture*: Ganymedes and Callisto) and the small Jupiter moon Amalthea (*upper left*), all on the same scale. (**b**) Below its ice crust, Jupiter moon Europa has a 15 km deep ocean which contains more water than all oceans on Earth together. It is not impossible that here life has originated. Recent evidence indicates that also Ganyemedes and Callisto have large salt water oceans below their surfaces

Fig. 2.12 Crater-covered icy moon Phoebe of Saturn has an irregular shape

space-crafts that closely passed these planets, particularly the two Voyagers. Many of the smaller moons are probably captured asteroids or ice dwarfs. Their surfaces are in most cases full of impact craters (see Fig. 2.12), due to collisions with numerous smaller bodies, which were moving between the planets in the early days of the solar system and were gradually swept up by the larger bodies. Apart from these moons, all of the four largest planets have ring systems. Only in the case of Saturn the rings are so large and rich in material that we can easily see them from Earth with a small telescope. The ring systems of the three other large planets are much more tenuous and therefore hard to observe from Earth. Those of Jupiter and Neptune were discovered by the Voyager space-crafts when they passed these planets at close range. The rings of Uranus were discovered in 1977 when this planet closely passed by a bright star, which was eclipsed by the rings. The ring systems consist of small icy bodies with sizes ranging from centimetres to meters, that orbit the planets like satellites.

It is interesting to notice that two moons in the solar system, Ganymede and Titan, are larger than the planet Mercury. Titan with its dense atmosphere is even more planet-like than Mercury, which lacks an atmosphere. It is strange to realize that astrologers—horoscope drawers—totally neglect these two large planet-like moons in their deliberations, but on the other hand, assign important powers to the minute planet Mercury and even to the tiny ice-dwarf Pluto. We all know, of course, that astrology is a romantic superstition, similar to the "reading" of the curling of intestines of sheep for predicting the future, as practised by the Babylonians, the same ones that invented astrology. The fact that astrologers totally neglect Ganymede and Titan in their horoscope drawing, underlines how senseless this activity is.

Especially interesting is the moon Europa of Jupiter. It is covered by an ocean with a depth of some 15 km, consisting of salt water, of which the few top kilometres are frozen (Fig. 2.11). That the water is salt has been deduced from the presence of a measurable magnetic field in Europa. Such a field can be generated only by a conducting liquid, like Earth's liquid metal core. Europa is, however, too small to have such a liquid metal core, and the only alternative is that its ocean is electrically conducting, which requires it to be salty. The presence of a large amount of liquid water makes that, in principle, life may have developed here, since we know that on Earth life started in the oceans, where it remained and developed for over three billion years, before it started to spread also over land some 400 million years ago. Possibly also Jupiter's moons Ganymede and Calisto have salt-water oceans under their thick ice mantles.

The Age of the Solar System

In the foregoing we already mentioned several times that the age of the solar system is about 4.6 billion years. This is the age of the oldest meteorites that have been found on Earth.

A meteor is the fiery track that briefly appears in the sky when a small stone or rock from space enters Earth's atmosphere. Earth moves in its orbit around the sun with a speed of about 30 km/s and rocks orbiting the sun in our neighbourhood easily have a velocity with respect to earth of several tens of km/s. When such a stone enters Earth's atmosphere, it moves much faster even than the Space Shuttle or the space-capsules of astronauts returning from the International Space Station, which have a speed of 8 km/s, and have a heat shield that becomes glowing hot when entering Earth's atmosphere, due to friction caused by its entrance velocity of some 28 times the sound velocity. Similarly, the stone, with its much higher speed, gets extremely hot due to the friction against the air and starts to glow, and to evaporate. The hot vapours and the heated air glow brightly, producing a fiery track along the sky (Fig. 2.13), which is called a *meteor*. If after crossing the atmosphere still some piece of the rock is left, and found on the ground, one calls this object a *meteorite* (Fig. 2.14). Small stones, of a few inches, burn up completely. The resulting meteor dust slowly rains down on the Earth's surface, and is found in the ice-caps of Greenland and Antarctica, as well as on the ocean floors. In order to reach the ground a stone from space must be at least a few pounds or, by accident, happen to have a very low velocity with respect to Earth when it enters the atmosphere.

Fig. 2.13 Meteor track on the sky. The long tracks of the background stars show that the camera which made this picture was opened for a long time. A meteor or "falling star" is produced by a small cosmic stone that enters the atmosphere with a speed of tens of kilometres per second. Due to the large friction against air it gets heated such that at an altitude of about 100 km it starts to glow and burns, which produces a short-lasting track of light

Fig. 2.14 Holsinger meteorite is the largest recovered piece of the about 50-m size asteroid that produced Arizona's Meteor Crater (Fig. 2.9b)

Radioactive Age Determination

Figure 2.15 illustrates the age determination of rocks by using the radioactive decay of two of the isotopes of the element uranium and one isotope of thorium. The isotopes uranium-235 and uranium-238 decay with a half-life of 0.7 and 4.5 billion years, respectively, into the isotopes of lead with masses 207 and 206, respectively. Isotopes of the same element have the same number of positively charged protons in the nucleus (see Appendix B) and, because the atom is electrically neutral, have the same number of negatively charged electrons circling around this nucleus. The chemical properties of an atom species are determined by the number of electrons in the atom. For this reason the two isotopes of uranium have exactly the same chemical properties, and the same holds for the two isotopes of lead. In addition to the protons, the nucleus also contains neutrons, which have practically the same mass as a proton, but are electrically neutral. Every Uranium nucleus has 92 protons. In addition, the nucleus of Uranium-235 has 143 neutrons, and the nucleus of Uranium-238 has 146 neutrons (Appendix B gives more details about the structure of atoms). The above-mentioned half-lives mean that after this time half of the nuclei of the isotope have decayed into the corresponding isotope of lead. If one assumes that when the rock was born it did not contain any lead, then after 700 million years half of the atoms of Uranium-235 have turned into lead-207. After 1400 million years, only one quarter of the original atoms of Uranium-235 are left, and there will be three times as many

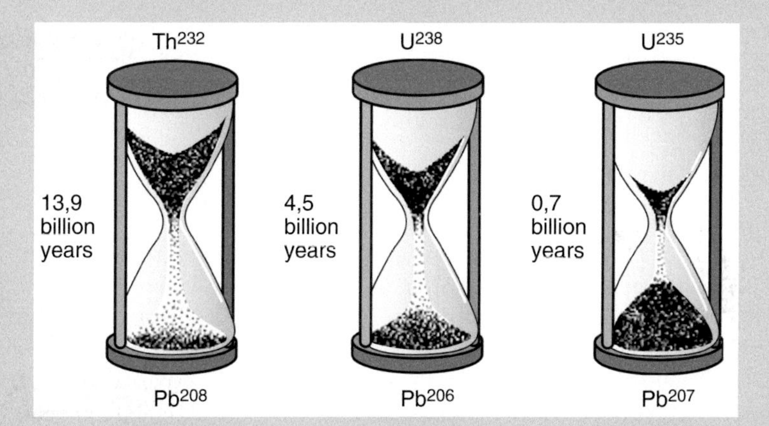

Fig. 2.15 For determining ages of rocks one uses the half-life of radio-active elements. This is the time in which half of the original number of atoms of this element or isotope has decayed into another element. Depicted here are the decay of the isotopes 232-thorium, 235-uranium and 238-uranium into respectively the isotopes 208, 206 and 207 of the element lead (Pb). After one half-life, half of the atoms of the radioactive element are left, after two half-lives one quarter of the atoms is left, after three half-lives one eighth, etc.

(continued)

(continued)

lead-207 atoms in the rock. After 2100 million years, only one eighths of the Uranium-235 atoms will be left and there will be seven times more lead-207 atoms, etc. One therefore sees that the ratio of the number of atoms of Uranium-235 and of lead-207 in the rock gives one the age of the rock. The method can also be applied if at its formation the rock did already contain some lead-207. In that case everything becomes somewhat more complicated, but by making use of elements with different half lives one can still determine the age of the rock.

The age of rocks, both of meteorites and of the Earth's crust, is determined by studying the decay of radioactive elements present in the rock, as explained in the box. This research has shown that the oldest meteorites and the oldest lunar rocks brought to Earth by the Apollo astronauts, have ages from 4.60 to 4.65 billion years. Since nowhere in the solar system older rocks have been found, it is now generally assumed that the solar system formed about 4.65 billion years ago. We will see later that the age of the universe is about 13.8 billion years and that our Milky Way galaxy already existed at least some 13 billion years ago. This implies that the universe already existed over nine billion years when the sun and its family of planets was born in an interstellar cloud in a corner of our Milky Way. The crust of Earth, and also the dark basalt planes on the moon, are younger than 4.6 billion years. On Earth no rocks older than 4.3 billion years have been found. Apparently, in the beginning Earth's crust was completely hot and liquid, and the first crustal rocks solidified only 4.3 billion years ago. These oldest rocks are found in Canada, near the Hudson Bay.

The Great Bombardment

The large dark areas that form the moon's "face" or the "man in the moon" (Fig. 2.8) are called *maria* (plural of mare = sea) or "lunar seas", since in the past astronomers thought these were seas. Study of the rocks which the Apollo astronauts brought back to Earth show that these almost crater-less "seas" consist of basalt that solidified 3.8–4.0 billion years ago. The basalt lavas that flowed out over the Mare Imbrium (diameter 1120 km) and Mare Orientalis (327 km) solidified 3.8 billion years ago. The more lightly coloured strongly cratered highlands of the moon are somewhat older: about four billion years. Also on the planet Mercury large impact basins have been discovered, similar to the maria of the moon, with diameters of 800–1500 km. Also these originated at very early times, some 600 million years after the solar system formed. Apparently around that time large numbers of asteroids—presumably originating in the asteroid belt between Mars and Jupiter—were migrating towards the inner parts of the solar system. When they passed the Earth-Moon system and Mercury many of them collided with these celestial bodies. Most of these asteroids were not larger than about 10 km, and produced the enormous numbers of craters on the lunar highlands and on Mercury. However, the impacts of the largest ones, objects with diameters of 50–100 km or even larger, were so powerful that they penetrated the already solidified crusts of the moon and Mercury, causing lava from the interior to flow out and form the large basalt maria. Also somewhat later, between 4.0 and 3.8 billion years ago, a number of times lava still flowed out over the maria, possibly caused by crustal quakes produced by smaller impacts. The traces of these later lava flows can still be well observed on several of the maria.

It is unavoidable that at the same time also Earth was hit by similar impacts. Vulcanism, erosion and plate tectonics have, however, completely erased the traces of this early impact episode. It is thought that the material that nowadays forms Earth's continents may largely be the left-over material of the asteroids that impacted Earth during this late heavy bombardment. The fact that the lunar maria show very few impact craters demonstrates that 3.8 billion years ago the heavy bombardment that produced the crater-rich parts of the surfaces of the moon and Mercury, was over. By this time it had become much more quiet in the inner solar system and after this Earth and the other four rocky planets were not any longer very frequently hit by asteroids.

Surprisingly, shortly after the end of the large bombardment there was already primitive life on Earth. The oldest fossils of single-celled organisms, found in rocks in Greenland and Australia (Fig. 2.16), have an age of between 3.5 and 3.8 billion

Fig. 2.16 *Left inset*: Fossil stromatolites (colonies of bacteria) in West-Australian rocks with an age of 3.5 billion years. *Below*: Present-day stromatolites in Shark-Bay, Western Australia. *Inset below*: One-billion years old string of fossil cyano-bacteria from the Bitter-Springs formation in Western Australia. *Size*: A few thousandths of a millimetre

Fig. 2.17 Impact spherules (tectites), small glassy spheres, from the Cretaceous-Tertiary boundary sediment layer on Haïti. This layer, with a thickness of half a metre, marks the end of the Cretaceous epoch (and the end of the epoch of the dinosaurs) and was produced by the impact of an about 10-km-size asteroid on the Yucatan peninsula of Mexico, close to the coast. The spherules are the cooled-down remains of the large amount of molten rocks from the Earth's crust that were launched by the impact into the atmosphere and rained down over a large part of the globe. The scale at the bottom of the picture is in millimetres

years. Still, in recent years it was found that in the Archeïcum, between 3.8 and 2.5 billion years ago, Earth was hit more often by asteroids than had been thought before. Geologists have found in rocks from this period many layers of several inches thickness that contain remnants of large impacts: drops of molten rocks that were spread worldwide by large impacts, and solidified as little glassy balls called *tectites* (Fig. 2.17). It was found that in these days on average one large impact occurred every 40 million years, and not every 100–200 million years as had been thought before. The largest of these impacts may have heated parts of the oceans to the boiling point and must have been very disastrous. It is now thought that these large impacts in these times may have prevented life to develop beyond the stage of simple bacteria.[2]

Possible Causes of the Late Heavy Asteroid Bombardment

Present thinking is that the cause of the late heavy bombardment by asteroids, 600 million years after the formation of the solar system, has to be sought in events taking place in the region between Uranus and Neptune. It is thought that Jupiter and Saturn, which are by far the two heaviest planets, formed early, already 4.6 billion

[2] Science, Vol. 332, 302–303, 2011 (April 15).

years ago, by direct condensation of large amounts of hydrogen and helium present in the disk of gas and dust, that in these days surrounded the young proto-sun. This disk extended to beyond the present orbit of Neptune and computations show that the largest amounts of matter in the disk were indeed located close to the places were nowadays Jupiter and Saturn are found. Inside Jupiter's orbit many rocky objects resembling asteroids condensed out of the nebular disk, so-called *planetesimals*. As this part of the disk was close to the sun, solar heat caused ices of water, carbon-dioxide, etc. to melt and evaporate, such that only rocky objects remained. The biggest rocks with their gravity attracted smaller ones, thus sweeping clean the parts of the disk in their neighbourhood and growing in about 100 million years to the present four rocky inner planets. Close to Jupiter, in what now is the asteroid belt, Jupiter's gravitational influence was so strong that it disturbed the disk of rocks and prevented the formation of one more rocky planet. As a result, here were left millions of asteroids. Also, due to Jupiter's gravitational influence, the planet Mars, which is close to the asteroid belt, was prevented by Jupiter to grow into a larger-size planet like Earth or Venus, and remained relatively small.

Far from the sun, outside Saturn's orbit, the temperature was below −100 °C, so cold that that the planetesimals that condensed out of the disk here consisted of a mixture of ices and stones consisting of silicates, metals and oxygen. This disk of icy planetesimals stretched from Saturn's orbit to beyond Neptune's present-day orbit. Also here the larger planetesimals swept up the smaller ones and finally grew to the present planets Uranus and Neptune. Computer simulations show that the formation of these two planets was a very slow process that took between 500 and 1000 million years. The increasing gravity of these two growing fairly massive planets will have disturbed the orbits of Jupiter and Saturn, and caused their orbits to gradually expand at different rates. As a result there could temporarily have resulted a situation in which the orbits of Jupiter and Saturn were in "resonance", for example that Saturn's orbital period was exactly three times that of Jupiter. In such a situation, if one starts with a situation that Jupiter and Saturn are on a straight line with the sun (a so-called conjunction), then after Jupiter has three times orbited the sun, exactly the same situation will occur again, and again and again. In such a situation, when they are on a straight line with the sun, the gravitational actions of the two planets on the asteroid belt will be added, and will cause an extra disturbance of their orbits, again, and again, and again. Therefore, this "resonant" behaviour will push asteroids out of their orbits, and it is thought that such a resonant behaviour of Jupiter and Saturn has moved millions of asteroids out of their orbits and caused them to migrate towards the inner solar system, producing the late heavy bombardment, 600 million years after the formation of the sun. The culprits causing this bombardment must then, indirectly, have been the growing planets Uranus and Neptune! The further growth of these planets would gradually have moved Jupiter and Saturn out of their resonance again, terminating the bombardment, and leaving the asteroid belt in its present heavily depleted form. In any case, it appears from this that in the early days the solar system was far less stable than it is nowadays!

How did Earth Acquire Such a Large Moon?

Another mystery is the question of Earth's abnormally large moon. Only the giant planets in the solar system have some moons of a size similar to our moon, but those moons have a composition very different from that or ours. With the exception of Jupiter's volcanic moon Io—which has been outgassed by heating due to the large tides raised on this moon by Jupiter—the moons of the large planets consist largely of ices and water, mixed with rocky material while, in contrast, our moon consists entirely of rocky material.

Earth together with its moon forms a unique "double planet". Compared with Earth, however, our moon does have some strange properties. Its mean density is only 3.35 times that of water, much lower than the 5.4 times water density of Earth, and also much lower than the mean densities of the three other rocky planets (see table in Appendix A). The reason why the four rocky planets have such high mean densities is the presence of a heavy nickel-iron core, with a density of some ten times that of water. In the case of Earth, this core has a mass about one third the mass of the planet. This core is surrounded by Earth's mantle, in which the density gradually decreases outwards from 5.5 to 3.3. On top of this mantle floats the crust of still lighter material in the form of the continents and ocean floors. This crust has a thickness of only about 10 km (oceans) and 50 km (continents) and contains only about 1 % of Earth's mass. The crust consists of basalts and silicates of iron, magnesium and aluminium, with a density that decreases outwards from 3.3 to 2.8 times that of water. This layered composition, with a density that decreases outwards, must have been established long ago when Earth was still red-hot and liquid. Like in a blast furnace, in the molten rocky ores the heavy nickel-iron sinks to the bottom, while the lighter silicates float to the top. The outer mantle of Earth has a depth of about 400 km and contains about 10 % of the mass of the planet: eight times the mass of the moon, and has a mean density of about 3.5, quite similar to that of the moon. Study of the lunar rocks brought to Earth by the Apollo astronauts shows that the moon's composition is globally similar to that of Earth's outer mantle: silicates of magnesium and iron, the only difference being that the lunar rocks contain far less volatile ingredients (chemically bound water, carbon-dioxide, etc.). Since the ratios of the oxygen isotopes ^{16}O, ^{17}O and ^{18}O in the moon and the outer mantle of Earth are very similar, and deviate strongly from those of the meteorites, which we know to originate from the asteroid belt between Mars and Jupiter, we know that the lunar material must have condensed out of the "solar nebula" (disk) at about the same distance from the sun as Earth. The existence of nickel-iron meteorites shows that in the very early solar system there have been more objects that were large enough to have gone through a molten liquid state and formed a nickel-iron core. This requires a size of these objects of at least several thousand kilometres. Later collisions between such large objects then destroyed them and the fragments of their nickel-iron cores from time to time fall on Earth as nickel-iron meteorites. The existence of these meteorites therefore shows that in the inner solar system, inside Jupiter's orbit, in the early phases of the solar system, 4.5–4.6 billion years ago, there must have been quite a number of planet-like objects. The great similarity in composition of the moon and Earth's outer mantle, in combination with a number of other facts (the large amount of "rotation" in the Earth-moon system, the fact that the moon's orbital plane does not coincide with Earth's equatorial plane) has led to the idea that the moon did not form together with Earth, but probably originated later, as a result of a collision of the very young Earth with a planet-like body with about the size of Mars: some ten times less massive than the present Earth. Such collisions can be well simulated on computers and these simulations show that the nickel-iron core of the colliding planet merges with Earth's nickel-iron core, while parts of the outer mantles of the Earth and the planet are ejected. Due to the violence of the collision the ejected matter is hot and liquid and the volatile parts of it evaporate. Most of the ejected matter is found to escape, but a small fraction—1–2 % of the mass of Earth—remains bound to Earth, and after cooling forms a disk of rocks orbiting Earth, resembling Saturn's rings. The largest of the rocks sweep up the smaller ones and finally coalesce into one larger body: the moon. This all must have happened in the first 50–100 million years of the formation of the inner planets.

This model for the formation of the moon is now generally accepted in the astronomical community. The absence of large moons for the three other rocky planets indicates that these did not experience a collision with large primordial planet-like body. Our Earth-moon system therefore had a unique formation history. This uniqueness may even have had important consequences for the evolution of life on Earth, and possibly even for the emergence of intelligent life, as will be discussed in Chap. 17.

Open star cluster NGC 265

Chapter 3
How Distant Are the Stars?

Space is big. You just won't believe how vastly, hugely, mind-bogglingly big it is.
I mean, you may think it's a long way down the road to the drug store, but that's just peanuts to space.

Douglas Adams, British author

Copernicus (1473–1543) and Galileï already knew that the only object in the solar system that emits light is the sun, and that Earth, moon and planets are cool objects that we see only because they are illuminated by the sun. Some, like our moon and Mercury are quite dark and reflect only of order 10 % or less of the sunlight. Others, like Venus, Earth and the four large outer planets, reflect a large fraction of the sunlight, between 30 and 70 %. Still, the total amount of sunlight reflected by the largest planet, Jupiter, is a billion times less than what the sun itself emits. Copernicus, who in his book *De Revolutionibus Orbium Celestium* (About the orbital revolution of celestial bodies) proposed in 1543 that the sun and not Earth is the centre of the

E. van den Heuvel, *The Amazing Unity of the Universe*,
Astronomers' Universe, DOI 10.1007/978-3-319-23543-1_3

solar system, already suspected that the stars are in fact "suns", objects that emit their own light. And that we see them only as faint specks of light in the night sky because their distances are enormously much larger than that of our sun. One can easily calculate that in order for us to see the light of the sun as faint as that of the brightest stars in the sky, the sun should be placed at a distance of about 3 *light years* (a *light year* is the distance that light travels in 1 year).

Four centuries later, Copernicus' suspicion was proved to be fully right. While Earth has a distance of only 8.3 *light minutes* from the sun, and the most distant planet, Neptune, only about 4 *light hours,* even the nearest stars have distances that should be counted in *light years*. One year contains about 31 million seconds. Since light travels about 300,000 km/s, 1 light year is about 9.46 trillion kilometres—a number of 13 digits. In the box *Powers of Ten* the different distance measures used in astronomy are indicated, and is explained how large numbers can be written as powers of ten.

Powers of Ten

In order to facilitate working with very large and very small numbers, one uses in science powers of ten. For example, one writes:

$100 = 10 \times 10 = 10^2$, and $1000 = 10 \times 10 \times 10 = 10^3$. One says then: ten to the power two, to the power three, etc. The number written at the upper right of 10 is called the *exponent* of ten. The exponent is equal to the number of times ten is multiplied with itself.

In this way, one million $= 10 \times 10 \times 10 \times 10 \times 10 \times 10 = 10^6$, one billion is 10^9, one trillion 10^{12}, etc. When multiplying two powers of ten, one simply has to add their exponents together: For example: $10^2 \times 10^3 = 10^5$.

With very small numbers one follows the same procedure: $0.1 = 10^{-1}$, $0.01 = 10^{-2}$, $0.001 = 10^{-3}$, etc. Also here, when multiplying powers of ten (in this case: negative powers) one adds the exponents.

In this way it becomes very easy to write very large and very small numbers. For example, the number of atoms in a kilogram of hydrogen is 6×10^{26}. This number is so large that if one would like to fully write it down one would get a 6 followed by 26 zeros.

Also with writing distances of stars one might proceed in this way, but this is not very practical, as the distances of even the nearest stars are tens of trillions of kilometres. In order to still be able to handle these very large distances, astronomers have introduced the *light year*: this is the *distance* travelled by light in 1 year.

One year on average counts 365.2422 days. One day is 24 h of 3600 s. By multiplying these numbers one obtains the number of seconds in 1 year: 3.15569×10^7 s. By multiplying this number with the velocity of light (299,792 km/s) one finds the number of kilometres in a light year: 9.4601×10^{12} km.

(continued)

(continued)

Another much used astronomical distance measure is the parsec. This is the distance of a star with a parallax of 1 s of arc. One parsec is 3.2615 light years.

Examples of very small but very important numbers in physics are:

- Planck's constant: $h = 6.626 \times 10^{-34}$ J s
- Boltzmann's constant: $k = 1.3807 \times 10^{-23}$ J/K

These will appear in later chapters.

The Parallax Method

Not long after Copernicus published his theory (in 1543), Danish nobleman-astronomer Tycho Brahe (1546–1601, Fig. 3.1), in the last quarter of the sixteenth century attempted to measure the distances of stars. He did this by trying to measure the so-called *yearly parallax*, which nearby stars should show, as the result of the yearly motion of Earth around the sun. Copernicus had already predicted that, as a result of Earth's orbital motion, a nearby star should—relative to much more distant background stars—describe in a year a small ellipse on the sky, as schematically outlined in Fig. 3.2. In December such a nearby star is seen from Earth in a slightly different direction than in June, when Earth has moved to the opposite side of the sun. The phenomenon that we see an object in a different direction when we move to a different position, is called *parallax* (literally: *sight difference*). We humans use this parallax phenomenon with our two eyes, to see "depth"—that is: see differences in distances of objects. From our left eye, an object is seen in a slightly different direction than from our right eye (Fig. 3.3). Land surveyors use the parallax phenomenon to determine distances to remote objects such as church towers, without having to go there (see Fig. 3.3).

Tycho Brahe carried out his measurements with the naked eye, as in his time the telescope had not yet been invented. The smallest angular difference that one can measure with the naked eye is half a *minute of arc*, that is: (1/120)th of a *degree of arc* on the sky, where a degree of arc is about twice the diameter of the full moon. Even though his instruments for measuring angles on the sky with accuracy of half an arc minute were by far the most precise ones in his time, Brahe did not succeed in measuring the parallax of a star. He therefore rejected Copernicus' idea that Earth is orbiting the sun. Much later it was discovered that the parallaxes of stars are always smaller than a *second of arc*, which is (1/60)th of an arc minute. Tycho Brahe could therefore never have measured a stellar parallax with the naked eye.

It was the invention of the telescope, in 1608, which made it in principle possible to measure smaller angles on the sky, but it would take more than two centuries before telescopes had become so accurate that measuring stellar parallaxes could become reality. The first one who, thanks to improved precision technology for

Fig. 3.1 In the second half of the sixteenth century Danish nobleman-astronomer Tycho Brahe built his observatory Uranienborg on the island Ven, with instruments that allowed him to make extremely accurate naked-eye measurements of the sky positions of stars and planets. Thanks to his decades-long accurate measurements of the positions of Mars, Tycho's pupil Johannes Kepler was able to discover his three famous laws of the orbital motions of the planets

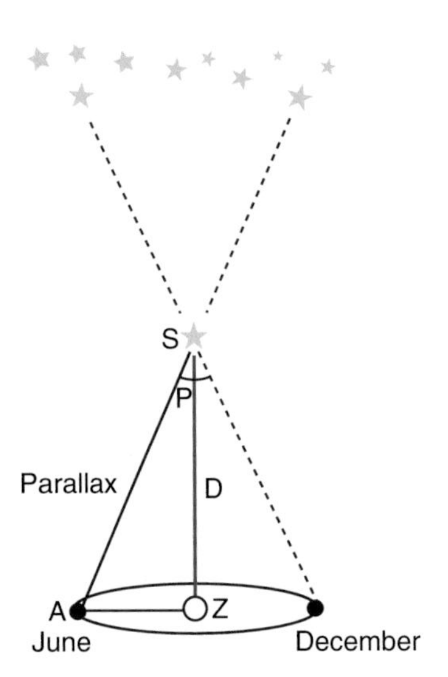

Fig. 3.2 The yearly motion of Earth (A) around the sun (Z) makes that a nearby star is observed to describe a small yearly ellipse on the sky relative to much more distant background stars. In the situation depicted here, with a nearby star in the direction perpendicular to the plane of Earth's orbit, this little ellipse is a circle. The measured angle ASZ=p is called the parallax of the star. Since the distance AZ is known, as well as the angle AZS, the distance D to the star can be calculated

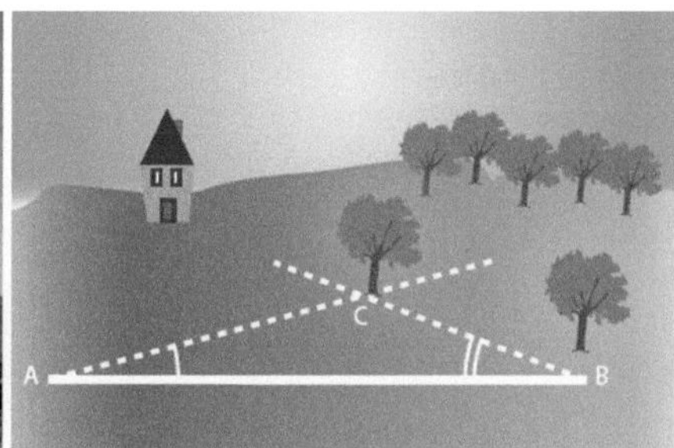

Fig. 3.3 *Left*: The parallax of your own finger: seen from your left eye you see the finger in a different direction with respect to the background than seen from your right eye. *Right*: The distances to the remote tree C from the points A and B, respectively, are AC and BC. These distances can be calculated by measuring at A the angle CAB between the directions to B and C, and at B the angle ABC between the directions to A and C, and then measuring the distance AB between A and B. Now two angles and one side (AB) of the triangle ABC are known and AC and BC can be calculated. This is the way in which prospectors measure distances to remote objects

making machines and instruments, succeeded in measuring a stellar parallax, was German astronomer Friedrich Wilhelm Bessel (1784–1846, Fig. 3.4). With this he gave in 1838 the first *direct* proof that Earth is moving around the sun. One century earlier British astronomer James Bradley had already found an *indirect* proof by discovering the so-called *aberration* of starlight, which is due to the finite velocity of light, as explained in Fig. 3.5.

In practice, the parallax method works as follows. As depicted in Fig. 3.2 in December we see a nearby star in a slightly different direction than in June. The difference in direction is the top angle 2P of the depicted triangle. (For simplicity we have taken here a star located in a direction perpendicular to the plane of Earth's orbit, but one can equally well do a similar calculation for other directions.) The angle P is called the parallax. In triangle AZS we know the distance AZ (distance sun-Earth), the angle P and the angle AZS; in the depicted case the latter angle is 90°, but also if the star is located in another direction, this angle is known. When two angles and one side of a triangle are known, trigonometry allows us to calculate all the other sides of the triangle. We thus can now calculate the distances AS and ZS. Since stars are very far away, the difference between these two sides in, in fact, negligible with respect to these two quantities themselves.

Fig. 3.4 German astronomer Friedrich Wilhelm Bessel was in 1838 the first to succeed in measuring the parallax of a star

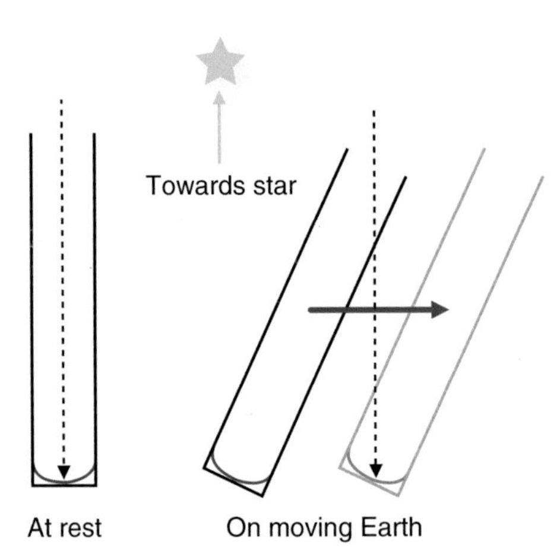

Fig. 3.5 Aberration of starlight. Earth moves in its orbit around the sun with a speed of about 30 km/s, and light has a speed of about 300,000 km/s. Because of the finite speed of light one should, in order to capture the light of a star in a telescope, tilt the telescope a little bit—by 20 s of arc—with respect to the direction of the star, into the forward direction of Earth's motion. Half a year later, the telescope should be tilted in the opposite direction. Due to this 'aberration' all stars seem to describe a yearly ellipse on the sky, with a semi-major axis of 20 s of arc, the same for all stars. This effect was discovered in the eighteenth century by English astronomer Bradley (1693–1762), who also gave the correct explanation for it

It was found that the parallaxes of stars are extremely small. The largest stellar parallaxes ever measured are 0.747 and 0.772 s of arc, for the bright binary star Alpha Centauri and its faint distant companion, Proxima Centauri, respectively. The southern star Alpha Centauri is the third brightest star of the sky. In a telescope one sees that it consists of two almost equally bright stars, close together, which are called Alpha Centauri A and B, and this system has a faint distant companion, Proxima Centauri, which moves in a wide orbit around Alpha once in several 100,000 years. This is the star that is closest to the solar system and its distance is 4.2 light years.

An arc second is one sixtieth of an arc minute. The parallaxes of Alpha and Proxima Centauri are therefore about (1/2400)th of the diameter of the full moon. For comparison: 1 arcsec is the angle which spans the thickness of a human hair, seen from a distance of 20 m (66 ft), or the diameter of a dinner plate seen from a distance of 40 mi. To accurately measure such a small angle is technically so difficult that only around 1820 the required combination of precision optics and fine-mechanical technology—for making accurate cam wheels and circular scales for measuring angles, etc.—had sufficiently advanced for achieving this goal. Thanks to the demands for making precision clocks and sextants, for accurate measurement of angles, required for navigation at sea, great progress had been made in these fields in the second half of the eighteenth century. In the first half of the nineteenth century German genius instrument maker Joseph Fraunhofer (1787–1826) in München was the first to succeed in building telescopes and *wire micrometers*[1] with highly accurate screws and scales, which allowed to measure angles with a precision of a fraction of an arc second. With a telescope that Fraunhofer especially built for him, Bessel was able in 1838 to measure the 0.3 arcsec parallax of the star 61 Cygni. He measured how in the course of a year the position of this star on the sky moved with respect to the positions of neighbouring faint stars (which he assumed to be far more distant). Bessel chose 61 Cygni because this star at that time was the star with the largest known *proper motion*. The proper motion of a star is the speed with which one observes the star to move with respect to neighbouring stars on the sky.

This proper motion is due to the fact that the positions of the so-called "fixed stars" are in reality not completely fixed, but change very slowly in the course of centuries, due to the real motions of the stars in space. With the naked eye one does not notice these motions, even if one observes the stars for decades or even a century.

---

[1]A *moving wire micrometer* is placed in the focal plane of the eyepiece of the telescope. It consists of a fixed cross of two perpendicular thin straight metal wires, a horizontal one and a vertical one, and one moving wire, parallel to the vertical wire. (The principle of the micrometre was invented around 1640 by British astronomer William Gascoigne, who used hairs for the wires). The moving wire is moved by an accurate screw, with a divided scale on which one can read off the angle by which the wire has been moved. The zero point on the scale corresponds to the position at which the moving wire coincides with the vertical wire. The angular distance between two stars is found by placing the horizontal wire over the two stars, and placing one of the two stars in the centre of the fixed cross. One then moves the moving wire onto the other star, and reads the angular distance between the stars on the scale attached to the micrometer screw.

This is the reason why it appears to us that the stars keep fixed positions with respect to each other, and the starry sky of the constellations never changes, even over timespans of millennia. However, if one adds the proper motions of the stars over thousands of years, one will observe that the shapes of the constellations gradually change (Fig. 3.6). The first one to discover the *proper motion* of a star was British astronomer Edmund Halley (1656–1742), of Halley's comet fame. Early in the eighteenth century he compared the positions of the bright stars Sirius, Procyon and Aldebaran relative to the fainter stars in their surroundings, with the positions which Greek astronomer Hipparchus of Nicaea had measured for them around 150 BC (Hipparcus' catalogue counted about 850 stars with a positional accuracy roughly similar to the size of the moon). Halley noticed that in 1900 years Sirius had moved with respect to neighbouring stars over a distance of about one-and-a-half times the diameter of the full moon. (Sirius had even moved measurably relative to the position that Tycho Brahe had measured at the end of the sixteenth century.)

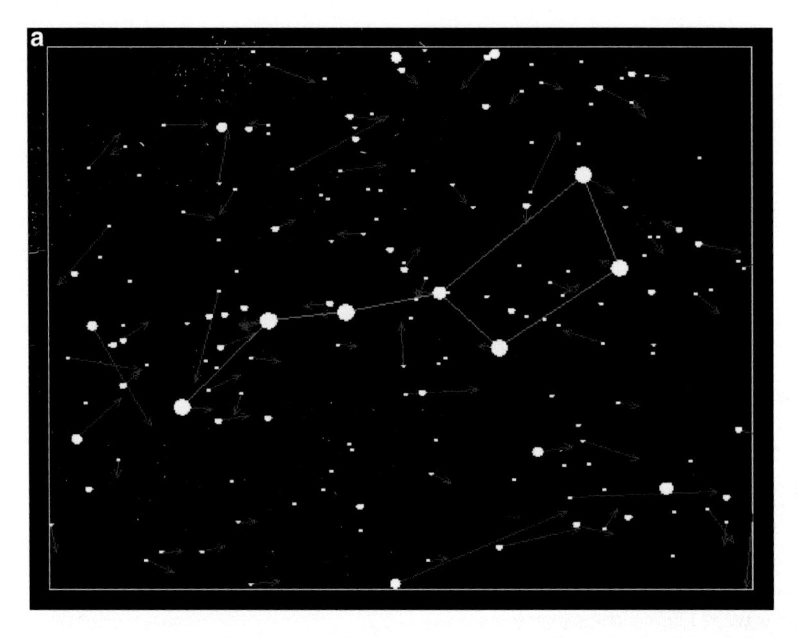

Fig. 3.6 (**a**) The constellation of the Big Dipper (Ursa Major) today. The lengths of the arrows indicate the distances which the stars travel in 50,000 years as a result of their proper motions. Stars β, γ, δ, ε, and ζ (see Fig. **b**) all have practically the same proper motions and move through space together. They form a loose star cluster with a common origin. Also the star Sirius on the other side of the sky and several tens of other stars belong to this *moving cluster*. *Bottom picture*: The Big Dipper 100,000 years ago and 100,000 years from now, respectively. (**b**) The distances of the stars of the Big Dipper in light years

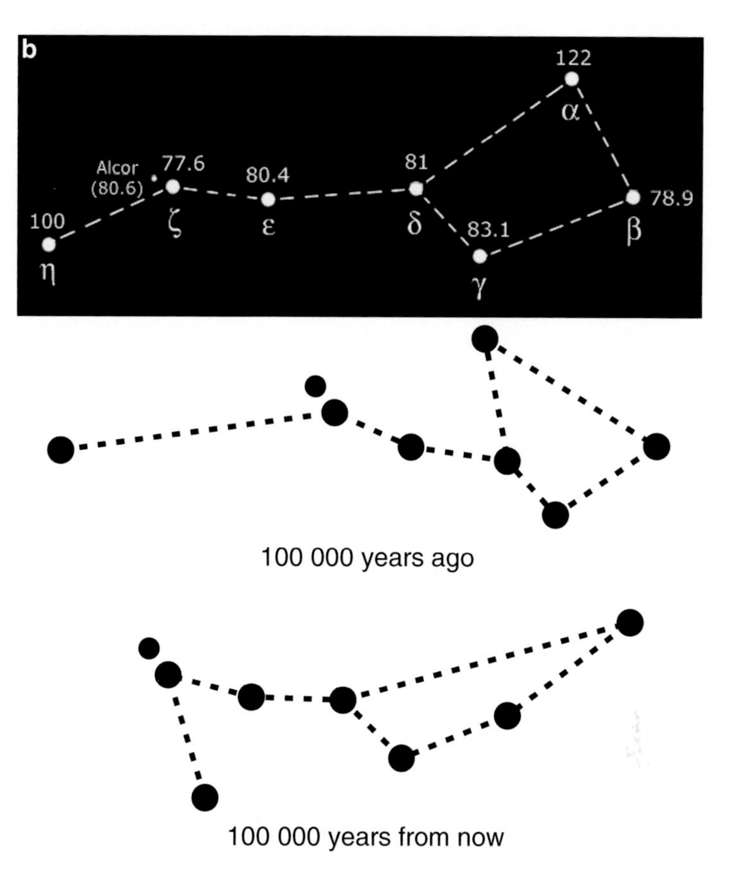

Fig. 3.6 (continued)

Friedrich Wilhelm Bessel (1784–1846)
Bessel, a self-made astronomer and mathematical genius (among many other things, he invented the Bessel functions, nowadays an essential mathematical tool of physicists and electrical engineers), began his career at age 15 as bookkeeper in a ship-broker office in the harbour of Bremen, where thanks to his great intelligence he rapidly rose in the office ranks. His interest in astronomy was triggered by books on astronomical navigation which he found in the office and which he—after working days from 8 in the morning till 8 in the evening, 6 days a week—devoured in his spare time. At age 20 he had already advanced far enough in mathematics to be able to calculate orbits of comets and other celestial objects. Through his interest in astronomy he met Bremen pharmacist/physician and amateur astronomer Wilhelm Olbers

(continued)

(continued)
(1758–1840), famous for discovering in 1802 and 1807 the second- and fourth-known asteroids Pallas and Vesta, and of a number of comets, and from formulating the important “Olbers’ paradox” (see Chap. 9). Olbers was so much impressed by Bessel’s talents that he recommended him for the position of assistant at the observatory of the wealthy magistrate and amateur astronomer Johann Schröter in Lilienthal. In 1809, after working in Lilienthal for a number of years, Bessel now aged 25, was invited by Prussian king Friedrich Wilhelm for the position of director of a new astronomical observatory that the king had ordered to be built in Köningsberg in East Prussia (since 1945 called Kaliningrad, and now part of Russia). Bessel accepted the invitation and started in March 1810. He remained director of the observatory for the rest of his life. In 1812, when Napoleon with his Great Army on his way to Moscow passed Köningsberg, he was greatly astonished: “How is it possible that in these days the King of Prussia can still think of building a new observatory?”.

Moving Star Clusters
Figure 3.6 shows the well-known constellation the Big Dipper. The five ‘inner’ stars (Zeta, Epsilon, Delta, Gamma and Beta), which all have a distance of about 80 light years, move together through space: their proper motions are parallel to each other and have about the same value. These stars therefore appear to form one family with the same origin. It appears that Sirius, at the opposite side of the sky, moves in space in the same direction and with the same velocity, just as several tens of other, fainter stars, in different parts of the sky. Together these stars form a very widely-spaced ‘star cluster’, through which our sun is presently passing. Such a group of stars that, in contrast to high-density star clusters like the Pleiades (Fig. 3.7), cannot be directly recognized as a cluster, but only appears to form a unity when one measures the proper motions, is called a *moving cluster*. Another moving cluster, also partly visible with the naked eye, is the Hyades, in the constellation Taurus.

(continued)

(continued)

Fig. 3.7 The Pleiades—the 'Seven Sisters'—is a group of six stars visible with the naked eye in the constellation Taurus. They are the brightest members of an open star cluster composed of several hundreds of stars

The measured proper motions of stars are expressed in arc seconds per year. It appears reasonable to assume that, in first approximation, all stars move with more or less similar velocities with respect to each other in space. One expects then that the stars that show the largest proper motions are nearest to us. Bessel therefore assumed that the stars with the largest proper motions will also show the largest parallaxes. This was the reason why he chose the star 61 Cygni for his first attempt to measure a parallax, as this star had the at that time largest known proper motion, of 5.2 arcsec per year. The present record holder is Barnard's star, a faint red dwarf discovered by American astronomer E.E. Barnard (1857–1923), which moves 11 arcsec per year along the sky (Fig. 3.8).

Nowadays, the parallaxes of thousands of stars have been measured—with accuracies of one thousandth of an arc second per year—by the European Hipparcos satellite, launched in 1989. The successor of Hipparcos is ESA's GAIA spacecraft, launched in 2013, which at present is measuring proper motions and parallaxes of

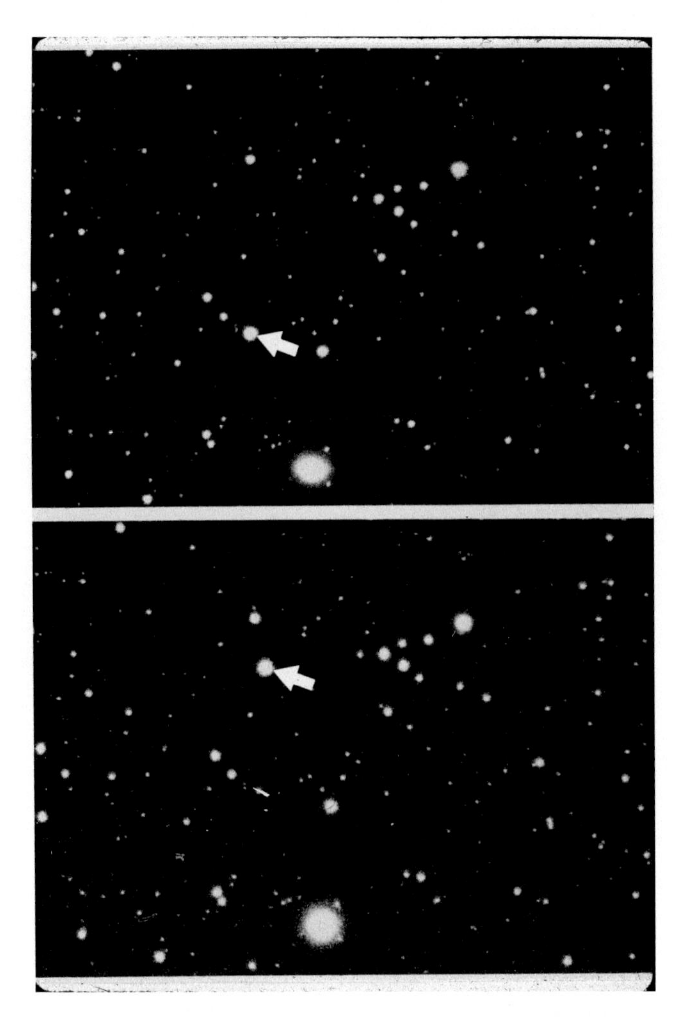

Fig. 3.8 Barnard's Star is the star with the largest known proper motion (11 arcsec per year) photographed in 1937 (*above*) and 1950 (*below*). One clearly notices its motion with respect to the background stars

stars with an accuracy of hundreds of times better than Hipparcos. The results from GAIA are expected to become available in the coming years. Hipparcos allowed to measure distances of stars out to about one thousand *parsecs*. One parsec (pc) is the distance of a star with a parallax of 1 arcsec, and is equal to 3.26 light years. So, Hipparcos was able to determine distances of stars out to 3260 light years. A distance of thousand parsec is called a kiloparsec (kpc) and a distance of a million parsecs is called a megaparsec (mpc).

The Distance to the Nearest Star as Compared to the Distance to the Sun

One way to realize how enormous the distances to the stars are, is by comparing the 4.2 light year distance of the nearest star to our distance to the sun. If one takes the sun to have the size of a grapefruit (4 in. diameter), Earth is as small as a pinhead ((1/25)th of an inch). On this scale the distance between Earth and Sun is 38 ft: about 110 suns (grapefruits) fit between sun and Earth. If in New York we put the grapefruit at the position of the Empire State building, the entire solar system fits in the central part of Manhattan, with the planet Neptune near the southern edge of Central Park. On this scale, the nearest grapefruit (Proxima Centauri, at 4.2 light years) is located 1400 miles away, in the Rocky Mountains, halfway to California. This shows how empty the space between the stars is. The most distant object launched by humans, the Voyager 1 spacecraft (launched September 5, 1977) has presently reached a distance not more than four times the distance of Neptune. On the scale with the sun as a grapefruit near the Empire State building, Voyager 1 has not yet reached the Northernmost point of Manhattan, not very far on its way to the nearest star, in the Rocky Mountains. With its speed of about 15 km/s it will need some 80,000 years to reach this star. This illustrates that with present-day technology space travel to other stars is no easy matter. We will return to this subject in Chap. 17.

Far Away = Long Ago

All stars that we can see at night with the naked eye—some 5000 on a very clear night—have distances between 4.2 and several thousands of light years. Figure 3.6b shows the distances in light years of the stars in the well-known constellation Big Dipper (Ursa Major), as measured by the Hipparcos satellite. The two outermost stars are located at 100 and 122 light years, respectively, while the five inner stars all have distances of about 80 light years. As the light of these inner five stars needs about 80 years to reach Earth, we see them as they were 80 years ago. And someone located near any of these five stars will see Earth as it was about 80 years ago, that is: in the mid-1930s. That person will see Earth during the 1930s economic crisis, and F.D. Roosevelt being elected in the USA, and the rise of Nazi power in Germany.

In the winter constellation of Orion (Fig. 3.9) we see even more distant stars. Betelgeuze, the brightest one, is 420 light years away, and Rigel, the second-brightest one, 770 light years. From Rigel one see Earth as it was in the year 1245: one just sees the crusaders in Europe saddle their horses to go to Jerusalem. And from the star Alnilam, in the middle of Orion's Belt, 2000 light years away, one just sees the Roman legions entering Britain and founding London.

We see here the consequences of the fact that the velocity of light is finite. This means that we never can see distant objects as they are *now*, but always as they *were in the past*: the further away, the longer ago! An object at a distance of a billion light years we see as it was a billion years ago, and a galaxy at a distance of ten billion

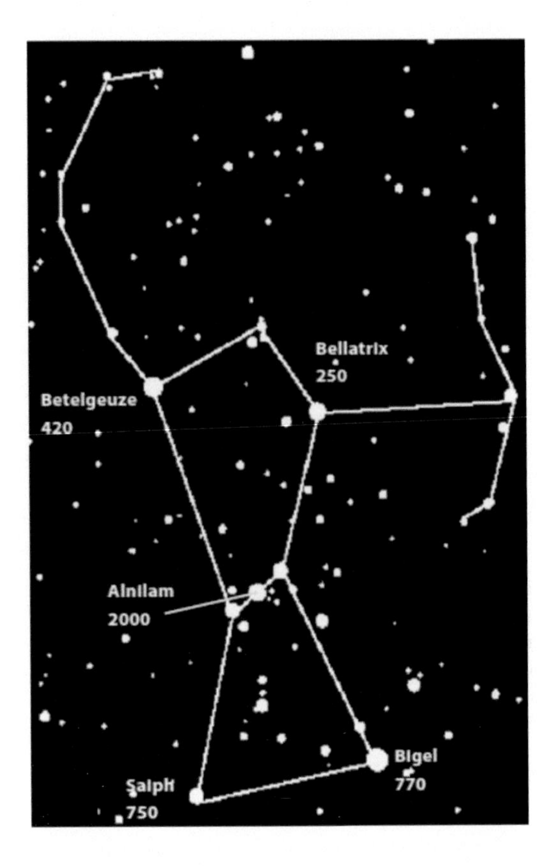

Fig. 3.9 The constellation of Orion with distances of some bright members indicated in light years

light years: as it was ten billion years ago.[2] When a biologist wishes to know how plants and animals have evolved in the past, he/she will have to dig deep in the earth to find fossils from these times. For an astronomer, life is much simpler: if we want to know how galaxies looked like 200 million years ago, we simply look at galaxies at a distance of 200 million light years. And if we wish to know how galaxies looked 5 or 10 billion years ago, we look for galaxies at a distance of five or ten billion light years, respectively. All our "fossils" still are visible in the sky. In fact, the entire history of the universe is visible on the sky. In astronomy holds literally:

$$Far\ Away = Long\ Ago$$

Of course, the more distant a star or galaxy is, the fainter the light that we receive from it (see Fig. 3.10). To see very distant galaxies, that tell us about the earliest history of the universe, we therefore need very large telescopes. This is the reason why for astronomers no telescope is, in fact, large enough, and why they keep building larger and larger telescopes!

[2] At very large distances this is no longer completely true, due to the expansion of the universe. We come back to this in Chap. 6 and later.

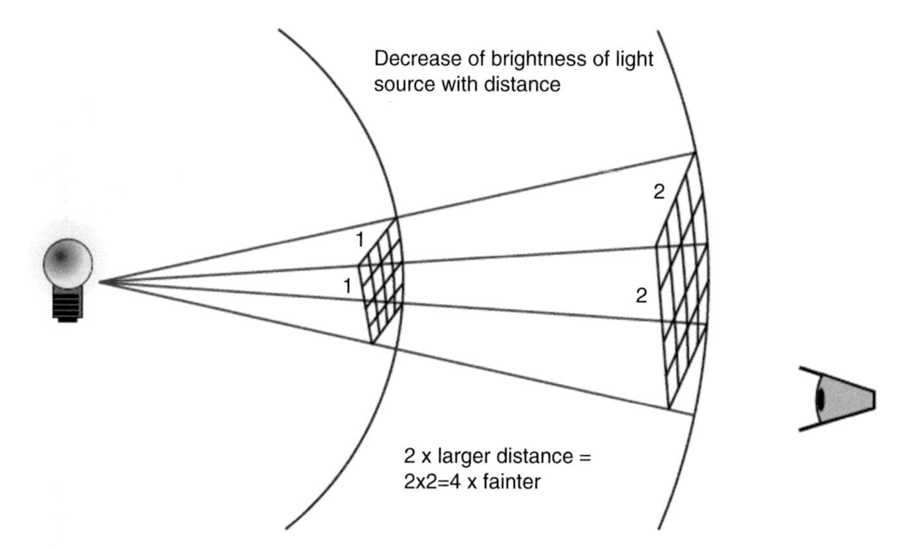

Fig. 3.10 Decrease of observed brightness of a light source with increasing distance. The light from the source spreads evenly in all directions, and passes through surfaces of spheres of increasing radius. As the surface area of a sphere is proportional to the square of its radius, the amount of light caught by 1 m^2 of a sphere (or by the pupil of your eye) is inversely proportional to the square of the distance to the light source. The observed brightness of two identical light sources, of which one is twice as distant as the other, therefore differs by a factor of 4

Open star cluster Pismis 24 and the emission nebula NGC 6357

Chapter 4
The Discovery of the Structure of Our Milky Way Galaxy

The road from Earth to the stars is not smooth

Lucius Annaeus Seneca (4 BC–65 AD)—Roman Philosopher

Between the about 5000 stars that we can see with the naked eye on a dark moonless night, stretches the faint-glowing band of the Milky Way (Fig. 4.1), which was recognized by man since time immemorial. Its name originates from ancient Greek mythology. Persian scholar Abu Rachman Biruni (973–1048)—inventor of spherical trigonometry and author of over 100 books in all fields of science, from mathematics, physics and astronomy, to geography and ethnography—was the first to propose that the Milky Way is a collection of uncountable numbers of faint stars. This suggestion was confirmed in the fall of 1609 when Galilei pointed his

E. van den Heuvel, *The Amazing Unity of the Universe*,
Astronomers' Universe, DOI 10.1007/978-3-319-23543-1_4

Fig. 4.1 Wide-angle picture of the Milky Way taken in Northern Chile. The picture covers half of the sky and stretches from the European Southern Observatory's Very Large Telescope on mountain Paranal, at *left*, to the VISTA Telescope—also European—at the *right*. The brightest regions, interrupted by dark dust lanes, give a glimpse of the lens-shaped central part of the Milky Way system, which we observe from the side. At the *right*, above the mountain of the VISTA telescope, one notices the Andromeda Nebula (M31) as a small elongated speck of light. At a distance of about 2.5 million light years this is the nearest large neighbour galaxy of our Milky Way, just visible to the naked eye

telescope—invented in the Netherlands 1 year earlier—at the Milky Way. The fact that the Milky Way stars are so faint implies that—on average—they are far more distant than the *individual* stars that we can see with the naked eye. Around 1750, English instrument builder and amateur astronomer Thomas Wright (1711–1786) proposed in his book *An Original Theory or New Hypothesis of the Universe* that the sun and the nearby stars are part of an almost flat disk, "resembling a millstone", of millions of stars, which in the plane of the disk extends much farther than in the direction perpendicular to the disk, and which surrounds us on all sides. In those days this was a purely theoretical thought, as one was not yet able to determine the distances of the stars. The real mapping of the Milky Way system started near the end of the eighteenth century and took almost two centuries. Step by step it proceeded as follows.

The Work of William Herschel

The first attempt to map the Milky Way system was carried out in the last decades of the eighteenth century by musician and astronomer William Herschel (1738–1822, Fig. 4.2). Herschel belongs to the most important astronomers of all time, but in his youth nobody would have predicted this. Herschel was born in the city-state of Hannover in Germany. His father was musician in the army of this small kingdom which in these days belonged to the English royal family, which descended from the Hannover royal house. William, then still named Friedrich Wilhelm, and his brothers all were educated to become musician-soldiers in the army music-corps. When the king of England needed soldiers William, 15 years old and hobo player, was moved with his army to London, and became acquainted with England,

Fig. 4.2 William Herschel, discoverer of planet Uranus, was the first to attempt to map the Milky Way stellar system. As depicted in this portrait, Herschel also discovered the existence of infrared radiation: he discovered that in the solar spectrum beyond the colour red there still is (invisible) radiation that is able to make the temperature of a thermometer rise

a country he came to like. However, at the outbreak of the 7-year war between Germany, England and France, the army of William, then 17 years old, was moved back to Germany to fight the French. Herschel was very unfit for the battlefield and managed at the age of 19 to be dismissed from the army. He returned to London, where he became music teacher and musician. He played a great variety of instruments: apart from the hobo: also the violin, clavicord and organ—and also was active as a composer of beautiful baroque music. His reputation as a music teacher rapidly grew as did his name as a conductor. At the age of 30 he had reached the position of conductor of the symphony orchestra and chorus of the distinguished Southern-English beach town Bath. In his spare time he taught himself English, Italian, Latin and mathematics. In 1770 his brother Alexander, an excellent cellist, came to England to live with him and in 1772 he convinced his sister Caroline (1750–1848) to also come to England. Later, she would for almost 200 years hold the record of the largest number of comets discovered by a woman. William taught her English and mathematics and thanks to his singing lessons she became an excellent soprano in his chorus in Bath.

William's father had already an interest in astronomy, and had taught him the constellations. When William was 35 he built simple telescopes with lenses (so-called refractors) to study the stars, but these were disappointing, due to the poor quality of the lenses in those days. He then rented a small reflecting telescope which he saw offered in a shop window. He was deeply impressed by what this telescope was able to show him of the moon and planets. As he had insufficient money to buy a larger reflecting telescope, he decided to start building a reflecting telescope himself. This was the first of long series of telescopes, with increasing mirror sizes, culminating in the largest telescope of his time, with a 3.5 ft diameter mirror. Telescope mirrors in these days were made of metal ("speculum", a kind of bronze), which Herschel cast himself. After this, he carefully ground and polished the mirror into a parabolic shape. In this work he was assisted by his sister Caroline, who also was his housekeeper (William married only at the age of 50). Later also his brother Alexander assisted him.

On March 13, 1781 he made a spectacular discovery: he discovered a new planet, which later was to be named Uranus. Since antiquity, only five planets had been known, visible to the naked eye: Mercury, Venus, Mars, Jupiter and Saturn. Uranus is not visible to the naked eye, and is the first planet discovered with a telescope. Herschel saw in his telescope an object with a diameter larger than a star—even in the largest telescopes stars just remain miniscule points of unmeasurably small sizes—and he first thought that he had discovered a comet, because in the course of several nights it clearly moved with respect to the surrounding stars. However, on closer inspection this object appeared to move much slower than a comet, and moved also exactly in the plane of the ecliptic, where the planets describe their paths on the sky, something that comets almost never do. Since in his telescope the object also was perfectly round, he realized that he had discovered a new planet. He made this discovery with a telescope with a half-foot diameter mirror, and a 7 ft focal length.

This discovery immediately made Herschel a celebrity, and English king George III, who had a great interest in astronomy (the same king against whom the Americans fought their war of independence) granted him a lifelong salary of 200 pound sterling per year, such that from then on Herschel could devote all his time to astronomy.

He built larger and larger telescopes, and with one of these, with a one-and-a-half foot diameter mirror and 20 ft focal length, he started to map the Milky Way. To this end he selected 683 fields, each with a diameter of 15 arcmin (half the diameter of the full moon), evenly distributed over the sky. Since he did not know the distances to the stars, he made the assumption that the stars are uniformly distributed in space—like the trees in a forest that have been planted at equal distances from each other. If in such a forest one is close to the edge of the forest, one will see in one's field of view fewer trees in the direction of this edge than in other directions. But if one is far from the edge of the forest, one will see many trees in one's field of view in the direction of the forest's edge. So: the number of trees one sees in one's field of view in a certain direction is a measure for the *distance* to the edge of the forest in that direction. Herschel therefore reasoned that the *number of stars* that one sees in one's field of view is a measure for the *distance to the edge* of the Milky Way in the direction of this field. Sitting behind his telescope during many nights, he carefully counted the numbers of stars that he saw in each of his 683 fields. He dictated these numbers to his sister Caroline, who at some distance was sitting with a candle behind a screen, noting down the numbers. With the assumption of a uniform star distribution in space, the number of stars seen in a certain direction is proportional to the distance to the edge of the Milky Way to the third power. The counted numbers

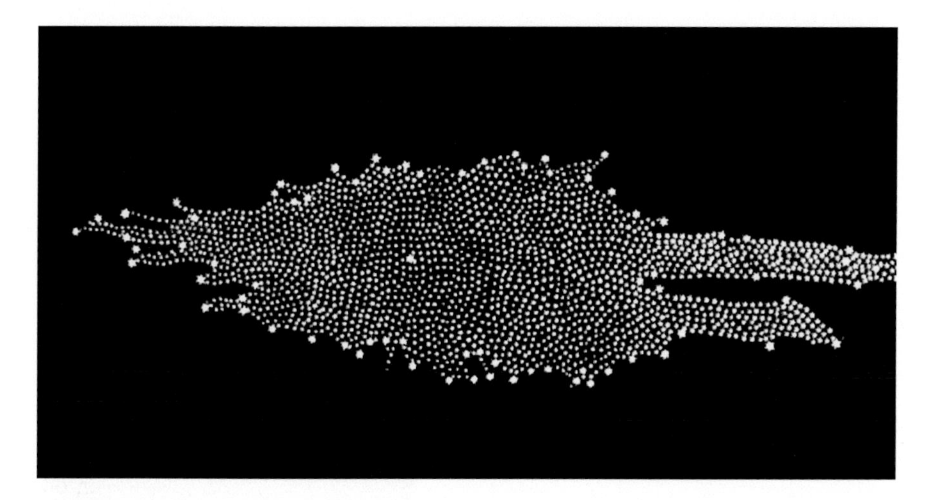

Fig. 4.3 Cross-section of the Milky Way, perpendicular to the Milky Way plane in the direction of Cygnus, as determined by William Herschel. The star slightly *left* of the centre is the position of the sun. The splitting of the Milky Way in Cygnus (*right*), due to the presence of a dust lane, is also clearly visible to the naked eye

of stars therefore gave him in each direction the distance to the boundary of the Milky Way in that direction. In this way Herschel was able for the first time to make a three-dimensional map of the Milky Way system. He found the system to be a flattened disk of stars—with about the shape of a thick pancake or pizza—with the sun near its centre. Figure 4.3 depicts a cut through this system in a plane perpendicular to the disk, in the direction of the constellation Cygnus. The split at the right-hand side is due to the well-known dark band in Cygnus, where to the naked eye the Milky Way seems to split, and hardly any stars are seen.

After this great project, Herschel made many more discoveries. He discovered, for example, that there exists infrared radiation ("heat radiation"), invisible to our eyes (see Fig. 4.2). Also, he was the first to discover the orbital motion of a double star system: the visual binary star Castor in the constellation Gemini.

For nearly a century, Herschel's map of the Milky Way remained the "standard model" of our Galaxy.

The Work of Kapteyn

Around 1885 the work of charting the Milky Way system was taken up again. This time by Dutch astronomer Jacobus Cornelius Kapteyn (1851–1922; Fig. 4.4), who applied the new technique of photography, which had been introduced in astronomy in the second half of the nineteenth century. In 1877 Kapteyn, at age 26, had been appointed professor of astronomy and theoretical mechanics at Groningen University, while at the same day his brother Willem, one-and- a half year his senior,

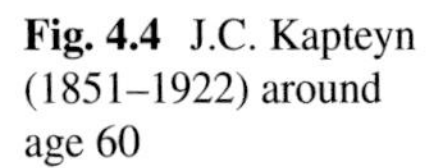
Fig. 4.4 J.C. Kapteyn (1851–1922) around age 60

was appointed professor of mathematics at Utrecht university, where they both had studied. Apart from a few hours lecturing per week, Kapteyn could spend the rest of his time doing research. However, there was a great problem: Groningen university had no observatory and not even a telescope, and Kapteyn's request to the Dutch government for money for a telescope (all Dutch universities are government-funded) was turned down, due to the advice of the directors of the observatories of Leiden and Utrecht, who were of the opinion that two observatories are more than sufficient for a small country like the Netherlands.

Kapteyn then came into contact with astronomer David Gill in Capetown, South Africa, who had just the opposite problem: he had an excellent telescope and took magnificent photographs of the Southern sky, but had no time or personnel to scientifically analyse this beautiful material. Kapteyn offered to analyse these photographs, that is: to carefully measure the position and brightness of each star, as such a high-precision survey of the Southern stars had not been done before. In this way the Southern sky would be charted at least as accurately as had been done for the stars in the Northern sky 30 years earlier by German astronomer Argenländer of Bonn—a pupil of Bessel—in the so-called *Bonner Durchmusterung*. In Argeländer's time astronomical photography did not yet exist, and he had done all his work visually behind the telescope.

Gill was delighted by Kapteyn's offer and this led to a collaboration of many decades, in which Gill took the photographs and Kapteyn—first alone, but later with a small army of assistants—carried out the measurements on the photographs. In astronomy such photographs are always taken on glass plates, because contrary to photographs on celluloid or paper, glass keeps its shape, such that the relative positions of the stars on the plate are not subject to deformations. Kapteyn started his work in two small rooms in the cellar of the Physiological Laboratory of Groningen University, offered to him by his friend Gerard Heymans, professor of physiology.

The result of this work was the publication in the years 1896–1900 of the *Cape Photographic Durchmusterung*, with accurate positions and brightness of 454,875 stars. This work made Kapteyn world famous, but it took till 1913 before the university finally was able to give him his own "Astronomical Laboratory".

Kapteyn realized that he could statistically determine stellar distances by measuring the proper motions of the stars, as explained in Chap. 3. If on average the stars have the same velocities in space, those with the largest proper motions will be nearest to us, and those with the smaller proper motions will be more distant. In order to determine proper motions, one should after a number of years again take a photograph of the same area of the sky, and again measure the positions of all the stars. This was done, and so the proper motions and statistical distances of groups of stars were obtained. This does not work for the faintest, most distant stars, as their proper motions are negligible. For these stars Kapteyn used other methods to estimate their distances, mainly based on their observed brightness: the amount of light we receive from a star decreases inversely proportional to the square of the distance, as explained in Fig. 3.10. A star that is twice as distant will be four times fainter, one that is three times more distant is nine times fainter, etc. If one then assumes that the sun is an "average" star, and other stars emit on average the same amount of light as the sun, the observed brightness of a star allows one to calculate its distance. Later it was found that intrinsic (true) brightness of stars is not the same, and there is a relation between the colour and the intrinsic brightness of the star: stars bluer than the sun are intrinsically brighter, redder stars fainter (the latter holds for most of the red stars, but the red giant stars are an exception). Using he colours, one can then determine their mean distances more accurately.

Based on these average distance measurements, made over decades by himself and his collaborators, Kapteyn around 1920 presented his famous *Kapteyn Model* of the Milky Way system. This model is depicted in Fig. 4.5 (as published in 1922 in the American *Astrophysical Journal*). As in Herschel's older model, in Kapteyn's model the sun is close to the center. Going outwards from the sun, the mean space density of stars (number of stars per cubic light-year) decreases in all directions,

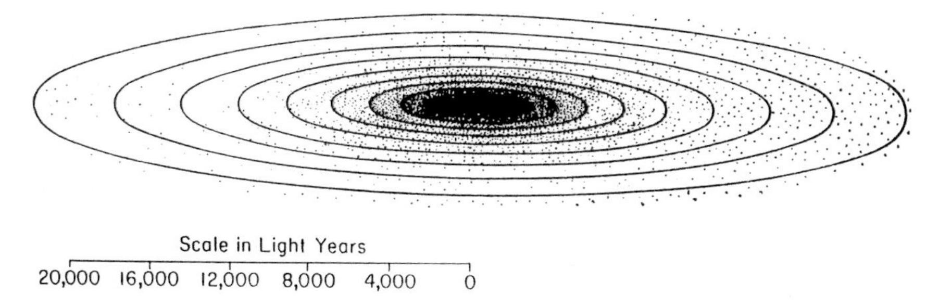

Fig. 4.5 Kapteyn's Milky Way model, published in the *Astrophysical Journal* in 1922. The drawing is a cross-section of this axial-symmetrical model in a plane perpendicular to the Milky Way plane. The sun is close to the centre, and in the Milky Way plane the system extends from the centre to about 20,000 light years

slowly in the central plane of the Milky Way, where the outer boundary is reached at a distance of about 20,000 light-years, and rapidly in the direction perpendicular to the plane where the boundary is reached at a distance of a few thousand light-years.

"Fog" in the Milky Way System

Since Copernicus we know that Earth is not in the center of the solar system: our planet does not have a special central position but is just one of eight planets that orbit the sun. For this reason it came as a surprise that in the early models of the Milky Way system, of Herschel and Kapteyn, our sun *did* occupy a *central*, and therefore very special, position. This appeared to make us again *very special* in the universe! Herschel and Kapteyn had assumed that the space between the stars is completely empty and therefore completely transparent, and thus that the light of the stars reaches us without having been subject to losses due to absorption or scattering by atoms, molecules or dust particles in the space between the stars. However, just around the time when Kapteyn presented his definitive model, some astronomers had started to doubt this hypothesis of complete transparency of the space between the stars. In many places on photographs of the Milky Way dark patches are visible, where the number of stars is much smaller than in their surroundings (Fig. 4.6). Astronomers were beginning to suspect that in these places there are dark clouds of particles that prevent the light from more distant stars from reaching us. Kapteyn himself was very much aware of this possibility, but since in his days there were no means for measuring the extinction of the star light by gas and dust, and thus for correcting for its effects, he was not able to take this into account in his model. The real proof for this interstellar extinction of star light by gas and dust was given only in the 1930s by Swiss-American astronomer Robert Trumpler, although German astronomer Johannes Hartmann had already in 1904 discovered interstellar absorption lines produced by atoms of Calcium, and in 1919 American astronomer Mary Lea Heger (after her marriage to later Lick Observatory director Donald Shane, her name became Mary Lea Shane) had discovered the interstellar absorption lines of the element Sodium in the spectra of distant stars. These discoveries already indicated that there are clouds of Calcium and Sodium atoms in the space between the stars.

Nowadays we know that in all directions in the plane of the Milky Way there is a "fog" of gas and dust that prevents us from observing stars more distant than about 20,000 light-years. Everybody knows that in fog one's sight is limited to a certain "sight distance", beyond which one can see nothing. If this sight distance is, for example, 300 ft, one cannot see anything beyond 300 ft, in all directions. So, in the fog, *everyone has the impression to be in the centre of the world.* If we walk away through the fog, this *sight circle* walks along with us: we all the time keep the impression that we are in the centre of the world. In the "interstellar fog" in the Milky Way this sight circle has a radius of 20,000 light-years in all directions, which led Kapteyn to the false impression that "we" (the sun) occupy a central position in the system.

Fig. 4.6 Dark and bright nebulae in the Milky Way in the region of the North-America Nebula in Cygnus. In the dark regions in the *lower-right* part of the figure far fewer stars are visible than in the bright regions in the *middle* and the *left* of this photograph: dark dust clouds in the *lower-right* block the light of the stars behind the nebula, while in the brighter parts the background stars can be seen out to large distances

Harlow Shapley and the System of Globular Star Clusters

Already in 1923 American astronomer Harlow Shapley (1885–1972, Fig. 4.7) had expressed his suspicion that the sun is not in the centre of the Milky Way system, and that the real centre is located some 50,000 light-years away in the direction of the constellation Sagittarius. He based this idea on the distribution in space of the so-called globular star clusters, shortly called *globular clusters*. Globular clusters are beautifully spherical collections of hundreds of thousands up to millions of stars (Fig. 4.8). They differ greatly from the so-called *open star clusters* like the Pleiades (Fig. 3.7), which have irregular shapes and in general contain only a few hundred to a few thousand stars. The many thousands of open star clusters are found in the band of the Milky Way, whereas the few hundred known globular clusters are mostly found just *outside* the band of the Milky Way. Shapley was able to determine the distances

Fig. 4.7 At right on this picture we see Harlow Shapley (below the lady with the hat). The picture was made in December 1923, during the meeting of the American Astronomical Society at Vassar College. In this picture also three young Dutch astronomers are visible: in the same row as the lady with the hat, the fifth from the *right* is Willem Luyten (1899–1994) and third from the *right* is Jan Oort (1900–1992), and somewhat lower, between Luyten and Oort: Piet van de Kamp (1901–1995). Luyten at that time already had obtained his PhD, Oort and van de Kamp not yet; they were working at Yale and Lick Observatory, respectively. Oort had been a student of Kapteyn, van de Kamp had studied in Utrecht with Nijland and had succeeded Oort as assistant of van Rhijn (Kapteyn's successor) in Groningen, before he left for the USA. Luyten did his bachelor studies in Amsterdam, and his masters and PhD in Leiden, with Hertzsprung. Luyten and van de Kamp both worked their entire further careers in the USA, while Oort returned to the Netherlands

of globular clusters by using a certain type of pulsating stars, called *Cepheids*. The brightness of these stars increases and decreases in very regular strictly periodic way (see Fig. 4.9). American astronomer Henriette Swan Leavitt (1868–1921) in 1908 had discovered that for these Cepheids there is a relation between their real *intrinsic* luminosity (the total amount of light emitted by the star per second) and the *length* of the pulsation period: the duration of one full pulsation. This relation is explained in Fig. 4.9. This means that by simply measuring the *length* of the pulsation period of the star, one knows how bright the star really is! By comparing this real luminosity of the star with how faint we observe the star, we can calculate the *distance* to the star. Since Cepheids are intrinsically very bright giant stars, these pulsating stars provide us with a very powerful way of measuring large stellar distances, and since these stars occur in globular clusters, they provide a nice way for measuring distances of these clusters. Shapley, a former journalist, working at Harvard University, measured the periods of the Cepheids in a large number of globular clusters, using the then new 60-in. reflecting telescope of Mount Wilson Observatory in California, where he worked at the

Fig. 4.8 Four characteristic globular star clusters. Clockwise, starting from the upper left one, their names are: M3, M5, M92 and M15. The letter M indicates that all of them are objects on the list of fuzzy objects collected by eighteenth century French astronomer Messier

invitation of George Ellery Hale, the founder of this observatory. When he plotted the distance of the globular clusters against the directions in which they are seen, he discovered that in space the clusters fill a more or less spherical volume, with a centre located in the plane of the Milky Way at a distance of about 50,000 light-years, in the direction of the constellation Sagittarius (Fig. 4.10). He therefore proposed the Milky Way system to be much larger than the Kapteyn system: about 200,000 light-years in diameter. Later it was found that Shapley had overestimated the distances, and that the sun is located about 30,000 light-years from the centre of the system of globular clusters, and the Milky Way system has a diameter of about 100,000 light-years. Shapley at that time thought that the Milky Way system was the entire universe.

Jan Oort Discovers the Motion of the Sun

In 1926 Shapley's model of the Milky Way system was confirmed in a completely independent way by a discovery of the then 26-year old Dutch astronomer Jan H. Oort (1900–1992), a former Groningen student of Kapteyn (see Figs. 4.7 and 4.11). Oort, who had just been appointed as staff member at Leiden Observatory by

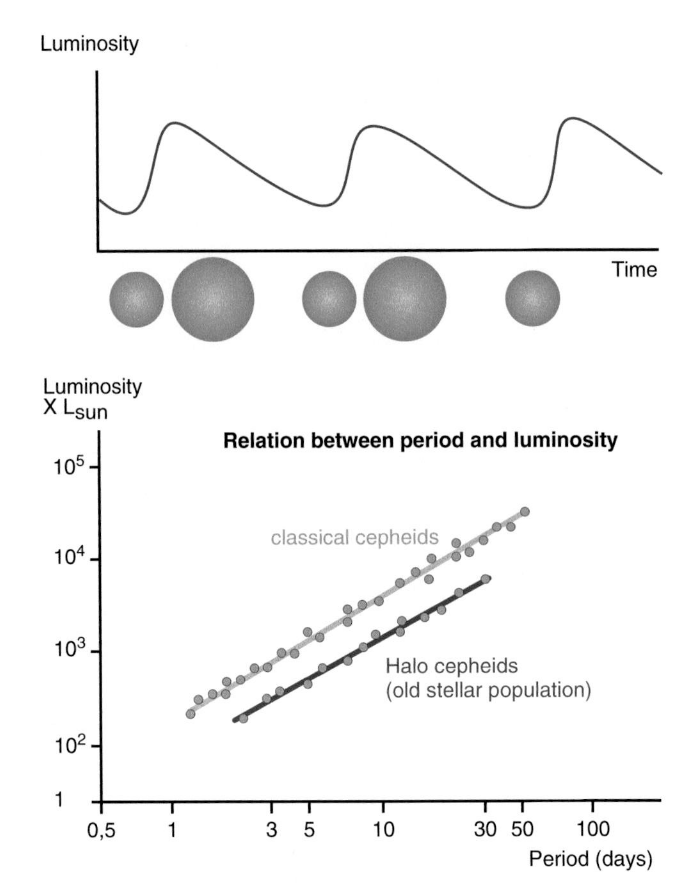

Fig. 4.9 *Upper frame*: characteristic light curve of a Cepheid-type pulsating star. The time between two consecutive light maxima is called the *pulsation period. Lower frame*: relation between luminosity of Cepheids (expressed in the luminosity of the sun L_{sun}) and their pulsation period. There appear to be two types of Cepheids: those in the disk of the Milky Way follow the upper relation, and those in the halo of the Milky Way (which includes the globular clusters) follow the lower relation. Cepheids of the latter type are also called W Virginis stars

its director Willem de Sitter (Kapteyn's first PhD student), discovered a systematic pattern in the space motions of stars in the sun's neighbourhood, which indicates that the sun and these stars move around a distant centre located in the direction of Sagittarius, just as Shapley had found for the globular clusters. Oort studied the so-called radial velocities and proper motions of nearby stars. The radial velocity is the speed with which a star moves towards us or away from us. It is measured by studying the shift in wavelength of lines in the spectra of stars, due to the so-called Doppler effect (explained in the next chapter, Fig. 5.9). Oort discovered that the stars we see in the direction of Sagittarius move in space in a direction just opposite to the direction of the motions of stars on the opposite side of the sky, 180° away

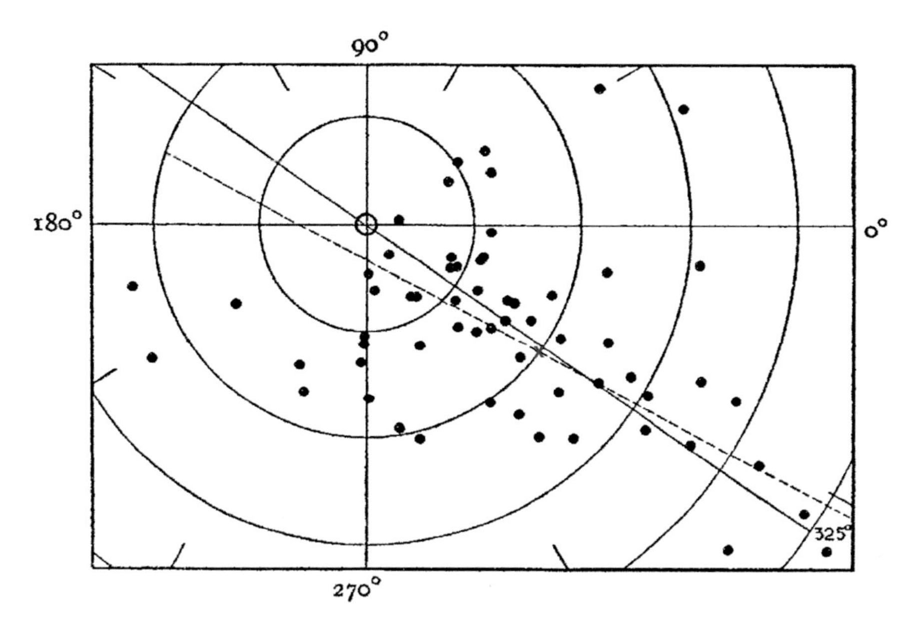

Fig. 4.10 Space distribution of globular clusters, projected on the plane of the Milky Way, as derived by Harlow Shapley. The black dots are the clusters. The sun is the centre of the circles. The centre of the distribution of the clusters is the red cross in the direction of Sagittarius

Fig. 4.11 Jan Oort and Hendrik van de Hulst at the reception for Oort's 40-year jubilee as staff member of Leiden University

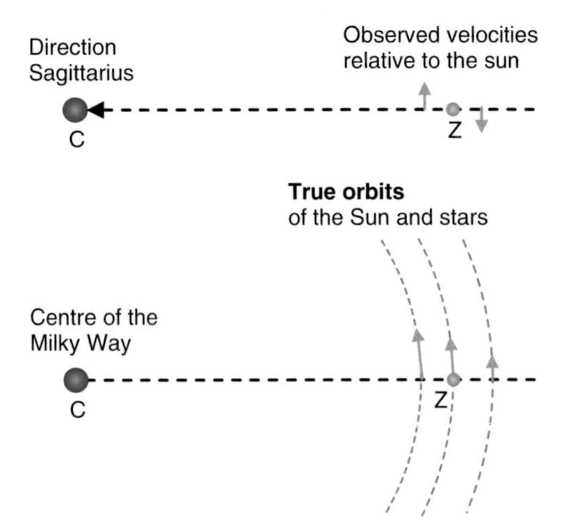

Fig. 4.12 *Upper frame*: motions of nearby stars in the Milky Way plane, relative to the sun, as discovered by Oort. The stars in the direction of the constellation Sagittarius move forward with respect to the sun, those in the direction opposite to Sagittarius appear to move backwards relative to the sun. *Lower frame*: the explanation given by Oort for these stellar motions relative to the sun: all stars, including the sun, describe orbits around a centre located at a large distance in the direction of Sagittarius. Like for the motions of the planets in the solar system, stars closer to the centre move faster in their orbits than stars further away from the centre

from Sagittarius, as depicted in the upper part of Fig. 4.12. At the same time, stars in directions perpendicular to the direction of Sagittarius appear to be standing still with respect to the sun. Oort interpreted these stellar motions as a *relative* effect, resulting from the fact that the sun itself is moving in the Milky Way plane with a speed of about 250 km/s perpendicular to the direction towards Sagittarius, while at the same time the stars in the direction of Sagittarius move *faster* than the sun, and the stars 180° away from Sagittarius move *slower* than the sun, as depicted in the lower part of Fig. 4.12.

The stars in the direction of Sagittarius are therefore overtaking the sun, while the sun itself overtakes stars in the opposite side of the sky, who stay behind with respect to us. This situation looks like the one in which one is driving in the middle lane of a three-lane freeway: the slower traffic in the right lane is overtaken by the traffic in the middle lane, while this traffic in turn is overtaken by cars in the fast left lane. The persons in the car in the middle lane see the traffic on their right moving backwards, and the ones in the left lane moving forward relative to themselves. Oort realized that the situation sketched in the lower part of Fig. 4.12 resembles that of the planets moving around the sun: planets closer to the centre move faster, and those farther from the centre move slower: Earth moves in its orbit slower than Venus but faster than Mars. Oort therefore concluded that all the stars in the Milky Way, including the sun, are describing orbits around a centre located far away in Sagittarius. Apparently, the bulk of the mass of the Milky Way system is located there, and forms a central mass concentration around which all other stars describe kepler-like orbits. From the measured velocities Oort derived that the distance to the galactic centre must be about 30,000 light-years. Oort's work confirmed the theoretical work of Swedish astronomer Bertil Lindblad, who somewhat earlier had proposed that the Milky Way system should show *differential rotation*, that means: the orbital velocities of the stars should decrease when going outwards from the centre, just as Oort discovered a year later.

Thus, since 1926 we know that the sun is not at rest in the centre of the Milky Way, but like billions of other stars describes an orbit around a distant centre located in the direction of Sagittarius. The systems of Herschel and Kapteyn represent only a small local part of the entire Milky Way system, which has a diameter of some 100,000 light-years and counts at least 100 billion stars. The sun completes one orbit around the centre in about 220 million years. In the 4.65 billion years that the solar system exists, it has completed some 21 orbits.

Radio Astronomy Shows Us the Entire Milky Way System

After the discoveries of Shapley and Oort the question arose: how do we find out how the rest of the Milky Way system, that is obscured from our view by interstellar dust and gas, looks like? That question could be answered only since the early 1950s, thanks to the new science of *radio astronomy*. Already before the second World War, American radio engineers Karl Jansky (1905–1950) and Grote Reber (1911–2002) had discovered radio waves coming from the sky. Radio waves are basically the same type of waves as light waves, the only difference being that their wavelengths are much longer. Both travel with the speed of light. Light waves have wavelengths between about 0.4 and 0.7 μm (1 μm is one-thousandth of a millimetre), while radio waves have wavelengths ranging from a few millimetres to several kilometers. Thanks to their long wavelengths, radio waves just bend around interstellar dust particles and raindrops and droplets of fog, such that interstellar clouds and fog are completely transparent for radio waves. This is the reason why ships and planes use radar (radio waves with wavelengths of centimetres) to see through clouds and fog.

Reber discovered in 1939 that the Milky Way itself is a source of radio waves. During the war Oort's university of Leiden was closed by the Germans, because it opposed the firing of Jewish professors. To escape from forced labour in Germany, Oort was in hiding at a farm in the central part of the Netherlands. There, in 1942, he wondered whether some atoms or molecules might be emitting radio waves of a fixed frequency, due to quantum jumps of electrons, just like spectral lines in the visible part of the light spectrum. If atoms in an interstellar cloud would emit waves of such a fixed frequency, one would be able to measure the velocity of this cloud by measuring the Doppler shift of the this frequency. (The police uses this frequency shift—so-called *Doppler-radar*—to measure the speed of cars and catch drivers for speeding). To answer Oort's question, Utrecht student Henk C. van de

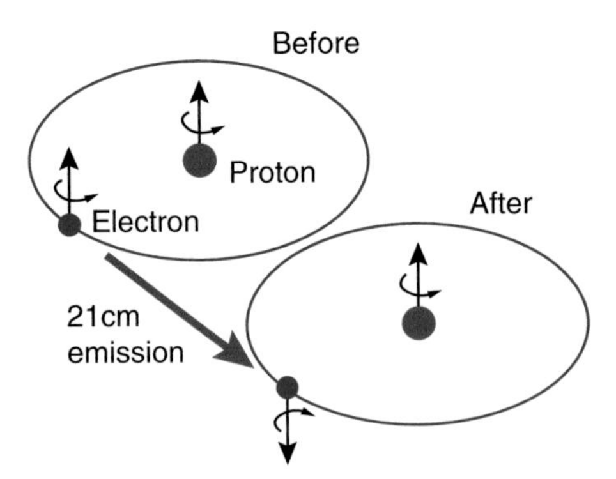

Fig. 4.13 A hydrogen atom emits radio waves when the spin of its electron reverses its direction. When the proton and electron turn in the same direction (*above*) the hydrogen atom has a slightly higher energy than when they turn in opposite directions (*below*). In 1944 Utrecht University student Hendrik van de Hulst calculated that the energy released at the reversal of the spin of the electron will be emitted in the form of radio waves with a wavelength of 21 cm. Since 70 % of all matter in the universe is hydrogen, this is the most important wavelength for radio studies of the universe

Hulst (1918–2000, Fig. 4.11) started in 1942 searching for energy transitions in atoms and molecules that might produce radio spectral lines. Hydrogen is the most common element in the universe, making up some 70 % of all mass. Van de Hulst discovered that atoms of hydrogen should emit radio waves with a wavelength of 21.2 cm. They do this when the direction of the rotation (so-called "spin") of the electron in the hydrogen atom is spontaneously reversed with respect to the spin of the proton-nucleus, as depicted in Fig. 4.13. Van de Hulst published his findings in the Dutch physics journal "Nederlands Tijdschrift Voor Natuurkunde" in 1944. In this article he predicted also a number of other radio lines, for example those of the OH-molecule, and of very high transitions in the hydrogen atom. (These were discovered several decades later with radio telescopes.)

Van de Hulst calculated that since hydrogen is the most abundant element in the universe, there are good chances to observe the 21.2 cm radio line of this element from interstellar clouds. Immediately after the war Oort and collaborators started building instruments for detecting this line. They used a former German radar dish of 7.5 m diameter, a so-called *Würtzburg antenna*, of which a number were left by the Germans at the Dutch coast. Van de Hulst's 1944 paper had been read also in the US and Australia, and there groups of physicists attempted to detect this line from the Milky Way. The first group to succeed in 1952 was that of Edward M. Purcell at Harvard University, a few months later followed by Oort, Muller and van de Hulst in the Netherlands, and the group of Christiansen in Australia. The results of the three groups were published that same year together in the same issue of the British journal Nature.

Already 2 years later, Oort and his collaborators published the first map of the locations of hydrogen clouds in the part of the Milky Way system observable from the Netherlands. The Southern part, not visible from the Netherlands, was filled in by the Australian colleagues. A few years later, the completion in 1956 of the 25-m

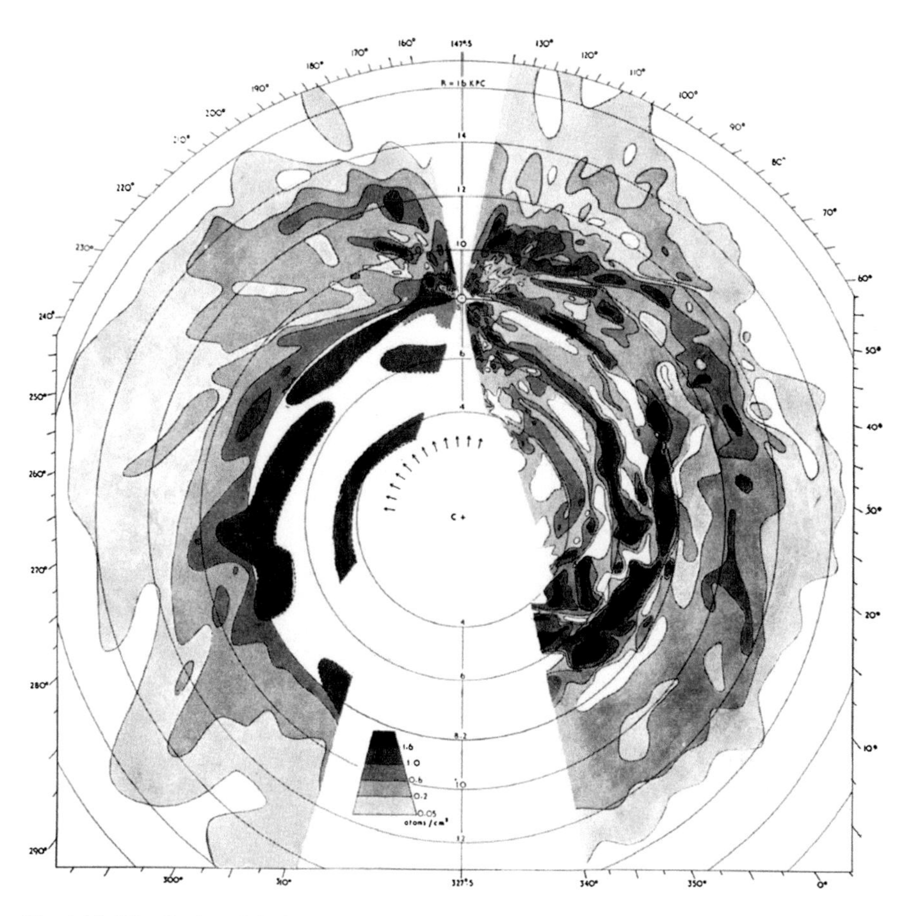

Fig. 4.14 Distribution of the locations of hydrogen clouds in the Milky way plane, derived in 1958 from hydrogen 21 cm line radio observations made with the Dwingeloo radio telescope in the Netherlands, in combination with radio telescopes in Australia. (It was not possible to determine the locations of the hydrogen clouds in the wedge-shaped white region, as in this region the clouds move perpendicular to the line of sight to the sun.) The centre of the picture is the centre of our Milky Way galaxy, and the small circle in the upper part of the figure is the location of the sun. The shades of darkness of the clouds indicates their density

diameter Dwingeloo radio telescope in the Northern part of the Netherlands allowed to present a much more refined map of the Milky Way system, depicted in Fig. 4.14.

In 1970 the Dwingeloo radio telescope was succeeded by the Westerbork Synthesis Radio Telescope (WSRT) near the village of Hooghalen in Northern Netherlands. This is an array of ten fixed and four movable parabolic radio dishes with a diameter of 25 m (82.5 ft), positioned in a 3 km (2 mi.) straight line (Fig. 4.15). This radio telescope, inaugurated by Dutch Queen Juliana in June 1970, was for 10 years the world's largest and most powerful radio telescope. In 1980 it lost its first place to the Very Large Array (VLA) in New Mexico, which consists of twice

Fig. 4.15 The Westerbork Synthesis Radio Telescope (WSRT) was inaugurated in 1970. It consists of 10 fixed-location and four movable dishes of 25 m diameter each, stretched along a baseline of 3 km (about 2 mi.). Although since 1980 larger radio arrays have been constructed in the USA and India, in certain wavelength ranges the WSRT still is the most sensitive radio telescope in the world

as many dishes of the same size. In a synthesis radio telescope the signals received by all dishes are combined, such that the telescope works as *one large telescope* with a mirror area equal to the sum of the areas of the individual dishes. The largest and most sensitive synthesis radio telescope in the world nowadays is the Giant Meter-wavelength Radio Telescope (GMRT) in India, near Pune, which consists of 28 dishes with a diameter of 45 m each. Its total collecting area is about 30,000 square meters (about 6.6 acres).

During the 1970s the WSRT pioneered the studies of the structure of other galaxies and of galaxy clusters, and here made breakthrough contributions, for example, to the discovery of the Dark Matter content of galaxies (see Chap. 13). Thanks to continuous upgrades of focal plane instruments, receivers, computers and software the WSRT still is one of the most powerful radio telescopes in the world and continues to make many new discoveries, largely in extragalactic studies, but also in the studies of Milky Way radio sources, such as star-forming regions, pulsars (neutron stars) and X-ray sources.

Figure 4.16 depicts schematically the structure of the Milky Way system derived from radio observations, combined with observations in visible and infrared light. The lower frame of this figure depicts the spatial positions of the sun, globular star clusters and a variety of other objects.

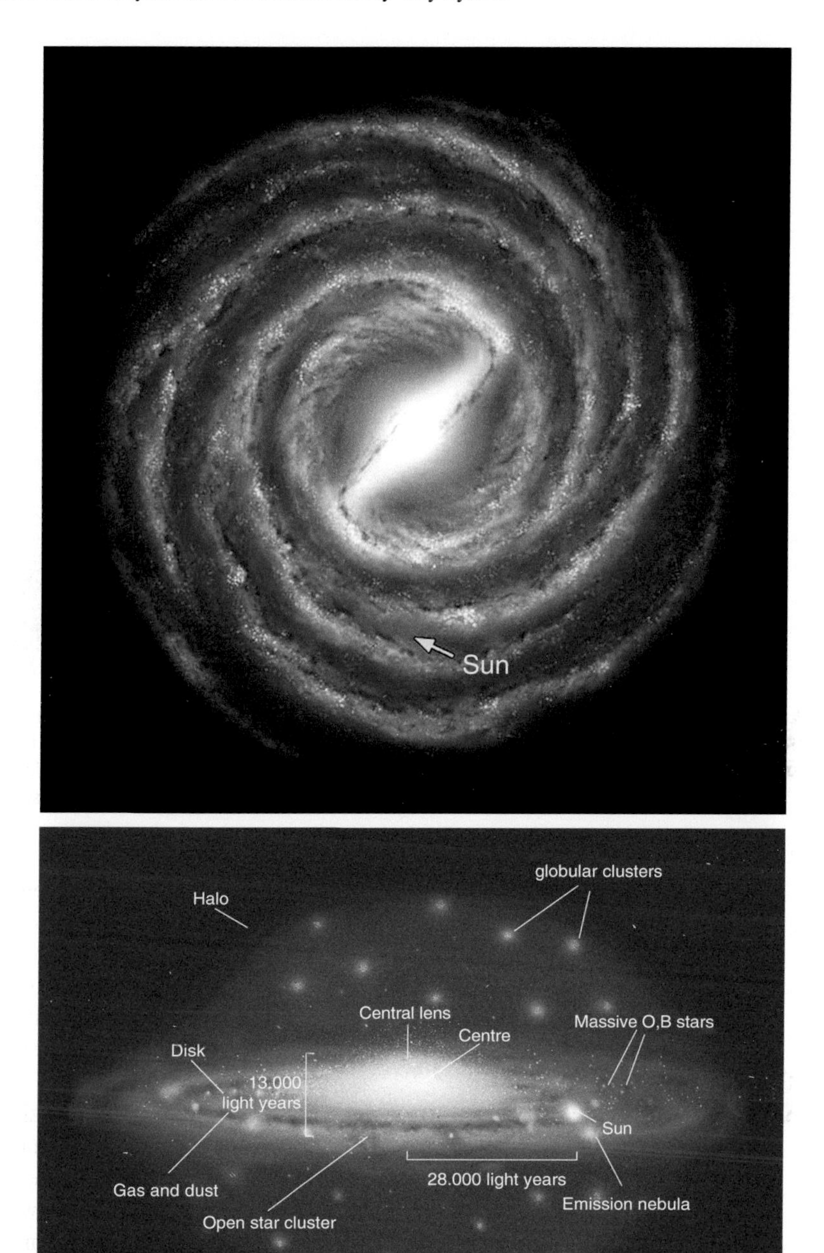

Fig. 4.16 *Upper figure*: Spiral structure of the Milky Way derived from radio and infrared observations. The lens-shaped inner part of our Galaxy has been found to be a tri-axial ellipsoid. This means that our Galaxy belongs to the category of barred spiral galaxies. *Lower figure*: Schematic model of our Milky Way system, in which the various components of the system, including the globular clusters, are indicated. (For clarity the brightness of the sun is highly exaggerated)

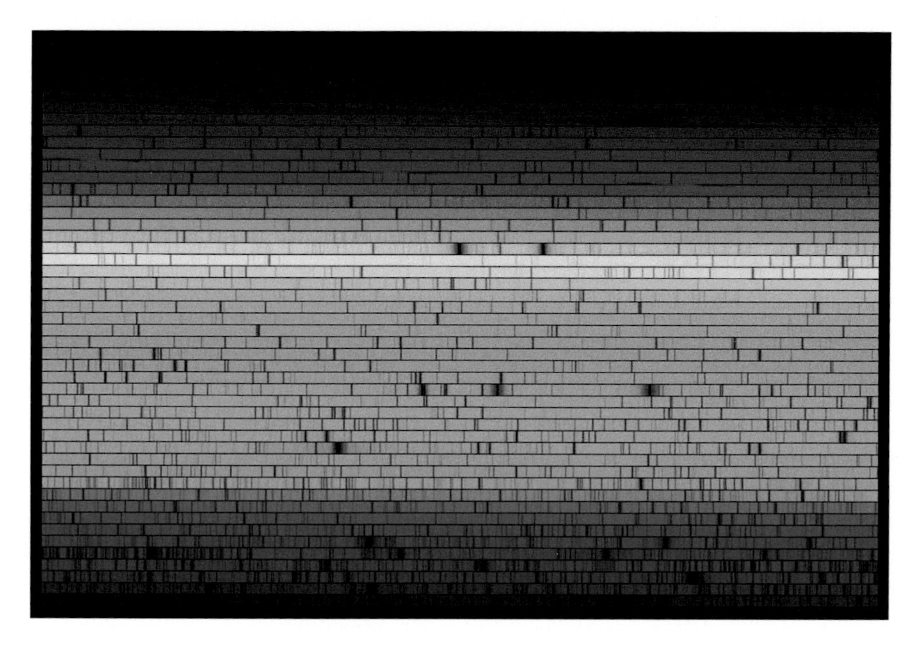

Spectrum

Chapter 5
The Chemical Composition of the Sun and Stars

There is nothing so absurd that some philosopher has not already said it

Marcus Tullius Cicero (106–43 BC), Roman philosopher

French philosopher Auguste Comte (1798–1857), founder of positivism and of the science of sociology, wrote about astronomy in his 1835 book about the sciences: "We will never be able to determine the chemical composition of the stars, nor their density and temperature." Only 25 years later German physicists Gustav Kirchhoff (1824–1887) and Robert Bunsen (1811–1899) discovered that by the analysis of the light of the sun and stars it is very well possible to determine of what substances these objects are composed. And nowadays also the densities and temperatures of stars and other celestial bodies can be determined with high accuracy.

This was not the first time that a philosopher went completely wrong in the field of astronomy. Another famous example is German philosopher Georg Hegel (1770–1831), the teacher of Karl Marx. In his Ph.D. thesis, Hegel made fun of the European astronomers, who near the end of the eighteenth century had set up a collaboration

E. van den Heuvel, *The Amazing Unity of the Universe*,
Astronomers' Universe, DOI 10.1007/978-3-319-23543-1_5

to carry out a systematic search for a planet between Mars and Jupiter. The reason for this campaign, in which each observatory was assigned a part of the Zodiac for searching for this planet, was that the sizes of the orbits of the planets form a geometrical series, called "Bode's law"; in this series, however, a planet between Mars and Jupiter is missing. Until the discovery in 1781 of the planet Uranus, astronomers had not paid much attention to this "law", and had thought that it might be just a coincidence. However, since also Uranus was found to fit exactly in this geometrical series, it became clear to the astronomers that that series seems to have a real significance, such that the absence of a planet between Mars and Jupiter became a serious matter.

Hegel, however, "proved" in his Ph.D. thesis "by logical reasoning" that no planet can exist between Mars and Jupiter. The printing ink of his thesis had hardly dried when on the first of January 1801 (the first day of the nineteenth century!), Sicilian priest-astronomer Father Guiseppe Piazzi discovered the first asteroid Ceres, which is orbiting between Mars and Jupiter at exactly the distance predicted by Bodes "law". Soon after that, also other asteroids were discovered, all orbiting at about the same distance from the sun. Hegel's blunder still gives astronomers a lot of fun, and makes them sceptical about other philosophical stuff he wrote, such as that the Prussian State is the highest form a state can attain.

Kirchhoff and Bunsen's Discovery

Kirchhoff and Bunsen made their discovery around 1860 by comparing the laboratory spectra of flames with the solar spectrum. A spectrum is the rainbow-like band of colours that appears when white light, such as that of the sun, has traversed a glass prism (Fig. 5.1). Already in the seventeenth century Isaac Newton had studied this band of colours, and shortly after the year 1800 English physician William Wollaston (1766–1828) discovered that when one uses a prism that spreads the colours very widely, dark lines become visible in the colour bands of the solar spectrum (Fig. 5.2). Around the same time it was discovered that light is a wave

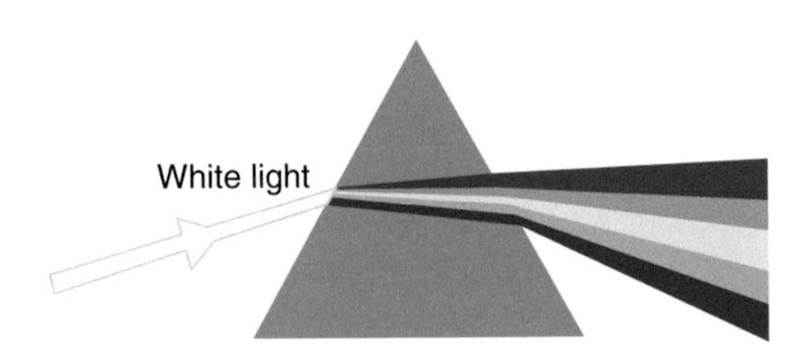

Fig. 5.1 Refraction by a prism causes white light to be spread out into a rainbow-like band of colours. Light is a wave phenomenon: *violet* and *blue* light have shorter wavelengths than *green* and *red* light. The shorter the wavelength, the stronger the refraction by the glass of the prism

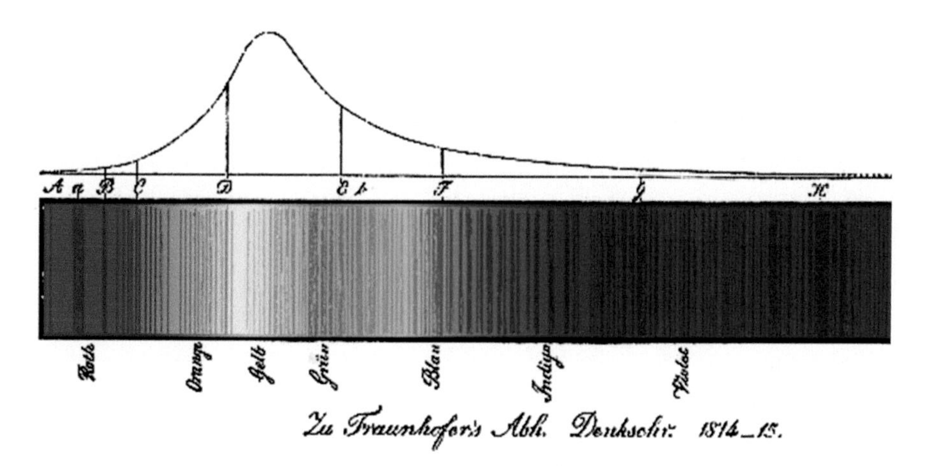

Fig. 5.2 Fraunhofer's solar spectrum in which the strongest dark lines are indicated by capital letters

phenomenon, and that the colour is a measure for the wavelength of the light: violet light has the shortest wavelength, and red—at the other end of the spectrum—the longest. Somewhat later, German optical instrument building genius Joseph von Fraunhofer, whom we already encountered in Chap. 3, measured the wavelengths of the strongest dark lines in the solar spectrum and named them with the letters of the alphabet: A, B, C, D1, D2,...etc. But nobody knew what these dark lines meant. The great breakthrough came in 1859/1860 when Kirchhoff and Bunsen discovered that if a small amount of a certain chemical substance is brought into a flame, bright lines appear in the spectrum of this flame, which are characteristic for only this substance, and different for any other substance. For example, when kitchen-salt (sodium-chloride) is added to the flame, two bright yellow-orange lines appear in the spectrum, that coincide in wavelength with Fraunhofer's dark lines D1 and D2 in the solar spectrum (upper frame of Fig. 5.3).

From this they concluded that the sun contains sodium (Na), and they found the same for a number of other elements, for example, the Fraunhofer H and K lines are due to ions of Calcium (Ca+). Also, Kirhhoff and Bunsen discovered that when one shines white light *through* a flame that contains kitchen salt and then takes the spectrum of this light, two dark lines appear, which exactly coincide with the dark D1 and D2 lines in the solar spectrum (lower frame of Fig. 5.3). From this they concluded that the white solar light must originate deep inside the sun, and that in the atmosphere of the sun there are atoms of sodium and other elements, which absorb the light on its way outwards through the solar atmosphere, and thus produce the dark lines.

Each of the elements in the periodic table has its own spectrum of lines, located at fixed wavelengths, that are characteristic for *just that element*, and different from those of the other elements (Fig. 5.4).

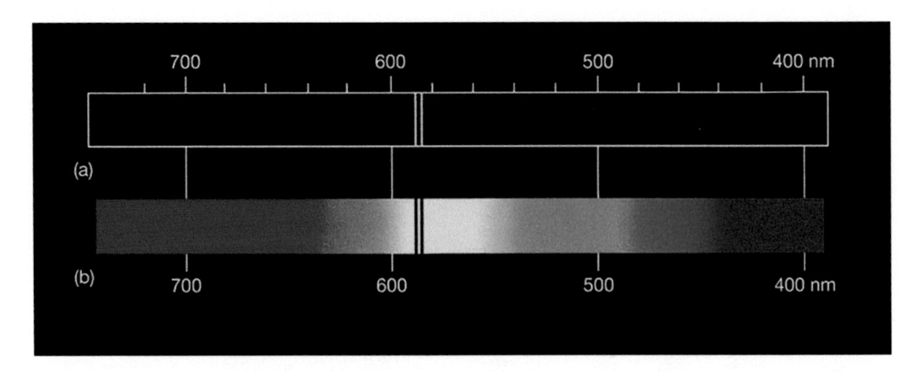

Fig. 5.3 *Upper*: bright emission lines in the spectrum of a flame into which sodium has been injected. *Lower*: In the spectrum of *white* light that shines through a flame with sodium, two dark absorption lines appear at the same wavelengths as of the two emission lines in the upper spectrum

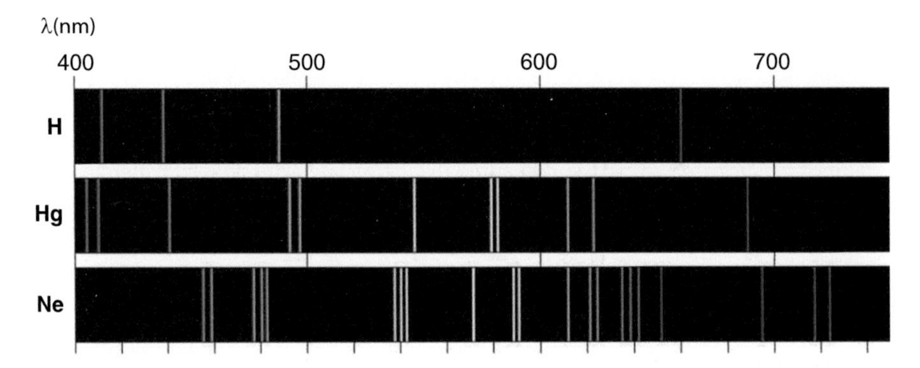

Fig. 5.4 Emission spectra of hydrogen (H), mercury (Hg) and neon (Ne). The wavelength scale is in nanometres: 1 nm is one millionth of a millimetre (and one billionth of a metre)

Under the guidance of professor Marcel Minnaert (1893–1970; Fig. 5.5) of Utrecht Observatory, thousands of lines in the visible solar spectrum were identified in the period between 1930 and 1960. In this way more than 60 of the 90 stable elements that are present on Earth were found in the sun. The remaining elements have no lines in the visible spectrum, but only in the ultraviolet or the infrared, types of light that are blocked by the Earth's atmosphere. After the start of space exploration, these elements were identified in the ultraviolet and infrared spectrum of the sun, by using spectrographs in satellites. It thus was found that all elements known on Earth are present in the sun. The element helium is a special case, as it was first found on the sun before it was found on Earth. This came about in the second half of the nineteenth century, when during a solar eclipse in India, British astronomer Norman Lockyer (1836–1920) and French astronomer Jules Janssen (1824–1907) discovered lines of an unknown element in the spectrum of solar prominences (structures

Fig. 5.5 Utrecht Professor M.G.J. Minnaert (1893–1970), who discovered the method for determining the amounts of the different elements in stars from the strengths of the absorption lines in their spectra. Independently, this method was also discovered by Harvard University's Donald Menzel. Minnaert also introduced the concept of *Equivalent Width*, which is a quantitative measure for the strength of a spectral line

Fig. 5.6 The spectra of the bright stars Sirius and Canopus both show strong absorption lines of hydrogen (see also Fig. 5.4)

in the solar atmosphere, which during a total solar eclipse stick out beyond the edge of the moon). They called the unknown element *helium* (after the Greek sun god Helios). Later, this element, a noble gas, was also discovered on Earth: it is very light and used for filling balloons, and for cooling objects to a very low temperature, where helium becomes liquid. We thus see that the sun contains all elements known on Earth, no less and no more.

In the second half of the nineteenth century, English amateur astronomer William Huggins (1824–1910) discovered that also the spectra of stars contain the same lines. As an example, Fig. 5.6 shows the spectra of the bright stars Sirius and

Canopus, which clearly show the lines of hydrogen. From the 1930s on, physicists in their laboratories started to make radioactive elements which, due to their limited lifetimes, are not found on Earth in nature. Examples are the *trans-uranic* elements, with atomic numbers beyond 92, Plutonium (which is used in nuclear weapons), Californium and Berkelium, as well as the elements Prometheum (element 61) and Technetium (element 43). The latter two elements occupy positions in the "normal" periodic table of the 92 elements through Uranium (element 92). However, in this table till the 1930s there were vacant places ("holes") at their positions, as they never had been found on Earth in nature. They appear to have no long-lived isotopes, and even if they had been around when Earth was born 4.65 billion years ago, they would since long have decayed. With the exception of Technetium, none of these unstable elements has been found in stars. The longest-lived isotope of Technetium has a half-life of 2.6 million years. In the 1950s lines of this element were found in the spectra of a certain type of red giant stars, the so-called S-stars. These stars appear to be able to produce Technetium by neutron-producing fusion reactions deep in their interior. On Earth short-lived isotopes of Technetium are nowadays continuously produced in nuclear reactors, for use in medical applications.

Formation of Spectral Lines: Quantum Jumps of Electrons in Atoms

In the early decades of the twentieth century New-Zealand-born experimental physicist Ernest Rutherford (1871–1937) and Danish theorist Niels Bohr (1885–1962) unravelled the structure of atoms. We since know that an atom consists of an electrically-positive charged nucleus, around which negatively charged electrons describe their orbits (Fig. 5.7). The nucleus consists of positively charged *protons* and almost equally massive *neutrons*, which have no electric charge. A proton and a neutron are both about 1840 times more massive than an electron. The negative charge of the electron has the same 'absolute' (intrinsic) value as the positive charge of the proton. The number of electrons orbiting the nucleus is equal to the number of protons in the nucleus, such that the atom as a whole has no electric charge. Bohr discovered that for an electron in an atom only discrete orbits are allowed, and no orbits in between. Each orbit corresponds to a certain fixed energy of the electron, which is different for the different orbits. When an electron drops from a higher to a lower orbit, the difference in energy E_2-E_1 between these orbits (numbered 1 and 2) is radiated away as light of the frequency ν_{12} given by $h\nu_{12}=(E_2-E_1)$, where h is Planck's constant (see first box in Chap. 3). The result is an emission line in the spectrum with a wavelength λ_{12} corresponding to this frequency, given by: $\lambda_{12}=c/\nu_{12}$. Also the reverse occurs. When white light (white meaning: with radiation of all kinds of wavelengths) falls on an electron in an orbit in the atom with energy E_1, the electron will be able to take up the energy corresponding to the wavelength λ_{12} and

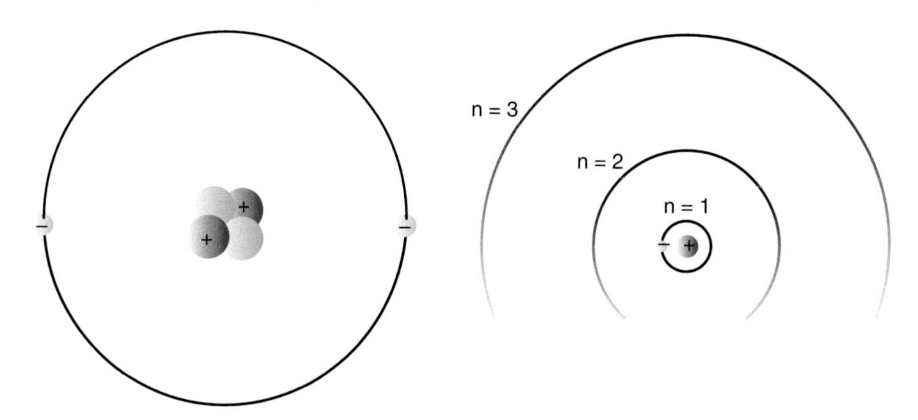

Fig. 5.7 *Left*: According to the Rutherford model, an atom consists of a small nucleus and electrons moving in wide orbits around this nucleus. This is depicted here for the atoms of helium (further explanation in the text). *Right*: according to Niels Bohr, the electrons are located in quantized orbits, with energies that are higher when the orbits are wider. Each quantized orbit corresponds to a quantized energy level of the electron: electrons can jump only from one such orbit to another one, transitions to energies between the quantized levels are not possible

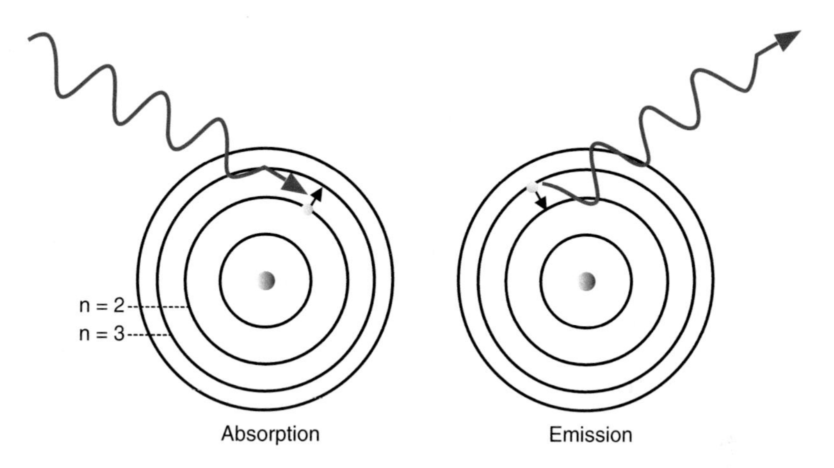

Fig. 5.8 When light shines on an atom, the atom can absorb an energy quantum, such that the electron jumps from a lower to a higher energy level. When the electron falls back from a higher level to a lower energy level, a light quantum of the same energy and wavelength is emitted

jump to the orbit with energy level E_2. In that case the light of this wavelength disappears from the white light, and a *dark line* ('absorption line') will appear at this wavelength in the colour band of the spectrum of the white light (Fig. 5.8). This is how the dark *Fraunhofer lines* in the solar spectrum are formed.

Astronomical Applications of the Doppler Effect

When a star moves towards us or away from us, the wavelength of every line in its spectrum shifts slightly towards the blue or the red, respectively. This is due to the so-called Doppler effect, discovered by Czech physicist Christian Doppler, who described this effect for the first time in 1842. Everyone knows this effect for the case of sound. When a fire truck with its loud siren is approaching us with high speed, we hear a higher sound than then when it is standing still, and after the truck has passed us we suddenly hear a lower sound, as explained in Fig. 5.9. Phrased differently: when the truck is approaching us, the wavelengths of the sound that we hear are shorter than the wavelengths emitted by the truck, while when it is moving away from us, the wavelengths that we hear are longer than those emitted by the truck. Since light also is a wave phenomenon, the same effect occurs for light waves. When a star approaches us, the wavelengths of the lines which we observe in its spectrum are slightly shorter than the true wavelengths emitted by the star, whereas when the star is moving away from us, the wavelengths that we observe for the lines are slightly longer than those emitted by the star. In other words: when the star approaches us, the lines in its spectrum are shifted towards the blue, and when it moves away from us, they are shifted towards the red. Astronomers then say that the lines show a *blueshift* or a *redshift*, respectively.

The shift in wavelength $\Delta\lambda$ divided by the true wavelength λ of the spectral line equals the velocity v of the star divided by the velocity of light c : $\Delta\lambda/\lambda = v/c$. This relation holds as long as the velocity v is much smaller than c. By measuring the wavelength shift $\Delta\lambda$ we therefore can determine the velocity v of the star or galaxy (as the true wavelength λ of the line and the velocity of light c are known). The thus measured velocity towards us or away from us of the star or galaxy is called its *radial velocity*. Throughout the year every star shows a regularly changing shift of the wavelengths of its spectral lines. This is due to the motion of Earth around the sun with a velocity of about 30 km/s. If one wishes to find the real velocity of a star towards or away from the sun, this Doppler effect due to the motion of Earth around the sun must be subtracted from its measured radial velocity.

(continued)

(continued)

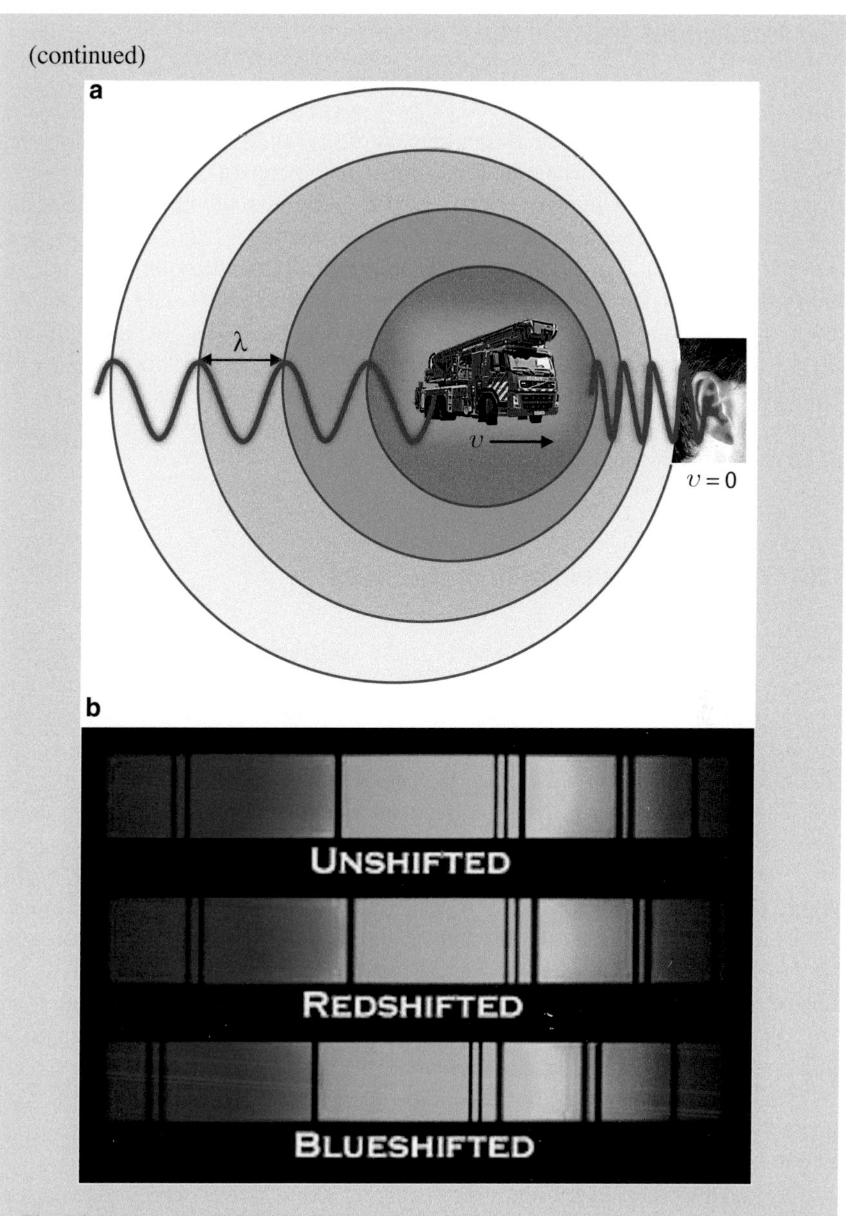

Fig. 5.9 (**a**) Due to the Doppler effect the sound of an approaching car has a higher tone (shorter wavelength) than the sound of a car that is moving away from the observer. (**b**) When the wavelength of light increases, the light becomes redder, when the wavelength decreases, it becomes bluer. As a result, for stars or galaxies that move away from us, the lines in their spectra are shifted towards the red, whereas when they move towards us, the lines are shifted towards the *blue*

The Miraculous Unity of the Universe

The fact that in the spectra of stars and galaxies out to the most distant corners of the universe one only observes the lines of the elements familiar to us from Earth and the sun, shows that everywhere in the universe matter consists of exactly the same kinds of atoms. It thus appears that matter everywhere in the universe obeys the same laws of atomic and nuclear physics that were discovered in laboratory experiments here on Earth. It is this great cosmic unity that allows us to study the universe and to unravel the birth and evolution of the sun, stars, planets and galaxies, by simply applying the laws of physics discovered here on Earth. Nowadays, every physicist seems to assume this as self-evident (one may call this "the arrogance of the physicist"), but without the painstaking and careful work of a large number of astronomers studying the spectra of the stars and galaxies during the past one-and-a-half century, we would not have known this.

The Chemical Composition of the Stars

The presence of the lines of an element in the spectrum of star shows us that that element is present in the star, but it does not yet tell us *how much* of that element is present. One would think, at first glance, that the strength (thickness) of the absorption line in the spectrum can be used as a measure of the amount of the element present in the atmosphere of the star. However, this has been found not true in practice. The situation is much more complex, and it turns out that the strength of an absorption line is determined by a large number of factors, the main ones being the temperature and pressure in the stellar atmosphere, and the precise values of the atomic parameters characteristic for the energy levels forming the line. The *amount* of the element present in the stellar atmosphere plays only a minor role in determining the strength of the absorption line. For example, the enormous strength of the ionized Calcium H and K lines in the solar spectrum might suggest that there is an enormous amount of Calcium present in the sun, but it turns out this strength is largely due to the temperature and pressure in the solar atmosphere, combined with the special parameters of the corresponding energy levels of the Calcium ion. The amount of Calcium in the sun turns out to be very modest, and in relative measure with respect to other elements not different from that on Earth.

There is, therefore, a large amount of complex atomic physics required to derive, from the strength of an absorption line, the amount of the element present in the stellar atmosphere. In 1925, British-born Harvard astronomer Cecilia Payne-Gaposchkin (1900–1979) discovered, from a study of the properties of the lines of hydrogen, helium and other elements in stellar spectra, that stars consist for the largest part—some 98 % of their mass—of hydrogen and helium, and that the other some 90 elements together make up at most a few per cent of the mass of the stars.

The method for deriving the relative amounts of these other elements from the strengths of their spectral lines we thank largely to two persons: Marcel Minnaert of Utrecht University, born in Bruges in Flandren and Harvard astronomer Donald

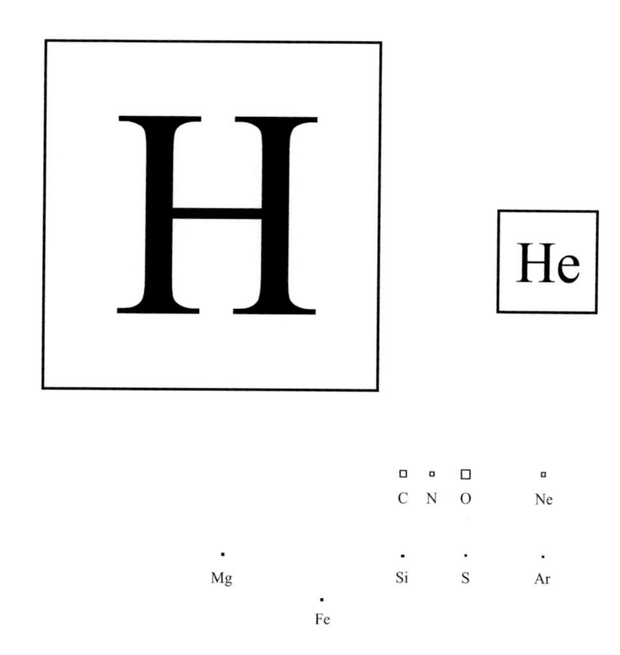

Fig. 5.10 The 'astronomical periodic system' (after Ben McCall) is a schematic representation of the relative amounts of the different elements in the universe. Hydrogen and Helium dominate everything, and are followed only at a distance by the lighter elements oxygen (O), carbon (C), nitrogen (N) and neon (Ne)

Mentzel (1901–1976). In the 1930s they independently developed the so-called "curve of growth" method (the name was coined by Minnaert and illustrates his background as a biologist, later turned physicist—with a doctor degrees in both these sciences). The curve of growth shows how for a stellar atmosphere with a given temperature and pressure, the strength of an absorption line of an element increases when the number of atoms or ions of that element per unit volume (e.g., per cubic metre) in the stellar atmosphere is increased. The theory of the curve of growth incorporates a lot of atomic physics. Much of this physics for stellar atmospheres was developed in the 1930s by German astrophysicist Albrecht Unsöld of Kiel, with whom Minnaert closely collaborated. Measuring the strength of the line, and then comparing it to the theoretical curve of growth allows one to read from this curve the number of atoms per unit volume in the stellar atmosphere. The curve of growth method was originally developed by Minnaert and Menzel for the sun, but later applied by many astronomers to stars.

As Minnaert's last master student, in 1961 I applied this method to study the turbulent motions in the solar atmosphere (turbulence also affects the strength of absorption lines).

Present-day work on the determination of the chemical constitution of stars, the interstellar medium and galaxies is all based on this fundamental research of Payne-Gaposchkin, Minnaert, Unsöld and Mentzel. This research has shown that everywhere in the universe matter consists for some 98 % of its mass (or more), of hydrogen and helium (some 70 % hydrogen and 28 % helium). All heavier elements together constitute at most a few per cent of the mass in the universe. Figure 5.10 schematically represents this "cosmic composition" of matter. After Hydrogen and Helium, the most abundant elements are Oxygen, Carbon and Nitrogen, followed by Argon, Magnesium, Silicon, Sulphur, Phosphorus and Iron.

Galaxy NGC 1672

Chapter 6
Other Galaxies and the Discovery of the Expansion of the Universe

In order to know where something is, one first has to find it

Johann Wolfgang von Goethe (1749–1832), German writer and philosopher

In the eighteenth century French comet hunter Charles Messier completed the first catalogue of nebulous objects in the sky. Seen through a small telescope, these hazy cloudlets look much like a comet. They gave Messier much trouble in his hunt for comets, so he decided to once and for all make a list of them such that they could no longer confuse him. An object in his catalogue is indicated by capital letter M followed by a number. M1 is the Crab Nebula in the constellation Taurus: the remnant of a supernova (an exploding star), and M42 is the Orion Nebula, a gaseous nebula in this constellation. One also finds in this catalogue globular star clusters, such as M3, M4, M5, M10, M13, M15 and M92, and open star clusters such as M6, M18, M21 and M67. Some of the "nebulae" in Messier's catalogue have, seen through a

E. van den Heuvel, *The Amazing Unity of the Universe*,
Astronomers' Universe, DOI 10.1007/978-3-319-23543-1_6

Fig. 6.1 The Andromeda Nebula (M31), at a distance of 2.5 million light years, is the nearest large neighbour galaxy of our Milky Way system. All the stars on this picture are foreground stars of our own Milky Way, the Andromeda galaxy is far behind them and individual stars cannot be recognized in it

larger telescope, a spiral shape, for example M31 in the constellation Andromeda (Fig. 6.1), M33 in the Triangle and M51 in Canes Venatici (Fig. 6.3a). The Andromeda Nebula M31 is the only spiral nebula visible to the naked eye. The first one to mention this hazy object, in the year 964 in his "Book of the fixed stars" was Persian astronomer Abd-al Rahman Al Sufi, who described the nebula as a *little cloud*. This "little cloud" is the very central bright part of the nebula, and one must have good eyes to see it. It is the most distant object visible to the naked eye.

For almost two centuries since the discovery in the eighteenth century of the spiral structure of these nebulae, it was unclear what they really are. William Herschel first thought that they are other galaxies like our own, but later he changed his mind and thought they are gaseous nebulae in our own Milky Way. Already before him, German philosopher Immanuel Kant (1724–1804) and British minister-geologist John Mitchell (1724–1793) had proposed that they are "island universes" resembling our Milky Way system. Both had read the book of Thomas Wright about our Milky Way, in which he had proposed that many faint nebular spots in the sky are very distant milky way systems. Kant put forward his "island universe" hypothesis in his book *General Natural History and Theory of the Celestial Bodies*.

The debate about what these nebulae really are lasted until 1923. In that year American astronomer Edwin Hubble (Fig. 6.2) pointed the 100-inch reflecting telescope on Mount Wilson at the Andromeda Nebula. This giant telescope with a parabolic-shaped glass mirror of 100 in. diameter and a total weight of 300 t had been completed in 1917 and remained the world's largest telescope till after the second World War. It had been funded by Los Angeles business man John D. Hooker

Fig. 6.2 *Right*: Between 1917 and 1948 Mount Wilson's Hooker telescope with its primary mirror of 100 in. (250 cm) diameter was the world largest telescope. It was designed and built under the guidance of George Ellery Hale. With this instrument Edwin Hubble (1889–1953; *Left-hand picture*) discovered the existence of other galaxies than our own, as well as the expansion of the universe

and carries his name. During the First World War Hubble, who after completing a degree in physics had obtained a law degree as Rhodes Scholar in Oxford University in England, served as army captain in France. As a result he later was often referred to as "the captain". After the war he came to Mount Wilson at the invitation of George Ellery Hale, who had taken the initiative for building the Mount Wilson Observatory, and had raised the funds for successfully realizing this project.

With this new telescope Hubble discovered that the Andromeda Nebula is not a gaseous nebula but consists of stars. He discovered in this nebula the same pulsating Cepheid stars which had allowed Shapley to measure the distances of globular clusters (see Fig. 4.9). Cepheids are giant stars which emit per second over 10,000 times more light than our sun, and therefore are observable out to very large distances. The Cepheids which Hubble found in the Andromeda Nebula are extremely faint—much fainter than those in the globular clusters. This means that they are much further away than the globular clusters—the most distant of which, according to Shapley, were 100,000 (one hundred thousand) light-years away. Since the pulsation period of the Cepheid directly gives one the real (intrinsic) luminosity of this star, one finds its distance by comparing its observed brightness with its intrinsic luminosity, as explained in Chap. 4. In this way Hubble determined the distance of the Andromeda Nebula to be about half a million (500,000) light-years, which implies that it is located far outside our Milky Way system. As on the sky the Andromeda Nebula has a longest diameter of 3° (six times the diameter of the full moon), it followed that the nebula must have a diameter of about 30,000 light-years—slightly larger than the Kapteyn model of the Milky Way. Hubble therefore concluded that the Andromeda Nebula is a milky way system resembling our own Milky Way, and that the same must be true for the other "spiral nebulae", such as M33 and M51 (Fig. 6.3a).

Much later, after the second World War, it was discovered that there are two types of Cepheids which, at the same pulsation period, differ considerably in their real (intrinsic) brightness. The ones that can just be seen in the Andromeda Nebula

Fig. 6.3 (**a**) Spiral galaxy M51 in Canes Venatici at a distance of 35 million light years gives a good impression of how our Milky Way Galaxy might look from a distance, if we look perpendicular to the Milky Way plane. (**b**) Spiral galaxy NGC4565, at about the same distance as M51, is seen edge-on. This gives a good impression of how our Galaxy would look if seen by a distant observer in the Milky Way plane. One clearly notices the band of gas and dust in the plane of this galaxy, closely resembling the band of gas and dust in the plane of our Milky Way. (**c**) Spiral galaxy NGC 1232 at a distance of 65 million light years

Fig. 6.3 (continued)

are the brightest of these two types, but the ones Hubble knew in his time, and had used for his distance estimates, happened to be the faintest of the two types. As a result he had severely *underestimated* the distance to the Andromeda Nebula; the real distance happens to be about 2.5 million light-years, some five times larger than

Hubble had thought. This means that its diameter is about 150,000 light-years instead of 30,000 light-years. It contains about twice as many stars as our Galaxy, and its total mass, including stars, gas and "dark matter"(see Chap. 13) is estimated to be close to one trillion (10^{12}) solar masses.

An observer in the Andromeda Nebula looking at our own Galaxy will at this moment observe the light that left our Galaxy 2.5 million years ago. With a super-telescope he/she will see from there that on Earth our apelike ancestors in Africa are just starting to pick up rocks to make primitive stone-age tools.

Distances to Other Galaxies

After he discovered that the Andromeda Nebula is a galaxy like our own, Hubble started a systematic study of the spiral nebulae and other galaxies. He measured their distances by using a variety of objects of which he knew the real (intrinsic) luminosity: globular star clusters, pulsating stars, and exploding stars called *novae* and later also *supernovae*. The last ones are dealt with in more detail in Chap. 13; they can during several weeks reach a luminosity a billion times that of our sun—comparable to the luminosity of an entire galaxy. They therefore can be used to determine distances of galaxies up to several billion light-years. Novae can at maximum brightness be some 20,000 times more luminous than our sun, and can be used to determine distances out to some five million light-years.

The different methods for distance determination show that the spiral galaxy M51 (Fig. 6.3a) has a distance of about 35 million light-years, just like the galaxy NGC4565 (Fig. 6.3b) in the constellation Coma Berenices (The Hair of Berenice). This galaxy we observe precisely edge-on, such that we clearly see the layer of gas and dust located in the symmetry plane of the system, just as in the plane of our Milky Way. These two galaxies give a good impression of how our Galaxy would look when seen from a distance.

Hubble and his colleague Fritz Zwicky discovered also that galaxies tend to live in groups. These can be small, like Stephan's Quintet (Fig. 6.4) and the Local Group, formed by our Galaxy, the Magellanic Clouds, the Andromeda Nebula and its satellites M32 and NGC 205, the spiral M33 and a few tens other nearby galaxies. But they also can be very large, such as the Virgo Cluster, the nearest large cluster of galaxies that consists of thousands of galaxies. The centre of the Virgo Cluster is dominated by the giant elliptical galaxy M87 (Fig. 6.5), which contains some 50–100 times more stars than our Galaxy. It has some 14,000 globular clusters—70 times more than our Galaxy—which can be seen as small specks of light against the diffuse background of the system. M87 is located at a distance of about 65 million light-years. An observer in M87 who is looking at Earth through a super telescope sees the arrival of light sent from Earth 65 million years ago; he/she will see the last dinosaurs walking on Earth, just before the asteroid hit and killed them off 65 million years ago. The asteroid left a 200 km size crater in the Yucatan Peninsula in Mexico, which since has been covered by 2 km of sediments.

Fig. 6.4 Stephan's Quintet, a group of five galaxies

Fig. 6.5 The giant elliptical galaxy M87 in the centre of the Virgo Cluster, at a distance of some 65 million light years, has some 50–100 times the number of stars of our Galaxy. It has some 14,000 globular star clusters—a factor of 70 more than our Galaxy—visible on this photograph as small white spots

Fig. 6.6 The Coma Cluster of galaxies, at a distance of 320 million light years, consists of thousands of galaxies. Among these there are two central giant elliptical galaxies which closely resemble M87 (Fig. 6.5)

Other relatively rich clusters are the Coma Cluster at a distance of 320 million light-years (Fig. 6.6) and the Hercules Cluster at 600 million light-years (Fig. 6.7). An observer in the Coma Cluster sees Earth in the Carboniferous epoch, when the continents were covered by giant ferns, which conquered the lands since the Silurian times, 420 million years ago. From the Hercules Cluster one observes Earth as it was 600 million years ago, when in the oceans the first recognizable multi-cellular animals began to appear. During this *Cambrian explosion* of life numerous new types of multi-cellular organisms appeared, with hard skeletons which left fossils that are easily recognizable with the naked eye. (Already during some 600 million years preceding the *Cambrian explosion* minuscule multi-cellular organisms with sizes up to a fraction of a millimetre had appeared in the oceans. These consisted of soft materials and left fossils that can only be seen through a microscope.)

The most distant cluster of galaxies that Hubble could observe with his 100-in. Mount Wilson telescope after exposing photographic plates for hours, was the Corona Borealis Cluster at 1.1 billion light-years distance (Fig. 6.8). Viewing Earth from there one would see appear in our oceans the very first microscopic multi-cellular organisms, which already stored their genetic material (DNA) in a cellular nucleus, the so-called Eukaryotes, the same large class to which also we belong. (The development of life on Earth is dealt with in more detail in Chap. 11).

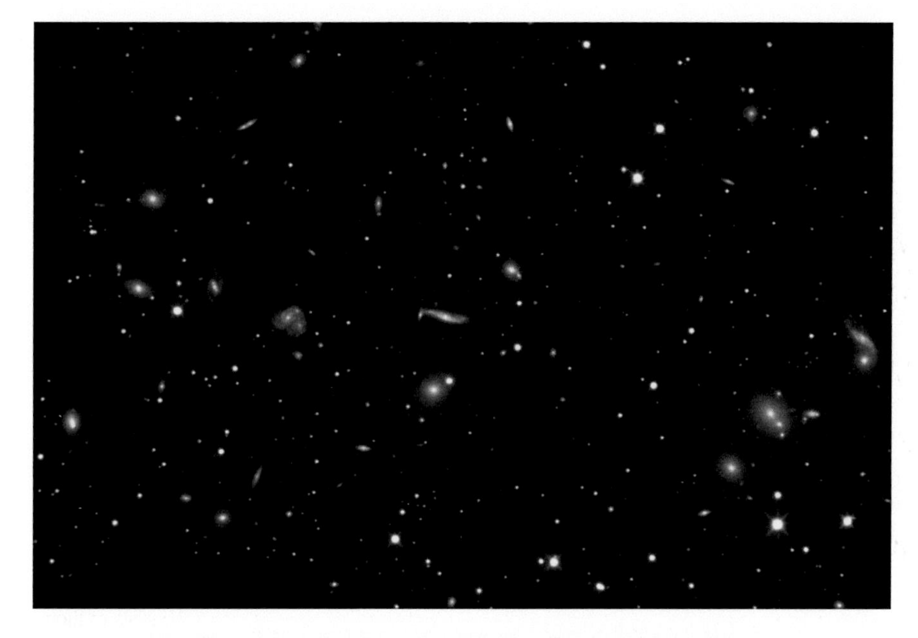

Fig. 6.7 The Hercules cluster of galaxies at a distance of 600 million light years

Fig. 6.8 Corona Borealis cluster of galaxies (Abell 2065) at a distance of 1.1 billion light years

The Most Distant Galaxies

The most distant galaxies that we can nowadays observe are so remote that we can look back in time more than 13 billion years. These are the faint red galaxies in the Hubble Ultra-Deep Field (HUDF), a very small area of the sky (1 % of the area of the full moon) which the Hubble Space Telescope observed for 10 full days to collect the light of the its faintest galaxies. This resulted in the "deepest" picture that has ever been made of a piece of the sky (Fig. 6.9). The faintest galaxies visible in the HUDF picture are more than ten times fainter than the faintest galaxies that can be pictured with the largest ground-based telescopes. This is due to the fact that the latter telescopes are troubled by the continuous random motions in the atmosphere, which cause the image of a star to continuously make random motions with amplitudes of order 1 arcsec. As a result the light of a star or galaxy is smeared out over a larger area on the picture, causing a "dilution" of the light of the object. This makes that even the largest telescopes on Earth, such as ESO's Very Large Telescope in Chile (four telescopes with mirrors of 8.5 m diameter; Fig. 6.10) and the Keck

Fig. 6.9 Hubble Ultra Deep Field. This picture of a very small piece of the sky (1 % of the area of the full moon) was exposed for more than an entire week by the Hubble Space Telescope. Thousands of galaxies are visible here. For some of them the light has been travelling to us for over 13 billion years. We see these galaxies as they were a few 100 million years after the Big Bang, when the universe was still very young

Fig. 6.10 The Very Large Telescope of the European Southern Observatory on mountain Paranal in Northern Chile. It consists of four telescopes with primary mirrors of 8.2 m diameter. Together, when the light of the four telescopes is combined, they form the largest optical telescope of the world, with an effective diameter of 16.4 m

Telescope on Hawaï (two telescopes with mirrors of 10 m diameter) cannot see galaxies as faint as those observable with the Hubble Space Telescope, which has a mirror of only 2.5 m diameter. The light of these galaxies, emitted over 13 billion years ago, was already on its way to us for 8.4 billion years when Earth and the solar system were born, 4.65 billion years ago.

As we have seen above, until 1923 it was thought that our Milky Way system with its diameter of 100,000 light-years, was the entire universe. With Hubble's 1923 discovery that the Andromeda Nebula is another galaxy like our own, followed by his discovery that there are millions of such galaxies, out to—in his times—the Corona Borealis cluster of galaxies at three billion light-years distance, the universe became 11 000 times larger! And in volume, Hubble made it 11 000 × 11 000 × 11 000 = 1.33×10^{12} times larger (1.33 trillion times). Not a small achievement!

The Expansion of the Universe

Between 1912 and 1925 American astronomer Vesto Slipher (1875–1969, Fig. 6.11) photographed the spectra of the 50 brightest spiral nebulae, using the 24-in. (60-cm) refracting telescope of Lowell Observatory in Flagstaff, Arizona. He

Fig. 6.11 Vesto Slipher (1875–1969) in 1905. Working at Lowell Observatory near Flagstaff, Arizona, he discovered between 1912 and 1925 that the lines in the spectra of all galaxies are shifted towards the red. (The only exception is the nearby Andromeda Nebula M31, which will collide with our galaxy in the distant future)

discovered that, with the sole exception of the Andromeda Nebula, the absorption lines in the spectra of all spiral galaxies are shifted to the red by considerable amounts. Small blue and red shifts of the lines are a well-known phenomenon in the spectra of the stars in our Milky Way. They are due to the motions of the stars towards us or away from us with velocities of a few tens to at most a few hundreds of km/s, as explained in Fig. 5.7. While these so-called *radial velocities* of the stars seldom exceed a few hundred km/s, the redshifts of the spectra of the spiral galaxies measured by Slipher indicated that these nebulae are flying away from us with speeds ranging from a few hundred km/s for the spiral M51 and M101, to over 1000 km/s for the galaxies in the Virgo Cluster. The very nearby Andromeda Nebula is the only exception among the large galaxies: it approaches us with a speed of about 300 km/s, and some 2.5 billion years from now it will collide with our Milky Way system. This is less worrisome than it might sound, as the stars in each of these galaxies are so distant from each other, that the stars from the Andromeda Nebula will just pass between the stars of our Galaxy, and not collide with them. On the other hand, the clouds of interstellar gas and dust of the two galaxies will collide and the resulting compression of these clouds will undoubtedly lead to the formation of large numbers of new stars—a so-called *starburst*—such as we see in other colliding galaxies (see Figs. 9.6, 9.7 and 9.8).

In 1921 Slipher announced in the New York Times the discovery of a spiral nebula that is flying away with a record velocity of 1800 km/s, and he suggested that this nebula must in the past have been very close to our Milky Way system. In hindsight, this was the first indication that something like a Big Bang must have taken place in the past. Already prior to this discovery, Slipher had suggested that the redshifts of the nebulae increase with increasing distance. This idea was based on estimates of the relative distances of the spiral nebulae, derived from their angular sizes on the sky. Slipher could not really measure these distances, as the light-gathering power of his 24-in. refractor was very much smaller than that of the 100-in. reflector on Mount Wilson. As a result, Slipher was unable to distinguish in the spiral nebulae individual stars and other objects of known intrinsic brightness (e.g., Cepheid variables), which are needed for distance determinations. Hubble in 1923 with the 100-in. Mount Wilson telescope was the first to be able to find such objects in the Andromeda Nebula, and thus could for the first time measure distances of galaxies.

In 1924 German astronomer Wilhelm Wirth tried to estimate the distances of galaxies by assuming all spiral galaxies have the same size, and he found that with this (in hindsight not very realistic) assumption, that on average the velocities of galaxies increase with their distance. That seemed to confirm the model of the universe proposed in 1917 by Dutch astronomer Willem de Sitter. De Sitter's 1917 model, which was based on Einstein's General Theory of Relativity (see Chap. 8), predicted that the spectra of distant galaxies should be redshifted and that the redshifts increase with distance.

From 1924 on, Hubble with his assistant Milton Humason—who formerly rented out donkeys and guided the public in the nature park surrounding Mount Wilson—started to photograph the spectra of the galaxies which he had discovered. He confirmed Slipher's discovery that the spectra of all nearby spiral galaxies are redshifted. They also made estimates of the distances of the galaxies, using various types of "standard candles", such as Cepheid variables, novae and globular clusters.

In 1927 Belgian priest Georges Lemaitre used these distance estimates—which had been presented by Hubble in talks at astronomical meetings—and combined these with the velocities measured by Slipher. He found that the velocities increased proportional to the distances. On this basis, Lemaitre in 1927 suggested that the universe is expanding. (His paper, written in French in an obscure journal was, however, not read by anyone, such that later the discovery of the expansion of the universe was ascribed to Hubble. We will come back to this question in Chap. 8).

In 1928 the world conference of astronomers—the General Assembly of the International Astronomical Union—was held in Leiden in the Netherlands. Hubble's wife Grace noticed in her diary that at this conference de Sitter urged Hubble to verify if his theoretical prediction that the redshifts of galaxies should increase in proportion to the distance is correct (see the biography "Edwin Hubble" by Gale E. Christiansen, *Farrar, Strauss and Giroux*, New York 1995, p. 198). After returning to California Hubble went to work to see whether de Sitter's prediction was right. In 1928 the redshifts of 46 nebulae were available to Hubble and Humason, 39 of which had been measured by Slipher, and 7 by themselves. They had been able to measure the distances of 24 of the galaxies measured by Slipher. In 1929 Hubble published the relation between velocity and distance for these 24 galaxies, shown in Fig. 6.12. According to this figure the velocity increases by 170–180 km/s per million light-years increase in distance. The value of this increase of velocity with distance has since been called the *Hubble constant*, and the finding that the

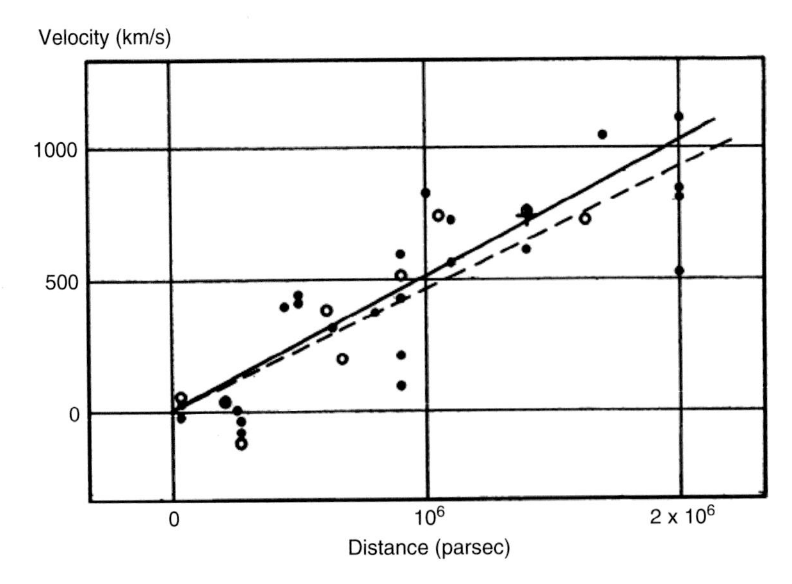

Fig. 6.12 The first 'Hubble diagram', as published by Hubble in 1929. In this diagram the velocities of galaxies are plotted against their distances in parsec (1 parsec is 3.26 light years). The black dots are velocities measured by Slipher, and open circles are velocities measured by Hubble and Humason. The latter authors estimated the distances of the galaxies. Hubble discovered with this diagram 'Hubble's law': galaxies fly away from us with velocities that increase with distance. (In recent years it was discovered that Belgian priest Georges Lemaître already in 1927 had discovered this same 'law', and published it, in French language, in a little-read scientific journal; see the text of this chapter)

velocity increases with distance is called *Hubble's law*. The discovery of the relation between redshift and distance was a worldwide sensation which made Hubble an instant celebrity.

(It should be noticed here that in his above-mentioned paper of 1927 Lemaitre had obtained the same value for the Hubble constant, not surprising, because he had used the same distance data, and Slipher's redshifts, see Chap. 8).

In order to establish with certainty that the discovered relation between redshift and distance keeps holding also at large distances, Hubble decided to ask Humason to photograph spectra of faint and therefore very distant galaxies. Photographing the spectra of very faint galaxies requires very long exposure times of the photographic plates, often stretching over several consecutive nights. And indeed: every new spectrum further confirmed the increase of redshift with distance. In 1931 Hubble and Humason published the new measurements. The largest redshift measured was for a galaxy in a galaxy cluster in the constellation Leo at a distance which (at the time) was measured to be 104 million light-years. This galaxy was measured to move away from us with the colossal velocity of 20,000 km/s, corresponding to a *Hubble constant* of 190 km/s per million light-years. Later it was found that Hubble had greatly underestimated the distance to this galaxy. It is about ten times more distant, at about one billion light-years. All Hubble's distance estimates from these

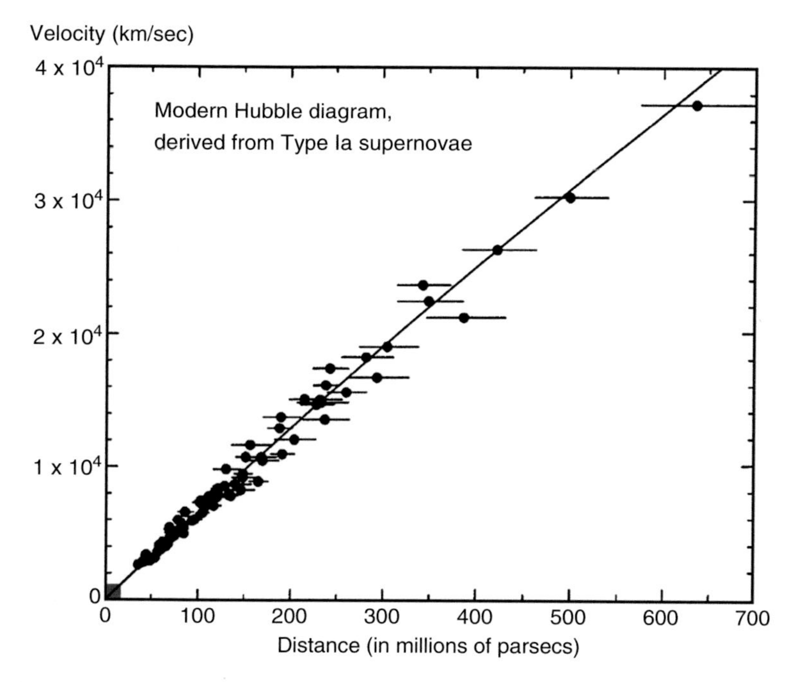

Fig. 6.13 A modern version of the 'Hubble diagram'; the distances of the galaxies were determined by using the so-called Type Ia supernovae (see Chap. 13). It has been found that in Hubble's original 'Hubble diagram' of Fig. 6.12—indicated here by the small red rectangle in the lower left corner—the distances were underestimated by about a factor of ten, such that the 'Hubble constant' was overestimated by about a factor of ten

times came out far too small, since at that time the real luminosities (intrinsic brightness) of the "standard candles" which he used were greatly underestimated. The best present-day determination of the *Hubble constant* is about 20 km/s per million light-years. (Later it was found that the Hubble constant is not really constant but changes in the course of time, as is explained in Appendix C). Figure 6.13 shows a recent plot of the Hubble relation between velocity and distance for galaxies with well-established distances.

Is Our Galaxy at the Centre of the Universe? The Raisin-Bread Model of an Expanding Universe

Hubble's discovery means that the further away a galaxy is, the faster it speeds away from us. As depicted in Fig. 6.14 it therefore seems as though our Galaxy occupies a very special position in the universe: we are in the centre and all galaxies are moving away from us, with velocities that increase with the distance. Does this mean that, after all, we are in the centre of the universe?

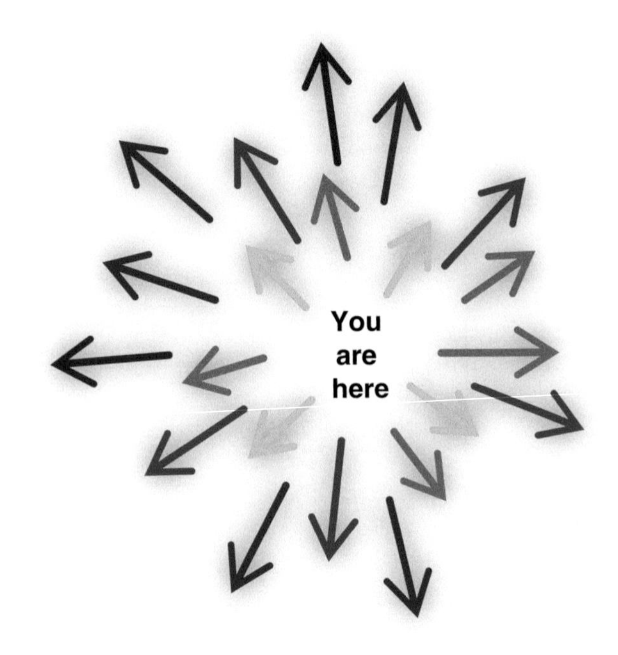

Fig. 6.14 Hubble's law implies that all galaxies fly away from us with velocities that increase with their distances. This gives the appearance that we are in the centre of the universe (everybody wants to fly away from us) but this is only seemingly so: it is due to the expansion of the universe, see Fig. 6.15

Since Copernicus we know that Earth does not occupy a special (central) position in the solar system. And since Shapley and Oort we know that the sun does not occupy a special (central) position in the Milky Way system: it is just one of some 200 billion stars orbiting the distant centre of our Galaxy, located in the direction of Sagittarius. The principle that our position in the universe is in no way special, is called the *Copernican Principle*. Considering this principle, it would be very surprising if our Galaxy would suddenly occupy a central position in the universe of galaxies, as Fig. 6.14 seems to suggest.

Indeed there appears to be a much more plausible explanation of Hubble's law, namely that the universe is expanding. This is most easily explained by picturing the universe as a raisin bread (Fig. 6.15). If one takes a ball of dough with yeast and raisins in it, this ball will, due to the action of the yeast, start expanding. In 1 h its size becomes twice as large, as depicted in the right-hand side of Fig. 6.15. This right-hand picture shows how the positions of the raisins numbers 1, 2 and 3 have changed during the expansion. Relative to raisin 1, raisin 2 has doubled its original distance from 1 to 2 inches, while raisin 3 has doubled its original distance from 2 to 4 inches. This shows that *in the same time* of 1 h, the originally more distant raisin number 3 has moved relative to raisin 1 *over a twice as large distance* as raisin number 2, which originally was twice closer to number 1 than number 3. As this happened *in the same time*, this means than raisin number 3 moved away from raisin number 1 with a *two times larger velocity* than raisin number 2. So, during the expansion of the dough, the further away a raisin is from another raisin, the faster it moves away from that raisin. So, due to the expansion of the dough, every raisin sees the other raisins speeding away from it, with velocities that are larger, the

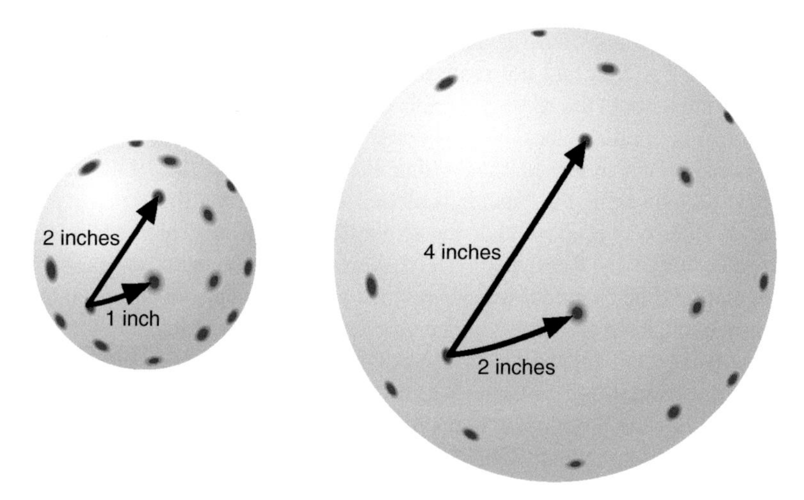

Fig. 6.15 Explanation of Hubble's law: the universe is expanding. To demonstrate this we use here the example of a rising ball of raisin-bread dough (the 'universe'). The ball of dough starts small, as depicted at left, and because yeast has been added to it, after 1 h the ball has expanded by a factor of two. We consider now the motions of the different raisins ('galaxies') with respect to raisin number 1 in the lower left of the ball. After an hour, raisin number 2, who originally had a distance of 1 inch with respect to raisin 1, is now located at a distance of 2 inches from number 1, while raisin number 3, which originally was at a distance of 2 inches from number 1, has now moved to a distance of 4 inches from number 1. So, raisin 3, which originally had a twice as large distance to number 1 than raisin 2, has now, *in the same time of 1 h*, moved over a twice as large distance than raisin number 2. Since velocity is the distance covered per unit time, we see that raisin number 3 moved away from raisin number 1 with *twice as large a speed* than raisin number 2. Thus, raisin number 1 sees other raisins moving away from it with speeds that are larger when their original distance to him was larger. This is exactly like the Hubble law! As raisin number 1 was chosen arbitrarily, one could have chosen any other raisin to start with and would again have seen that the larger a distance another raisin originally had with respect to it, the faster away it moved when the ball of dough expanded. So, *every raisin* notices other raisins to move away exactly according to the Hubble law. Thus, the Hubble law means that the universe is expanding

farther away the other raisin is. So, every raisin has the impression that it is in the centre: every other raisin is speeding away from it, with velocities that are larger, the more distant the other raisins are. This is exactly the same situation as was discovered by Hubble for galaxies relative to our Galaxy, as depicted in Fig. 6.14. We thus see that the simplest explanation for Hubble's law is that the universe is expanding! Every galaxy in the universe then observes the same phenomenon: it sees all other galaxies moving away from it, with speeds that increase with increasing distance. We thus see that our Galaxy by no means has a special central position in the universe, and that this is an illusion, created by the expansion of the universe. We have no central position, and actually, nowhere in the universe a centre can be distinguished.

As we mentioned earlier, this expansion of the universe was predicted already by Willem de Sitter in 1917 on the basis of a rather special model of the universe that started from Einstein's General Theory of Relativity. When he constructed his model, de Sitter at first had not realized that this model was expanding. He had found that in his model the redshift of galaxies should increase with increasing distance, but had not realized that this was due to expansion. That was discovered only 5 years later by British astronomer Arthur Eddington. Until the 1930s de Sitter's model was thought to be the "standard model" for explaining the expansion of the universe. It then was realized that the expansion could be explained by simpler models, discovered by Russian physicist Alexander Friedmann, as explained in Chap. 8, and de Sitter's model was abandoned as a rather special and peculiar one. Amazingly enough, in the past two decades de Sitter's model has made a remarkable come-back, both for the description of the earliest phases of the universe as well as for the present-day universe. We will come back to this in Chap. 13. In order to understand how de Sitter came to his model we first have to explore the nature of gravity and the ways in which Einstein changed our views of gravity, and how this, in turn, has influenced ideas about the development of the universe. This is the subject of the following chapter.

The discovery of gravity

Chapter 7
Gravity According to Galilei, Newton, Einstein and Mach

Discovering is seeing what everybody has seen and thinking what nobody has thought

Albert Szent Gyorgyi (1893–1986), American scientist

Introduction

As mentioned in Chap. 1, Galileo Galilei (1546–1642, Fig. 7.1) discovered that without visual information it is impossible to distinguish whether we are at rest or moving with a constant speed. Earth moves with a speed of about 30 km/s in its orbit around the sun, but we feel nothing of this motion. It appears to us that Earth is completely at rest. Galilei discovered that the only thing we can feel is a *change in velocity*: when a car accelerates, we are pressed against the back of our seats, and

E. van den Heuvel, *The Amazing Unity of the Universe*,
Astronomers' Universe, DOI 10.1007/978-3-319-23543-1_7

Fig. 7.1 Galileo Galilei

if it suddenly brakes, we shoot forward. The reason for these effects is that if one does not exert any force on an object, it wants to persist in moving on with the same speed. (This makes you shoot forward when your vehicle is braked.) This is Galilei's *law of inertia*. It was Newton's discovery that, in order to change the velocity of an object, a *force* is needed. To make a car accelerate, we need an engine that causes the speed of a car to increase. In order to make the speed decrease, we must hit the brakes, such that a decelerating force is exerted on the wheels. Newton (1643–1727, Fig. 7.2) has expressed this relation between force and acceleration in his famous law of inertia:

$$\text{Force} = \text{mass} \times \text{acceleration} : F = m.a$$

In this law, the *mass* (the amount of matter, expressed in kilograms) is the so-called *inertial mass*, that is: the mass with which an object *resists* against changing its motion (acceleration or braking). If one wishes to accelerate two objects with different masses to the same speed, one needs a larger force on the one with the larger mass. For example, to accelerate a car with a mass of 1000 kg to the same speed as a car of 500 kg, one needs a twice as large force. The two times heavier car resists, as it were, twice as strong against getting into motion: it is *twice as inert.*

The second great discovery of Newton was that of gravity. He realized that the force that holds the moon in its orbit around Earth must be an attractive force exerted by Earth. He made this discovery in his *annus mirabilis* (wonder year) 1665/1666 when he, 23 years old, was staying at his mother's farm in Whoolsthorpe, as Cambridge University had been closed because of a plague epidemic. In that year he discovered the law of gravity, developed a new branch of mathematics (differential and integral calculus) and formulated his colour theory of light. While sitting in the garden and observing the fall of an apple, he realized that this apple falls because Earth pulls at it with an attractive force, the same force that keeps the moon in its orbit (Fig. 7.3). Hundreds of thousands of years our ancestors have seen stones,

Fig. 7.2 Isaac Newton

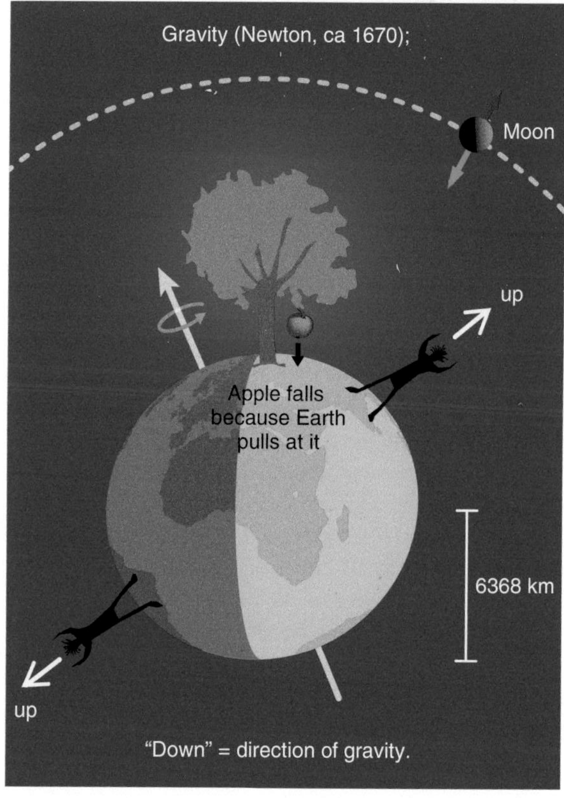

Fig. 7.3 An apple falls towards Earth because Earth is pulling at it with its attractive gravitational force. Newton realized that this is the same attractive force that keeps the moon in its orbit. What we call "down" is the direction of gravity, and what we call "up" is opposite to the direction of gravity. For our antipodes at the opposite side of Earth, "up" and "down" are just opposite to our "up" and "down". The centrifugal force that the Moon feels in its curved orbit (*red arrow*) is just compensated by the attractive gravitational force exerted by Earth (*blue arrow*) such the moon keeps going forward in its orbit (in vacuum there is no resistance to its forward velocity)

apples and other objects falling. It required a Newton, however, to realize that this falling is due to an attractive force exerted by Earth.

Thanks to the works of Johannes Kepler (1571–1630), Newton knew that the planets move around the sun in elliptic orbits with the sun located in one of the ellipse's two focal points (Kepler's first law). And also that in the elliptic orbit the planet moves fastest when it is closest to the sun, and slowest when it is farthest from the sun (Kepler's second law). And finally, that for all planets, the square of the period in which the planet orbits the sun, divided by the third power of the size (*semi-major axis*) of its ellipse, yields the same number (Kepler's third law). Newton realized that the same force that causes Earth to attract the moon and apples, causes the sun to keep the planets in their orbits. Using Kepler's third law, he was able to prove that the strength of the force of gravity exerted by a celestial body is proportional to the mass of this body, divided by the square of the distance to the centre of that body. So, the strength of the force of gravity decreases inversely proportional to the square of the distance from the body. For example, if one is twice as far away from the centre of Earth, the force of gravity is four times weaker, if one is three times farther away from the centre, gravity is nine times weaker, etc.

Newton was able to mathematically prove that with his law of gravity the orbits of celestial bodies (planets, comets, etc.) become conic sections, that is: ellipses, parabolas or hyperbolas. Ellipses are bound orbits, a parabola is a limiting case of an ellipse (as it were: an ellipse with an infinite semi-major axis), while hyperbolas are unbound orbits. He also showed that if one shoots away a cannon ball from a high mountain in a horizontal direction with a sufficiently large velocity, it may keep falling around the Earth and become an artificial satellite (Fig. 7.4). All these results which Newton discovered are still used nowadays, for example, for bringing satellites in orbit and to launch spacecrafts to other planets.

The Weakest Force of Nature

Although according to Newton's law of gravity all masses attract other masses, the gravitational attraction forces exerted by objects that surround us in our daily life are so extremely small that in practice we do not notice them. For example, two persons with a mass of 70 kg; at a distance of 1 yard from each other, exert a gravitational attraction force on each other of 30 billionth of a kilogram force (0.00003 g). This force is very much smaller than the resistance between the soles of our shoes and the floor, or between our pants and the seat of the chair on which we are sitting. We therefore will never fly towards each other due to the *gravitational attraction* which we exert on one another (but there may be other forces why people are attracted and move to each other!).

The force of gravity becomes noticeable only when the amount of matter is very large, such as in a planet or a star. Earth, with a mass of 6×10^{24} kg exerts a force of 70 kg (about 700 N; 1 kg force is 9.8 N) on a person with a mass of 70 kg (see Fig. 7.5). This is indeed a large force. The more mass a celestial body has, the

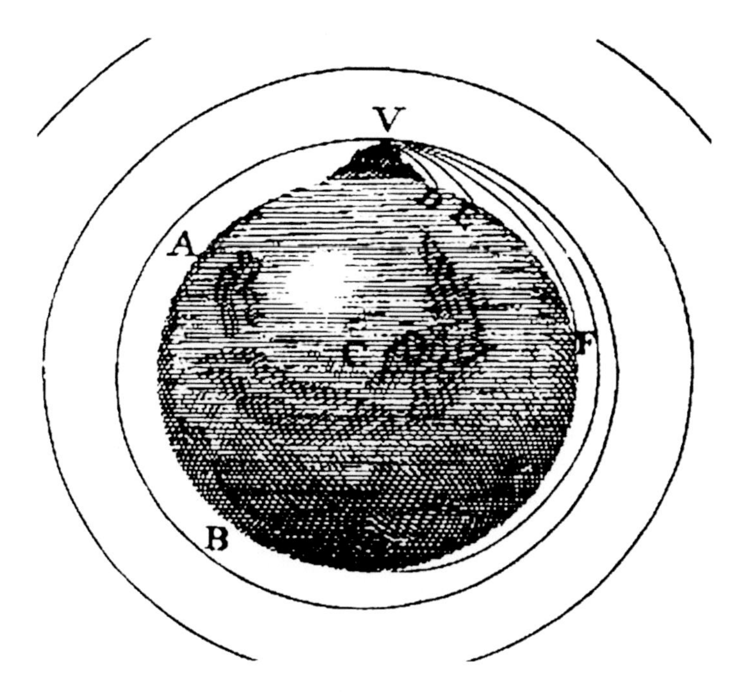

Fig. 7.4 This drawing from Newton's *Philosophiae Naturalis Principia Mathematica* of 1687 ("the Principia") illustrates that if cannonballs are launched in horizontal direction (parallel to the surface of Earth) with increasing velocities, they hit the ground at larger and larger distances: at large distances they are 'falling around Earth'. One sees that Newton realized that if the cannonball is launched with a very high speed from a high point (mountain top) it can keep falling around Earth in a closed orbit, and thus becomes a satellite of Earth. (In reality, a mountain top is not sufficient, because at its height there is still a lot of atmospheric air which produces a large resistance, causing the ball to slow down. Only at a height above some 300 km the air is so tenuous that the resistance becomes small enough to keep the ball falling around Earth)

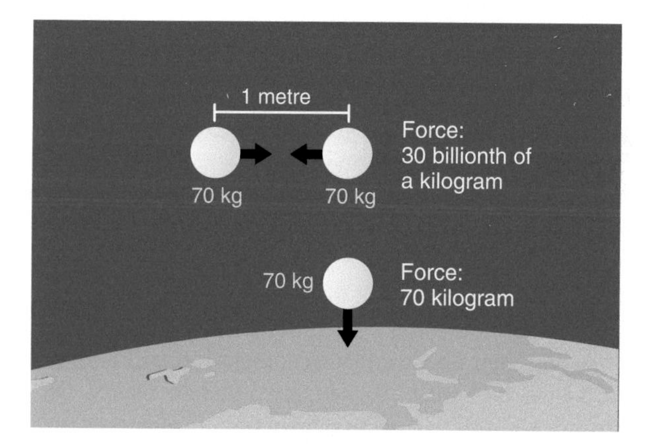

Fig. 7.5 Gravity is a very weak force: two masses of 70 kg at a distance of 1 m (some 3.3 ft) attract each other with a gravitational force of only 30 billionth of 1 kg force. However, when one has a large amount of mass, like that of planet Earth (6×10^{24} kg), an object with a mass of 70 kg (about the mass of a man) is attracted by Earth with a force of 70 kg. So, a person with this mass will standing on a scale observe to have a weight of 70 kg

larger the force of gravity that it exerts. And since mass is volume times specific density (in kilograms per cubic metre), the strength of the gravitational attraction is, in good approximation, determined by the *size* of the object. It turns out that when a celestial object is larger than about 600–800 km diameter (the precise value depends on its density), the gravitational force which it exerts on its molecules becomes stronger than the forces between its molecules and atoms (the so-called *van der Waals forces*). These are the forces which make that the objects in our daily surroundings, such as tables, desks, chairs, cars, trains, computers, our bodies, etc., stay in shape. If the own gravitational force exerted by the matter of a celestial object on itself becomes larger than the van der Waals forces, the gravitational forces will, in the course of time, cause the object to become *spherical*, such that the gravitational forces are distributed evenly and symmetrically throughout the body. This is the reason why all celestial objects larger than about 600–800 km have a spherical shape, while smaller bodies, such as the smaller asteroids and comets are not spherical (see Figs. 2.2, 2.11 and 2.12). In the latter objects, the gravity is so small that the forces between the atoms and molecules outweigh the gravitational forces, such that they keep their irregular shapes with which they were born.

Inertial Mass Versus Gravitational Mass

Perhaps Newton's most amazing finding about gravity was that the mass which determines the strength of the gravitational force exerted by a body—the so-called *gravitational mass*—is the same as its *inertial mass*, which determines the body's resistance against being accelerated. This is very strange, because the gravitational mass which produces the gravitational attraction exerted by the body is something like an electric charge, which determines the electric attraction (or repulsion) exerted by an electrically charged body (positively and negatively electrically charged bodies attract each other; bodies with the same charge repel each other). The gravitational mass is therefore, in fact, the "gravitational charge" of the body, which allows it to exert a gravitational force on objects surrounding it. There seems to be no reason why this "gravitational charge" would have anything to do with the inertial mass of the object, which causes the object to resist to being accelerated.

Newton himself already noticed that the equality of these two types of masses is very puzzling, but could not find an explanation for it. For more than 200 years scientists kept banging their heads about this strange equality of gravitational and inertial mass. It would last until 1915 before Einstein finally, with his General Theory of Relativity, was able to find an explanation for it.

Science is a Young Man's Game

Often Albert Einstein (1879–1955) is depicted as a friendly old man with a wild bush of beautiful grey hair. This gives the impression that his great contributions to physics and to our knowledge of the universe originated from this deeply-grooved old head. But nothing is less true: breakthrough contributions to theoretical physics and mathematics are seldom made by persons older than 35–40 years. Newton was 23 when he discovered the laws of inertia and gravity and invented differential and integral calculus, even though he did write all of this down only 20 years later in his great work the *Principia* (1687), after having been continuously pushed for 2 years by his friend Edmund Halley, who also paid for its publication out of his own pocket. Blaise Pascal (1623–1662) was only 16 years old when he wrote his famous essay on conical sections, in which he proved dozens of new mathematical theorems (Descartes first refused to believe this was the work of a 16-years old boy), at age 18 he invented the first calculating machine (all later calculating machines and computers followed the basic principles of this machine), at age 25 he discovered his famous law about the pressure of gases and liquids, after that he invented the theory for calculating probabilities, and age 31 he joined a monastery to devote the rest of his life to theology and philosophy of Jansenism, and made no more noticeable contributions to science. The great German mathematician Carl Friedrich Gauss (1777–1855)—with Archimedes and Newton considered to be one of the three greatest mathematicians of all time—already at age 16 had his first ideas about *non-euclidian geometry*, at age 18 invented the method of *least squares* and at age 20 proved the *fundamental theorem of algebra*. Between ages 18 and 21 he had his greatest ideas, laid down in his *Disquisitiones Arthmeticae* (1798), which was the basis for all his later mathematical work. At age 24 he discovered the method by which, on the basis of only three observations of a planet or comet, one can fully determine its orbit. In mathematical terms this was only one of his minor contributions, but for astronomy it proved to be of great importance, as it allowed Olbers to rediscover the first-known asteroid Ceres. Soon after its discovery on January 1, 1801, Ceres became lost because its discoverer, Father Piazzi, had fallen ill, and after that it had come too close to the sun to be observable. Lucky enough Piazzi had made observations of its position relative to nearby stars on three different dates, and Olbers, using the new method invented by Gauss, was able to find it back after it had moved away from the sun's neighbourhood. This triumph of Gauss' new method made him an instant celebrity, and he was offered the directorship of the Astronomical Observatory of Göttingen University, a position he held for the rest of his life. (His chair is still kept in the cellar of the old Observatory, and in 1968, after giving a talk at Göttingen Observatory, I was allowed to sit on it for a few moments, a great honour).

Einstein (Fig. 7.6) is no exception to the rule that "science is a young man's game". He made his most important contributions to physics between his 20th and 40th years, although he had been thinking on these fundamental problems for much longer, at least since his 16th year. In 1905 at age 26, he published four scientific

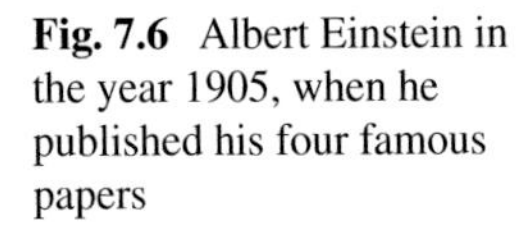

Fig. 7.6 Albert Einstein in the year 1905, when he published his four famous papers

papers which revolutionized our knowledge of physics. These papers were the results of his thinking in the six preceding years. Apart from the "wonder years" of Newton (1665/1666) and Einstein (1900–1905), there have in the history of science not been comparably fruitful periods of individual researchers.

In 1915 Einstein completed—after another 10 years of thinking—his General Theory of Relativity, which again was a revolution. It resulted in a new view of gravitation, which would prove to be of great importance for our view and understanding of the evolution of universe. But before we consider this theory in more detail, we first briefly consider two of his breakthrough contributions of 1905: his Special Theory of Relativity and the quantum theory of light. (For more details I may refer to the general literature on these subjects, such as Einstein's biography *Subtle is the Lord* by Abraham Pais).

The Special Theory of Relativity

The two 1905 articles of Einstein on the Special Theory were in fact the last of his four articles of that year: he sent them in September to the journal *Annalen der Physik*. In these articles Einstein showed that concepts such as absolute space and absolute time do not exist. Time runs different for you than for an observer that is moving with respect to you, and an object that is moving with respect to you has for you a different length than for an observer that is moving along with the object. This theory led to the famous formula $E = mc^2$, which states that mass (m) and energy (E) are equivalent. According to Einstein, the velocity of light (c) is a universal constant of nature that in every coordinate system, regardless of the velocity of this system, has the same value, also for an observer that is moving with respect to another coordinate system and is measuring the velocity of light emitted from a lamp in that other system. That means that the measured value of c gives us no information about whether we are moving or standing still: we always measure the same velocity of light. That last fact had been observed in the famous 1887 experiment of American physicists Michelson and Morley, and in dozens of later experiments since. Einstein was the first to realize that this discovery is something resembling Galilei's discovery that we cannot determine whether we are at rest or moving with a fixed (constant) velocity. Galilei's discovery in fact means that all physical phenomena ("laws of physics") remain the same if one goes from one system that moves with a fixed velocity, to another system moving with a different fixed velocity, and that we, as observers, cannot distinguish in which of these two systems we are. Since Michelson and Morley we know that, no matter at what (fixed) velocity we are moving, we always measure the same velocity of light c. This implies that the value of c is a universal constant of nature.

There is a story that around 1903 Einstein got the idea that time is a *relative* concept, when he was watching the clock of the city hall of the Swiss capital Bern. Einstein lived in this town from 1902 till 1909, when he worked as a clerk at the Swiss Patent Office, a job he got thanks to the father of his fellow student Marcel Grossman. Watching the clock on the city hall, Einstein realized that for a person travelling away from Bern with the speed of light, this clock will forever indicate the same time. Hence for that person the time in Bern is standing still, while the time on his own watch is still running forward. The person therefore sees a different time in Bern than for himself. At that moment Einstein realized that time is different for different observers, because observers in Bern see the city hall clock move forward. He thus realized that time depends on the way in which one is moving. He elaborated this idea in his Special Theory of Relativity. Special Relativity predicts that when an object moves with a high speed relative to us, we observe that its time (its clock) runs much slower than for a person that travels along with this object with the same speed. Also we see that the length of the object, in the direction of its motion, becomes shorter. This is the so-called *Lorentz-FitzGerald contraction*, which Dutch Nobel laureate Hendrik Anton Lorentz (1853–1928) and Irish physicist George Francis FitzGerald (1851–1901) had already calculated, based on the results of the Michelson-Morley experiment, without fully realizing its meaning in the way Einstein later discovered. Einstein predicted these two effects—which since have been experimentally verified many times—in his two 1905 articles on Special Relativity. He also found that when the velocity of an object increases, its mass increases, and derived, as a final bombshell, his famous formula for the equivalence of mass and energy.

The Quantum Explanation of the Photoelectric Effect

In his first 1905 paper, which he submitted on March 17th (3 days after his 26th birthday) to the *Annalen der Physik*, and was published in June, Einstein presented an explanation for the photoelectric effect by proposing that light is emitted in the form of small packages of energy. Five years earlier these so-called *quantums* had been proposed as a mathematical "trick" by German physicist Max Planck to explain the shape of the distribution of the energy of light and other electromagnetic radiation emitted by a hot object (a so-called *black body*) as a function of wavelength (see Fig. 9.2 and Appendix D). According to the mathematical equations used by Planck in his theory, the energy of such a *light-quantum* is proportional to the *frequency* ν of the light waves: $E = h.\nu$, where h is the so-called Planck constant (see box in Chap. 3). The frequency ν of the light (or other electromagnetic waves, from gamma rays to radio waves) is given by: $\nu = c/\lambda$, where λ is the wavelength and c the velocity of light.

Planck had thought that his quantums did not really exist, but were only needed as a mathematical "trick" to enable him to derive the correct formula for the black-body radiation energy distribution. But the photoelectric effect proved otherwise. This effect occurs when light shines on a metal: under certain circumstances this light may induce electrons to be emitted from the surface of the metal. This effect is used, for example, in photocells for theautomatic opening of doors of shops and elevators. When approaching such a door, your body blocks the light ray between a light source on one side of the entrance and a photocell at the other side. As a result the flow of electrons in the photocell is interrupted, which triggers a signal to be sent to an electromotor that causes the door to open.

Experiments prior to 1905 had shown a number of peculiar properties of the photoelectric effect, which Einstein was able to explain with his new theory of light quantums. (It had been found that electrons are emitted only if the frequency of the light exceeds a certain threshold value ν_t. Below that value, no matter how intense the beam of light that falls on the metal, no electrons escape. Einstein concluded from this that there is a threshold energy $E_t = h\nu_t$ required to make an electron escape from the metal. If the light quantums have an energy below E_t the electrons cannot escape; so light with frequencies below ν_t cannot make them escape).

Shortly after this, Danish theoretical physicist Niels Bohr proposed, as explained in Chap. 5, that electrons in an atom are allowed to occupy only a number of specific discrete energy levels (and no energies between these) and that when an electron drops from a higher energy level E_2 to a lower energy level E_1 a quantum of radiation with a frequency ν_{12} is emitted, given by $h\nu_{12} = E_2 – E_1$. And reversely, an electron can jump from the lower to the higher energy level by absorbing a light quantum of frequency ν_{12}.

This work of Einstein and Bohr, and the discovery of French prince Louis de Broglie (1892–1987) that particles also have a wave character, just as light waves have a particle character (the light quantums/particles are called photons), laid the foundation of quantum mechanics, the highly successful theory for the behaviour of atoms, electrons and molecules, and their interactions with electromagnetic radiation, and later extended to all elementary particles. Without this theory the modern technology of computers, mobile telephones, lasers, electron microscopes, etc., would not have been possible.

In 1921 Einstein was awarded the Physics Nobel Prize for his explanation of the photoelectric effect. In 1922 Bohr was awarded this prize for his atom model, and De Broglie received his prize in 1929 for discovering the wave character of elementary particles.

Einstein's General Theory of Relativity

Einstein's Special Theory of Relativity describes how observers which are moving with a *fixed* (constant) speed relative to each other, experience length, time and energy of an object in a different way. In the years after 1905 Einstein devoted his attention to the much more general problem of *accelerated* motions, for which, as Galilei and Newton had shown, forces come into play. One of the most important techniques Einstein used in his work is that of so-called *thought experiments.* He imagined a certain physical situation for which in his mind he carried out an experiment and then pondered about to which conclusions this experiment would lead him. Also Galilei used this technique for proving that in vacuum all objects fall with the same speed. Before him it was believed that, as Greek philosopher Aristoteles had stated some 2000 years earlier, heavier objects fall faster than light ones. Galilei, however, asked himself the following question: if a heavier object falls faster than a light one, what will happen if I glue together a light one and a heavy one? The combined object is heavier than the heavy one of the two, so it should fall *faster* than that one. But on the other hand, the lighter one falls slower than the heavy one, so it will—in the combined situation—slow down the speed of the heavy one, so the combined object will fall *slower* than the heavy one. Aristoteles' assumption that heavier objects fall faster than light ones therefore leads to a contradiction: on the one hand the combined object should fall faster than the heavier one of the two, and on the other hand it should fall slower. If an assumption leads to a contradiction, this assumption cannot be correct, so Galilei concluded that the assumption of Aristoteles is wrong, and all objects fall equally fast (in vacuum), regardless of their mass. It is important to notice that Galilei did not have to do a real experiment to come to this conclusion: the thought experiment was sufficient!

Einstein's thought experiment about accelerated motions led him to his so-called *principle of equivalence*, a principle that states that gravitation can in fact not be distinguished from accelerated motion. This idea came to him in 1907 when, seated behind his desk at the Swiss Patent Office in Bern, he realized that a free-falling person in a gravitational field is, in fact, weightless. The fall is uniformly accelerated but the falling person feels no forces at all. The uniform acceleration compensates exactly for the force of gravity, and therefore, Einstein reasoned, uniform acceleration and gravity must be equivalent to each other. Einstein called this "the happiest thought of my life".

This thought led him later to another thought experiment that showed that a light ray must be bent by a gravitational field, as depicted in Fig. 7.7. An observer who, as depicted in the lower left part of the picture, is located in a uniformly accelerated

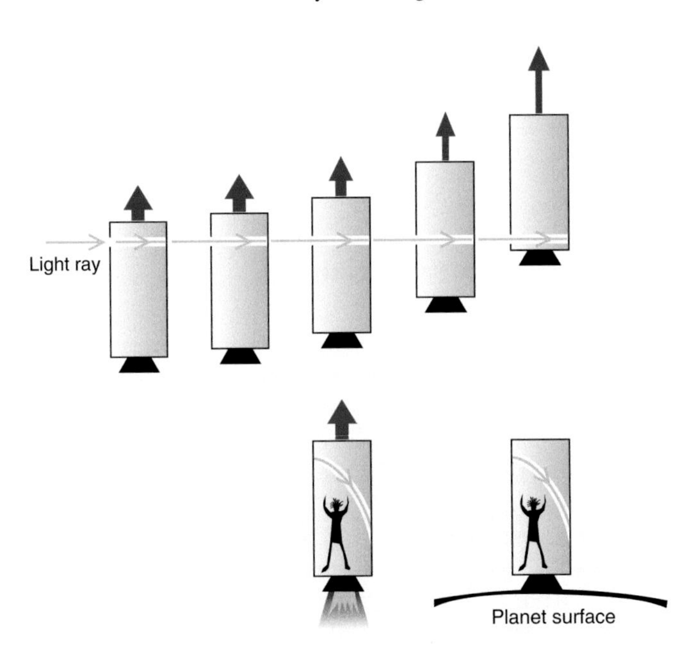

Fig. 7.7 A person in an accelerated rocket in free space, far away from any planet or star, is pushed towards the floor of his cabin, as depicted in the *bottom left* picture. If there are no windows in the rocket, this person cannot distinguish whether he/she is in an accelerated rocket in free space or if this rocket is standing still on the surface of a planet with gravity, as illustrated in the *bottom right* picture. This is the *equivalence principle*: gravity and acceleration are equivalent and cannot be distinguished from each other by any form of experiment carried out in the rocket. If one now makes a small hole in the top-left wall of the accelerated rocket (*upper left* picture), and shines a ray of light through it, then due to the accelerated motion of the rocket, this ray hits the opposite wall of the rocket just above the floor, so the ray describes a curved path in the accelerated rocket, as depicted in the *lower-left* picture. Using the equivalence principle, Einstein concluded that also in a rocket that stands still on the surface of a planet with gravity, a light ray shining through a hole in the *upper left* wall of the rocket, will describe exactly the same curved path as in the accelerated rocket. Therefore: a light ray is bent by gravity!

window-less rocket, in free space, far from all celestial bodies, will be pressed against the floor of the rocket cabin, and have the same feeling as when the rocket is at rest on the surface of a planet with gravity, just as depicted in the lower-right picture. As there are no windows in the rocket, he has absolutely no way to distinguish between these two situations: they are fully equivalent. All physical experiments must therefore proceed exactly the same in an accelerated rocket in free space and in a non-moving rocket standing on a planet with gravity. Einstein now reasoned as follows: make a small hole in the left-hand side of the moving rocket, through which one shines a light ray, as depicted in the upper frames of Fig. 7.7, for the accelerated rocket in free space. Due to the upward motion of the rocket, the light ray will describe a curved trajectory in the rocket and hit the opposite side of the rocket close to the rocket bottom, as depicted in the upper right picture. The observer inside the rocket will just see the light ray describe a curved trajectory, as depicted in the lower left picture. Now one does the same for the rocket that is standing still on a planet with a gravitational field, as depicted in the lower right figure: one makes a hole in the upper left side of the rocket and shines a light ray through it. As the observer inside the rocket has no way to distinguish between the two situations (being accelerated in free space or standing still on a planet with gravity), one will have that also in the situation of the lower right figure the light ray will describe a curved trajectory: one sees that gravity bends a light ray!

In fact, already Newton had calculated that light rays should be bent by gravitation. He thought that light consists of particles, whose trajectories are bent due to the gravity of a body. In 1911 Einstein still calculated this deflection of light rays in the same way as Newton, but when he later had reached the correct formulation of General Relativity, he found that this deflection is twice as large as it would be according to Newton. It would last until 1915 before he had formulated his final definitive form of his theory, and until 1919 before the deflection of light by the gravity of the sun was actually measured.

Until the publication of Einstein's new theory, in physics the idea had always been that three dimensional space can be represented by a rectangular system of coordinates, in which light rays describe straight paths. Einstein proposed now that light rays still exactly follow the shape of space, but that the bending of light rays that we observe is due to the fact that *space is curved* by the action of gravitation. Gravity bends space! This curvature of space concerns our three-dimensional space. Since our brains cannot easily imagine a picture of curved three-dimensional space, this curvature is often pictured for a space of one dimension less: that of a two-dimensional plane, like a flat sheet of rubber. A star or a planet with gravity is then depicted as a ball placed on this sheet of rubber. The ball with its weight deforms the two-dimensional space in its neighbourhood, making a depression in this sheet, as depicted in Fig. 7.8a. A celestial object with mass therefore literally produces a curvature of space in its vicinity, and a light ray must follow this curved space.

The Four Classical Predictions of General Relativity and Their Experimental Confirmation

The object with the largest mass in the solar system is the sun (330,000 times the mass of Earth), which therefore produces the largest curvature of space in this system. Since apart from the direct surroundings of the sun space in the solar system is flat (not curved) in very good approximation, the deviations from Newton's theory are everywhere in the solar system extremely small, except very close to the sun. Einstein applied his theory to the solar system and found that his theory predicts three small effects which would not occur if the older gravity theory of Newton were correct. These three effects have since been measured with high accuracy and have fully confirmed Einstein's theory, as we will describe now.

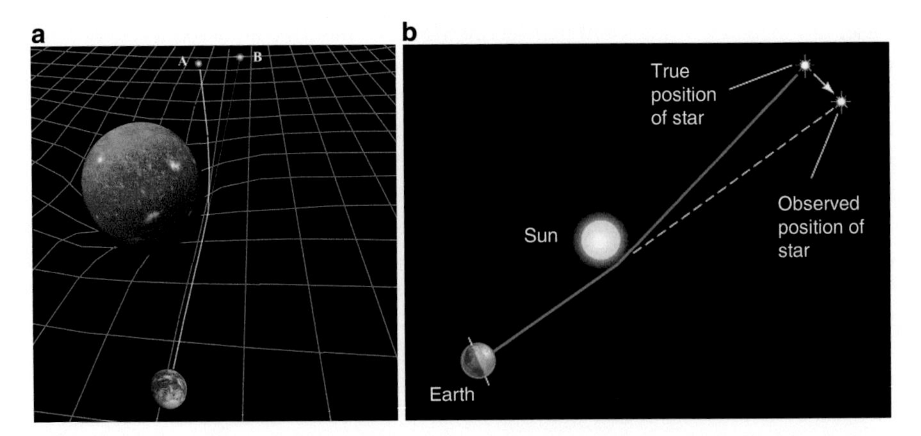

Fig. 7.8 (**a**) According to the General Theory of Relativity, three-dimensional space is slightly curved in the vicinity of a large mass, like the sun. This is depicted here for a space of one dimension lower, as the curvature of a two-dimensional plane. Without the presence of the large mass the coordinate system of this plane would have been perfectly rectangular. Now, however, close to the massive body the straight lines that indicate the shape of space have become slightly curved. Since light rays always follows the coordinate system of space, also light rays now become curved and light from a star at position A seems for an observer on Earth to come from the position B. (**b**) Due to the gravitational deflection of light, stars close to the edge of the sun appear to an observer on Earth to have been moved slightly outwards with respect to the position that one observes when the sun is not in this part of the sky, half a year later. The true deflection by 1.75 s of arc is much exaggerated here

Deflection of Light by the Sun

From the foregoing it is clear that a light ray from a star that passes very close to the edge of the sun will be deflected by the sun's gravity. The effect for an observer on Earth is that the star appears to be located slightly farther from the edge of the sun, than it really is, as sketched in Fig. 7.8b. One can only measure this effect when the star is still visible close to the edge of the sun. On Earth this is possible only during a total solar eclipse, when the sun is completely behind the disk if the moon for a few minutes. As the Earth moves around the sun in a year, a half year before or after the eclipse, the stars that were visible during the solar eclipse are now prominent in the night sky. By taking a picture of them and comparing the star positions with those in a picture taken during the total solar eclipse, one can measure whether the positions during the eclipse had been moved outwards from the sun by a small amount, as predicted by Einstein. In 1919 there was a total solar eclipse close, visible in Africa as well as Brazil. In two British eclipse expeditions that year the effect was indeed confirmed. One expedition, led by Cambridge astronomer Arthur Eddington, (Fig. 7.10) had chosen the island of Principe near the West-African coast, and the other expedition, organized by the Royal Greenwich Observatory, went to Sobral in Brazil. The apparent displacements of the stars near the sun's limb are only small, of order of 1 s of arc (see Chap. 3), but were clearly measurable.

Einstein was informed of this result by means of a telegram which he showed to one of his students. When she asked him what he would have said if the measurements had not confirmed his theory, his answer was: "I would have felt sorry for the Good Lord, because my theory is correct".

This confirmation by a British scientist of a theory of a scientist from a country with which, up till a year earlier, England had been in a deadly war, was front-page news in newspapers all over the world, and made Einstein an instant international celebrity. In addition, it enabled Eddington, who was a quaker and a pacifist, to demonstrate the absurdity of nationalism and war, and to show that science has no national boundaries.

According to American physicist John Archibald Wheeler, there is still another reason why this result made such an enormous impression: one thing of which people had always thought being certain was the simple structure of the three-dimensional space, which forms the background of our lives. Einstein had suddenly destroyed this certainty. If something as simple and evident as three-dimensional space suddenly turns out to be warped and deformable, what certainties are there left for us? (see John A. Wheeler: *"Geons, Black Holes and Quantum Foam"*, W.W. Norton and Co, NY, 2000, p. 230).

The Motion of Mercury's Perihelion

Einstein's theory predicts that, as a result of curvature of space, the orbits of the planets are no longer completely closed ellipses, as in Newton's theory, but have a so-called "rosette" shape (Fig. 7.9). This motion can be represented as that in an elliptic orbit of which the major axis is slowly turning around the sun in the orbital plane. The rate at which the axis turns around the sun is faster for planets that are closer to the sun. Therefore, this effect is largest for the planet Mercury, for which Einstein calculated that the major axis should turn 42.89 arcsec per century, a very small amount (less than three quarters of an arc minute per century!). In the nineteenth century astronomers had already measured that the major axis of Mercury's orbit, just like those of the other planets, is turning around in the orbital plane at a much larger rate. This is due to the fact that all planets attract each other according to Newton's law of gravity and in this way disturb each other's orbits. As a result the orbits are no longer perfect ellipses. These gravitational perturbations of the orbits can be accurately calculated with Newton's theory, and particularly French astronomer Urbain J.J. Leverrier (1811–1877) intensively carried out such calculations since the middle of the nineteenth century. He was very upset that he and his colleagues did not succeed to fully explain the rate of motion of the axis of Mercury's orbit: there remained an unexplained rotation of the axis by about 43 s of arc per century. As this rotation could not be explained in any way by them, it was proposed

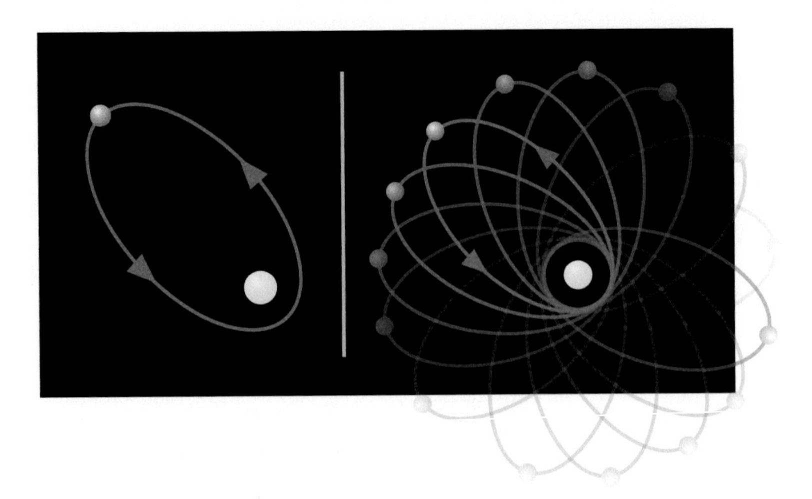

Fig. 7.9 Perihelion motion of the planet Mercury. General Relativity predicts that the major axis of the elliptic planetary orbits slowly rotate in the orbital plane. This effect is strongest for Mercury

that inside Mercury's orbit there is circling an unknown planet, which had already been given the name Vulcan, because of its closeness to the hot sun. Already since 1907 Einstein had been familiar with the existence of this unexplained deviation. When in November 1914 he completed the very complicated calculations of the effect of his new theory for the orbit of Mercury, and found the result to be 42.89 s of arc per century—within the measurement errors of a few tenths of an arc second equal to the observed effect, he experienced—as his biographer Abraham Pais writes—one of the strongest emotional experiences of his life. "For days I was beyond myself of excitement", he later told his friend Paul Eherenfest, who in Leiden had succeeded Lorentz as professor of theoretical physics (Fig. 7.10). Indeed, a greater triumph for a theoretical physicist is hard to imagine. This result also definitively convinced Einstein of the correctness of the prediction that the deflection of light near the limb of the sun must be twice the value given by Newtonian gravity. Apparently before this time he had not been fully convinced of this, as Abrahm Pais remarks.

The Gravitational Redshift of Light

Einstein's General Theory predicts that in a stronger gravitational field time runs slower than in a weaker field. Light waves are an oscillatory phenomenon with a frequency ν given by $\nu = c/\lambda$, where λ is the wavelength of the light, and c the velocity of light. Since time runs slower in a gravitational field, the frequency of the light oscillations will be slower in a gravitational field, which means that the wavelengths

Fig. 7.10 Einstein (1879–1955), Ehrenfest (1880–1933), de Sitter (1862–1934)—in back row—and Eddington (1882–1944) and Lorentz (1853–1928)—foreground—in Leiden in 1923. Picture taken in the house of de Sitter at Leiden Observatory

of the light that is emitted by an atom or molecule will be longer than in the absence of gravity. A longer wavelength means that spectral lines emitted by atoms and molecules in the gravitational field of the sun and stars will be shifted towards the red end of the spectrum, relative to the wavelength we measure here on Earth (where the gravitational field is very weak). Also this phenomenon, a very small effect in the case of the sun, has been confirmed by observations of the sun and stars, particularly: for white dwarf stars, which have very strong gravitational fields.

A Fourth Classical Effect: The Shapiro Time Delay

Apart from the above mentioned three *classical* solar system tests of General Relativity put forward by Einstein, there is a fourth test that was overlooked by him, but realized by American physicist Irwin I. Shapiro in 1964. This test is a direct consequence of the curvature of space near a body with a gravitational field. As depicted in Fig. 7.8a, a light ray from a star that passes close to the sun goes through a "dip" in space before reaching us. This means that apart from its direction being deflected by the gravitational field of the sun, the light also *takes longer* to reach us than if it had moved along a perfectly straight line. Normally, this effect goes unnoticed, but this is no longer the case if we are dealing, for example, with radio signals from spacecrafts that for us are located on the opposite side of the sun: in this case the signals reach us with a delay because of their travel through the curved space around the sun. Also if the source of the radio waves is located between the sun and us, this effect already occurs. The first detection of this effect was in 1964 by Shapiro, using radar signals reflected by the planet Venus. He predicted that the reflected signal would arrive on Earth 200 μs later than expected, which was fully confirmed by the observations. Since then this "Shapiro delay" has been measured with high accuracy for many interplanetary spacecrafts.

Navigation with GPS

Without realizing it, hundreds of millions of people every day are using Einstein's General Theory of Relativity. In the Global Positioning Satellites system (GPS), used for navigation of cars and airplanes and for finding one's position on Earth, Einstein's General Theory is built in the software. GPS satellites contain extremely accurate atomic clocks. For navigation these have to be compared with identical atomic clocks on Earth. For this comparison two relativistic effects come into play: (i) the weaker gravity of Earth at the altitude of the satellite causes the atomic clocks in the satellites to run 45 μs per day *faster* than the clocks on the ground (a microsecond is one millionth of a second), and (ii) due to the speeds of the satellites in their orbits, the clocks in the satellites run 7 μs per day *slower* than the clocks on the ground (this is a Special Relativistic effect). These two relativistic effects combined cause the clocks in the satellites to run 38 μs per day faster than the clocks on the ground. This looks like a very small effect. However, since this is a time difference *per day,* the time difference between the satellite clocks and the clocks on the ground increases with time and after a number of days becomes larger and larger. If one is unaware of this effect, and assumes that the clocks on the ground give the correct time for the satellites, one can make large errors with navigation and with finding one's place on Earth. The GPS system is primarily a military system, designed by the US Air Force for guiding rockets to their targets. If one would ignore the relativistic clock corrections, the rockets will miss their targets by a large margin. When the military engineers developed the GPS system, Colorado physicist Peter Bender drew their attention to the above-mentioned relativistic effects. As the engineers were not fully sure that this is really important, they included the relativistic effects in their software as an option that they could ignore or switch on. It then appeared that with the relativistic effects *off*, guiding of rockets to their targets was inaccurate, while with the effects *on*, rockets were nicely guided to their targets. This convinced everybody that these relativistic effects are real and important, and since that time they are built in standard in the GPS system. Everyone who uses GPS navigation in a car or plane therefore uses the fruits of Einstein's thinking. Who could in 1915 have dreamt this, about this esoteric pure scientific theory of curved spaces, which at the time seemed to have no relation to our everyday life? In the meantime General and Special Relativity thus have become an everyday engineering tool, and are indispensable for the billion dollar industry of navigation!

The Origin of Inertia

Einstein's insight that gravity and the forces experienced in accelerated motions are equivalent, has as a deeper meaning that the cause of the inertial forces that occur for accelerated motions, is to be found in gravity itself. According to modern insights, inertia is the result of the gravitational attraction of the entire universe, as had already been suggested in the second half of the nineteenth century by Austrian physicist and philosopher Ernst Mach (1838–1916). Einstein has remarked that in the development of his thinking about gravity, he was greatly influenced by Mach. Since its discovery by Newton, the amazing equality of gravitational mass and inertial mass has occupied the minds of countless scientists. As explained in the third paragraph of this chapter, at first glance the *inertial mass* of an object, which determines how large a force has to be applied to the object to accelerate it to a certain speed, has nothing to do with the *gravitational mass*, which determines how strong a gravitational force is exerted by the object. In his *Principia*, Newton had thought to have solved the problem of inertia, by proposing that there exists an *absolute space* which, also if there is no matter (no stars, planets, etc.) present in the universe, still remains noticeable and present. He thought that inertial forces occur as soon as an object is accelerated with respect to this absolute space. From the point of view of Mach's positivistic philosophy this absolute space, which can never be observed, is scientifically unacceptable. According to him, all science should ultimately be based on observations, and scientific theories, in order to be confirmed or rejected, should be testable by observations or experiments. In fact, already 20 years after Newton, the great Irish philosopher George Berkeley (1685–1753), also known as Bishop Berkeley, had put forward similar criticism of Newton's absolute space. According to him, the introduction of this concept was completely unnecessary, since he argued that the system of the stars fulfils the function of Newton's absolute space. This is because one can determine whether one is accelerated, by looking at the stars in the sky. Berkeley illustrated this by considering the acceleration experienced in a merry-go-round. Sitting in the merry-go-round and looking upwards to the stars, one notices that when one sees the starry sky rotating, one experiences inertial (centrifugal) forces, and when one does not see the starry sky rotating, one does not experience inertial forces. Therefore the stars are the background with respect to which one notices inertial forces: if one observes not to be accelerated with respect to the stars, one does not experience inertial forces, and if one observes to be accelerated with respect to the stars, one experiences inertial forces (see Dennis Sciama: *The Unity of the Universe*, Chapter VII, Faber and Faber, London 1959). In 1872, Mach went still one step further. To explain the mystery of the equality of inertial and gravitational mass, he proposed that inertia is *caused* by the joint gravitational attraction of all the stars in the universe. His proposal implies that as long as one moves with a uniform (fixed) velocity, one does not notice the gravitational attraction of the stars, but as soon as one changes speed, the attraction of the stars becomes noticeable, which we experience as an inertial force. The gravitational attraction exerted by all the stars in the universe combined opposes, as it were, our wish to change speed. Einstein was very intrigued by this idea, and his General Theory appears to confirm Mach's proposal. It appears that if one calculates, using General Relativity, the combined gravitational influence of the stars, inertia indeed originates from the combined attraction of galaxies, largely in the very distant universe. The contribution of nearby stars in our Milky Way Galaxy and nearby galaxies appears to be negligibly small, but the more distant the galaxies are, the larger their contribution. This is due to the fact that the number of galaxies—and therefore their mass—increases with the third power of the distance; as a result the galaxies at the outer boundaries of the observable universe contribute most (e.g. see I.R. Kenyon: *General Relativity*, Oxford Science Publications, Oxford, 1990, p.150).

It is amazing to realize that the forces we experience day to day in accelerating and braking cars, trains, airplanes and elevators, are due to the entire universe pulling at us. Can one imagine a better illustration of our intimate connection with the entire universe?

Can Newton's Theory of Gravity now go into the Garbage Bin?

It would be a grave mistake to think that now that we have the General Theory of Relativity, Newton's theory has become obsolete and can be discarded. This is not the way natural science works. Einstein's theory is an improvement and refinement with respect to Newton's theory, but for almost all gravitational forces and accelerations that we encounter in our daily lives the predictions of Newton's theory are still extremely accurate. As we have seen, in the solar system the General Relativistic deviations from Newton's theory are very small effects, and hard to measure. For this reason for almost all applications in the solar system, Newton's theory is still extremely well-suited, for example for the computation of the orbits of comets and planets and for launching spacecrafts to the moon and planets. Only when extreme precision is required, as with GPS navigation, or when gravity becomes very strong, as close to a neutron star or a black hole, the use of Einstein's General Relativity is required.

In physics, new theories, such as General Relativity or Quantum Mechanics, are refinements and improvements of the older ones, such as Newtonian gravity and classical mechanics, respectively. But this does not mean that Newtonian gravity or classical mechanics have become useless: they still remain extremely important and useful theories that in their domains of application (in the case of classical mechanics: the macroscopic domain) remain valid to high accuracy.

Postcards from Einstein to de Sitter in 1917

Chapter 8
Einstein, de Sitter, Friedmann, Lemaître and the Evolution of the Universe

The secret of science is to ask the right questions, and it is the choice of the problem more than anything else that marks the man of genius in the scientific world

Sir Henry Tizard (1885–1959)—British Chemist

As we saw in the last chapter, gravity is an extremely weak force, which becomes noticeable only when a large amount of mass is brought together, as in a star or a planet. Since the gravitational force is always attractive and—contrary to electric and magnetic forces—cannot be screened off, all stars, galaxies and galaxy clusters attract each other out to the largest distances. The gravitational attraction of a celestial body extends throughout the entire cosmos. This means that the combined gravity of all celestial objects together determines the structure and evolution of the universe. Newton already realized this. He asked himself the question what would have happened if originally all matter had been distributed completely smoothly (evenly) throughout the universe. He realized that if then, purely by accident due to random motions of particles, in some place the density of matter became slightly higher than elsewhere, such an over-dense region would exert more gravitational attraction to its surroundings than neighbouring regions, and that as a result matter from the neighbourhood would start moving towards the over-dense region, such that there the over-density would grow further. This would then cause a "snowball effect" of growth of that region, resulting in matter to be collected in a star or a planet. Newton thus realized that purely due to the workings of gravity an originally completely smooth distribution of matter would not have lasted forever, and would

E. van den Heuvel, *The Amazing Unity of the Universe*,
Astronomers' Universe, DOI 10.1007/978-3-319-23543-1_8

after some time have fragmented into regions of enhanced density, resulting in the formation of stars and planets. Our present ideas about the formation of stars and planets from interstellar clouds are still based on the occurrence of this type of *gravitational instability* (see further in Chap. 14 and Appendix E).

But how does this proceed for the entire universe and how does gravity act on this very large scale? This was the question which Albert Einstein asked himself in 1917. He realized that his General Theory, as a new theory of gravity, would be excellently suited for obtaining new insights about the structure of the universe.

The Cosmological Principle

In his computations on the structure of the universe Einstein assumed that the large-scale distribution of matter in the universe is *homogeneous,* which means that on average everywhere the density of matter is the same. And also that the universe is *isotropic*, which means that in every direction it looks the same (no preferred directions). These two assumptions together are called *the cosmological principle*. This principle means that if you are taken blindfolded to an arbitrary place in the universe, and open your eyes, you can never find out where you are and in what direction you are looking. In fact, this is an extension of the *Copernican Principle*, which states that in no way we occupy a preferred position in the universe. Of course, there will always be small local deviations from the average density—in galaxies and clusters of galaxies—but seen from a large distance, according to the cosmological principle, the universe is completely smooth, just like Earth, if seen from a large distance, looks like a smooth sphere, while seen in close-up its surface has small irregularities: mountains, valleys and oceans.

The outcome of Einstein's calculations about the universe was shocking to him. He found that as a result of the attractive force of gravity, the matter of the universe has a tendency to fall towards each other, so the universe as a whole would be shrinking. Einstein asked his astronomical colleagues if they saw any tendency of the world of the stars to be shrinking. In 1917 the astronomers knew only about the stars in our Milky Way system (in fact only a small part of it, the "Kapteyn system"), and they thought this was the entire universe. Since the Milky Way system shows no signs of contraction they told Einstein that there are no signs that the universe is contracting.

Also Einstein himself thought that the idea of a contracting universe was quite absurd. Already since antiquity, thinkers like Aristotle and Copernicus had been of the opinion that the universe is static and eternal, and that the starry world had always existed in its present form. This is a very natural thought and also Einstein thought that the universe must be static and unchanging. For this reason, when he saw that the computations with his new theory resulted in a contracting universe, he thought that there was something missing in his theory, in its simplest and most elegant form which he had just published. He therefore searched for a term that was allowed to be added to his equations without ruining its over-all structure, and which would prevent the universe from collapsing. The mathematical equations of Einstein's original General Relativity Theory did allow the addition of an extra term, which on the scale of the universe appears to work as a *repelling* kind of gravity, with a strength that increases with distance. Einstein found that with the addition of this term the universe could be prevented from collapsing. On the small scale of our solar system and even on the scale of the Milky Way system, the force produced by this term is so small that it goes unnoticed. Only on the scale of the entire universe it becomes noticeable.

This extra term for a repelling type of gravity was denoted by Einstein by the capital Greek letter Λ (lambda; at first he wrote it with a small letter λ), and he described it as a "so far unknown universal constant". Later the name *cosmological constant* was introduced for this term. By adding this Λ-term to his equations, Einstein had thought that he had obtained a static universe in a state of equilibrium, which does not shrink or expand, and in which the Λ-term exactly compensates for the attractive gravitational forces that would cause the universe to contract. The model which Einstein obtained in this way can be mathematically described as a three-dimensional sphere, curved in a four-dimensional space. Humans cannot visually imagine such a curved three-dimensional sphere, since we cannot visualize a four-dimensional space. Our visual imagination does not extend beyond three dimensions. In order to nevertheless get some idea of this model, we can compare it with an ordinary two-dimensional spherical surface, that is curved in three-dimensional space, for example: the surface of Earth, or the surface of a soccer ball. Such surfaces are two-dimensional because two coordinates are sufficient to characterize any point on the surface (on Earth: the coordinates of geographical latitude and longitude). That this spherical surface is curved in our three-dimensional space does not conflict with its two-dimensional character. Mathematically, such a spherical surface is the collection of all points that have the same distance to one point: the central point of the sphere. In the same way one can also define a one-dimensional "sphere" in a two-dimensional space: this is circle in a flat plane, as depicted in Fig. 8.1. It is easy to see that a circle in a two-dimensional plane or a sphere in three-dimensional space have a finite size (dimension). Nevertheless, on a smooth spherical surface, or along a circle in a two-dimensional plane, one can walk forever without ever meeting a boundary. After going around once, one returns to the same point, but one never meets a boundary. A spherical "space" is therefore finite but boundless.

A second important fact about spherical spaces is that nowhere on a spherical surface or on a circle one can find a centre. The centre of the one-dimensional circle is located in the two-dimensional flat plane in which this circle is located, but it is *not on* the circle itself. And the centre of the two-dimensional spherical surface is located in the three-dimensional space *inside* the spherical surface, but *not on* this surface itself (see Fig. 8.1a, b). Now we can understand that a three-dimensional spherical "surface" that is curved in a four-dimensional space, can have a finite size but nevertheless has no boundaries anywhere, and has no centre on this "surface". It is boundless but finite. In such a three-dimensional spherical space you can keep walking forever without ever meeting a boundary. But just like on a two-dimensional

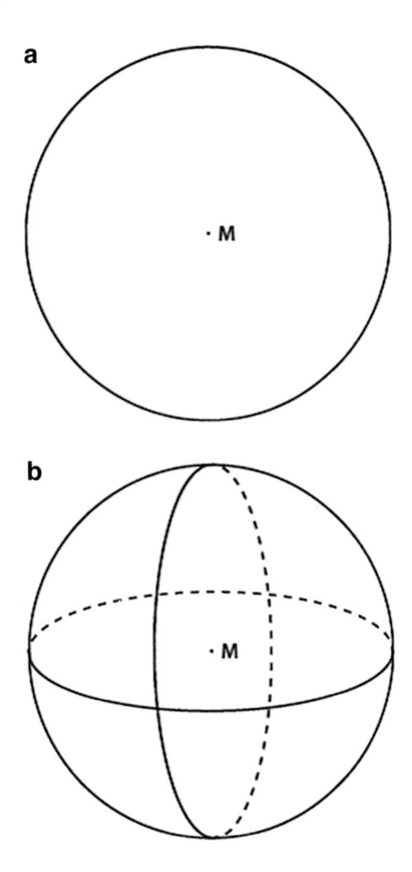

Fig. 8.1 (**a**) A one-dimensional 'sphere' is a circle in two-dimensional flat plane. The centre of this 'sphere' is *not* located *on* the sphere, but *in* the two-dimensional plane, in the centre of the circle. (**b**) A two-dimensional sphere is a super-thin spherical surface (like a Christmas ball) in three-dimensional space. The centre of this sphere is *not* located *on* the sphere, but *in* the three-dimensional space, in the centre of the sphere. In the same way a three-dimensional sphere—the Einstein model of the universe—has no centre in this spherical universe, but in the four-dimensional space-time

spherical surface, or along a circle, one will, after going around once, return to the same point. In Einstein's static equilibrium model of a three-dimensional spherical universe, obtained with his assumption of a cosmological constant Λ, a light ray will, after going around once, come back in its point of origin.

There was, however, a problem with this model of the universe, which Einstein had not considered. Although he had made sure that there was an equilibrium between the attractive gravity and the repulsive Λ-force, whose forces completely compensated each other on the scale of the universe, he had not checked whether this equilibrium was *stable*. A stable equilibrium situation of a system is such that if one disturbs the equilibrium, and then leaves it alone, the system by itself returns to the equilibrium situation. On the other hand, in the case of an unstable equilibrium, the system will, after a slight disturbance of the equilibrium, move further and further away from the equilibrium situation. An example of a stable equilibrium situation is that of a swing hanging from a horizontal beam, by two ropes. When it is at rest the ropes of the swing are perfectly vertical. If one moves the swing out of this equilibrium position and releases it, it will return by itself to the vertical position. It will overshoot through this position to the other side, but then move back to it, etc. Due to the drag of air it will move slower and slower and finally hang again vertically in its equilibrium position. On the other hand, an example of an unstable equilibrium is that of a pencil with a flat point, which one balances vertically on its point on a table. When a movement of air blows it slightly out of its equilibrium position, it will move further and further away from the vertical position and just fall on the table.

It has turned out that the equilibrium of Einstein's universe with the Λ-force is an unstable one, resembling the case of the pencil vertically balanced on its point. The slightest possible disturbance of this Einstein universe will cause it either to start shrinking (when the attractive gravity wins from the Λ-repulsion) or expanding (when the Λ-force wins) at an ever faster rate. The first one who discovered this by accident was Dutch astronomer Willem de Sitter (Fig. 7.10).

De Sitter's Discovery

When in 1916/1917 de Sitter started making calculations with Einstein's equations to which the Λ-term was added, he discovered something strange, namely that in such a universe the light of distant stars will become redder and redder: he found that the shift of the wavelength of the light towards the red increases in proportion to the distance: the further away an object, the more the wavelength of a spectral line is moved to the red (the wavelength increases). De Sitter's model was a strange one: since in the real universe the density of matter is very low (less than one atom per 33 cubic feet), he had ignored the presence of matter in the universe, such that there was just the repulsive gravity from the Λ-force. In this "empty" universe he found this strange effect of an increasing redshift of light with distance. De Sitter first thought that his universe, like that of Einstein, was static (not changing in time), but later he became aware that the redshift of the light of distant stars in his universe is due to the fact that his universe is actually expanding: the expansion produces a Doppler-effect which causes the redshift (the expansion of de Sitter's model was, in fact, discovered by Eddington, in 1922). De Sitter had made the first model of an expanding universe, although it was a rather peculiar one, as it contained no matter. In 1922, Eddington added matter to it, and it then became the Eddington-de Sitter model of an expanding universe.

Einstein's Λ-gravity in the de Sitter universe is just a repulsive force, which has nothing to do with matter. It is a force of the vacuum (empty space), which makes space expand. When the amount of space—the volume of the universe—increases, also the amount of this vacuum energy increases.

At first Einstein did not want to believe the results of de Sitter. This led to an extensive correspondence in 1917 between Einstein in Berlin and de Sitter in Leiden, the Netherlands. The latter country was neutral during the First World War which lasted from 1914 till 1918. Einstein's correspondence was conducted on post-cards, as during the war letters were opened by German censors, to check them for war-sensitive information. Einstein's postcards from this time were discovered in the archives of Leiden Observatory and are now in Leiden's Boerhave Museum. Figure 8.2a, b show two of these postcards, of April 14 and July 31, 1917. De Sitter,

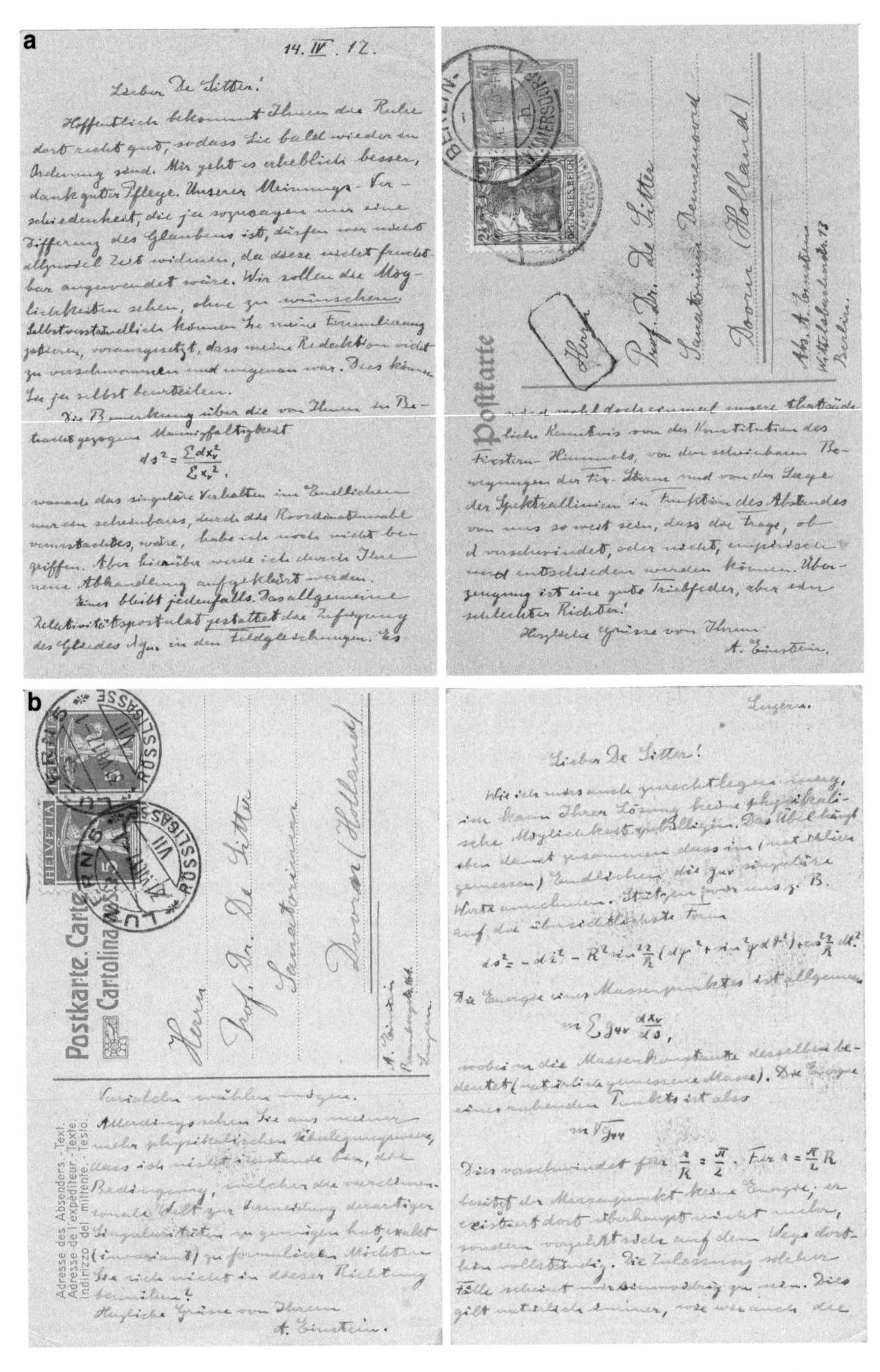

a

14. IV. 17.

Lieber De Sitter!

Hoffentlich bekommt Ihnen die Ruhe dort recht gut, sodass Sie bald wieder in Ordnung sind. Mir geht es erheblich besser, dank guter Pflege. Unserer Meinungs-Verschiedenheit, die ja sozusagen nur eine Differenz des Glaubens ist, dürfen wir nicht allzuviel Zeit widmen, da diese nicht fruchtbar angewendet wäre. Wir sollen die Möglichkeiten sehen, ohne zu wünschen. Selbstverständlich können Sie meine Formulierung zitieren, vorausgesetzt, dass meine Redaktion nicht zu verschwommen und ungenau war. Dies können Sie ja selbst beurteilen.

Die Bemerkung über die von Ihnen in Betracht gezogene Mannigfaltigkeit

$$ds^2 = \frac{\sum dx_\nu^2}{\sum x_\nu^2},$$

wonach das singuläre Verhalten im Endlichen nur ein scheinbares, durch die Koordinatenwahl vorgetäuschtes wäre, habe ich noch nicht begriffen. Aber hierüber werde ich durch Ihre neue Abhandlung aufgeklärt werden.

Eines bleibt jedenfalls. Das allgemeine Relativitätspostulat gestattet die Zufügung des Gliedes $\lambda g_{\mu\nu}$ in den Feldgleichungen. Es wird wohl doch einmal unsere thatsächliche Kenntnis von der Konstitution des Fixstern-Himmels, von den scheinbaren Bewegungen der Fix-Sterne und von der Lage der Spektrallinien in Funktion des Abstandes von uns so weit sein, dass die Frage, ob λ verschwindet, oder nicht, empirisch wird entschieden werden können. Überzeugung ist eine gute Triebfeder, aber ein schlechter Richter!

Herzliche Grüsse von Ihrem

A. Einstein.

Postkarte

Herrn Prof. Dr. De Sitter
Sanatorium Dennenoord
Doorn (Holland)

Abs. A. Einstein
Wittelsbacherstr. 13
Berlin.

b

Postkarte. Carte postale. Cartolina postale

HELVETIA

Herrn Prof. Dr. De Sitter
Sanatorium
Doorn (Holland)

A. Einstein
[illegible]
Luzern.

Adresse des Absenders. - Text.
Adresse de l'expéditeur. - Texte.
Indirizzo del mittente. - Testo.

Luzern.

Lieber De Sitter!

Wie ich mir noch zurechtgelegt [illegible], im Rahmen Ihrer Lösung keine physikalische Möglichkeit [illegible]. Das Übel hängt eben damit zusammen, dass im (mit [illegible] gemessen) Endlichen die $g_{\mu\nu}$ singuläre Werte annehmen. Stützen Sie [illegible] z. B. auf die übersichtlichste Form

$$ds^2 = -dr^2 - R^2 \sin^2\frac{r}{R}(d\psi^2 + \sin^2\psi\, d\vartheta^2) + \cos^2\frac{r}{R}\, dt^2.$$

Die Energie eines Massenpunktes ist allgemein

$$m \sum g_{4\nu} \frac{dx_\nu}{ds},$$

wobei m die Massenkonstante desselben bedeutet (natürlich gemessene Masse). Die Energie eines ruhenden Punktes ist also

$$m\sqrt{g_{44}}.$$

Dies verschwindet für $\frac{r}{R} = \frac{\pi}{2}$. Für $r = \frac{\pi}{2}R$ besitzt der Massenpunkt keine Energie; er existiert dort überhaupt nicht mehr, sondern vergeht [illegible] auf dem Wege dorthin vollständig. Die Zulassung solcher Fälle scheint mir bedenklich zu sein. Dies gilt natürlich immer, wie man auch die Variablen wählen mag.

Allerdings sehen Sie aus meiner mehr physikalischen Überlegungsweise, dass ich nicht imstande bin, die Bedingung, welcher die vierdimensionale Welt zur Vermeidung derartiger Singularitäten zu genügen hat, exakt (invariant) zu formulieren. Möchten Sie sich nicht in dieser Richtung bemühen?

Herzliche Grüsse von Ihrem

A. Einstein.

Fig. 8.2 (**a**) *Above*: postcard from Einstein to de Sitter of April 14, 1917, on which Einstein states, in the last two lines (*left*), that his General Relativity postulate allows to add a λ-term. (**b**) *Below*: postcard from Einstein to de Sitter of July 31, 1917

who suffered from tuberculosis, was at the time in the sanatorium "Dennenoord" in Doorn, a village in the Dutch countryside between Utrecht and Arnhem. On April 14, 1917 Einstein begins his writing with the wish that the rest in the sanatorium will do de Sitter well, such that he soon will be better again. He then states that it is going better with his own health thanks to being well cared for (he had suffered from intestinal problems, in the aftermath of leaving his first wife, and was now living with his cousin Elsa, whom he in 1919 married, and who cared for him very well). He then writes that his General Theory of Relativity does allow the addition of the cosmic repulsion, for which on this card he still uses the small Greek letter λ (lambda). Apparently, de Sitter at first had doubts whether the addition of this term was permitted. Einstein then expresses his expectation that at some time in the future our understanding of the stellar system will become well enough to establish whether the repulsive force which he introduced, really exists or not. On the postcard of July 31 Einstein raises objections against the occurrence of a singularity in de Sitter's model of the universe. In a singularity the values of a number of physical parameters, such as density and temperature, become infinitely large, and the dimensions infinitely small (formally: zero). (It later has turned out that this singularity really is there: it corresponds to the Big Bang; so Einstein's objections were not correct here.)

The problem with this discussion was that Einstein too much trusted the knowledge about the universe derived from astronomical observations at that time, which appeared to indicate that the universe is *static*, whereas his equations without the Λ-term produced a *dynamic* universe, that is contracting (or expanding, as was later discovered by Friedman). De Sitter, on the other hand, discovered that also a universe with a Λ-term is not static or stable, but does expand. To quote Indian astrophysicist Jayant Narlikar: "Einstein's model of the universe had matter but no motion, and de Sitter's universe had motion but no matter".

In any case, neither Einstein nor de Sitter had succeeded in finding a *general solution* of Einstein's equations. That honour was reserved for Russian physicist Alexander Friedman, who discovered that Einstein's original equations, without the Λ-term, allowed contracting as well as expanding universes, and that for obtaining an expanding universe the Λ-term is not needed.

Until the 1930s, Friedman's work went largely unnoticed, and just around the time when Friedman found his solutions, in the mid-1920s, de Sitter's model became famous, thanks to the work of Eddington. This author ascribed the redshifts of galaxies observed by Slipher to the expansion of the universe predicted by de Sitter's model. From this moment on, until the 1930s, the astronomical community worldwide believed that the expansion of the universe was produced by Einstein's Λ-term for repulsive gravity.

Friedman's Solutions

Physicist Alexander Friedman (1888–1925, Fig. 8.3) studied and worked in Leningrad (now: Sankt Petersburg), where before the First World War the later Leiden professor Paul Eherenfest (Fig. 7.10) was one of his teachers. Friedman's

Fig. 8.3 Russian physicist Alexander Friedman (1888–1925) discovered the general solutions of Einstein's equations for the evolution of the universe

father was a ballet dancer who later became a composer, and his mother was a concert pianist. Friedman, however, chose for an entirely different career: he was an outstanding theoretical and experimental physicist and mathematician. His scientific work, in the service of the young Soviet Union, was focused largely on meteorology. During the First World War he founded the navigation service of the Russian Air Force, at that time still in the service of the Czar. He was an excellent pilot and developed successful bombing strategies. Later he founded the meteorological service of the Soviet Union and made balloon flights high into the atmosphere to make meteorological measurements. After one of these flights, in August 1925, when he had reached an altitude of 7100 m (22,000 ft) he landed far out in the countryside, where he was helped by hospitable local farmers. The hygienic situation there was, however, very poor and he contracted typhoid fever. After his return to Leningrad, he died on September 16 of that year.

Friedman was an outstanding and versatile scientist who, next to his experimental work, had completely acquainted himself with the new theories of quantum mechanics and relativity, on which he taught his students and wrote textbooks. He taught numerous classes in all areas of physics, and in his meteorological research he collaborated with scientists in Norway and Germany—countries in which he had received part of his scientific education. Also, in the spring of 1924 he visited my country, the Netherlands where he participated in "The first International Congress for Applied Mechanics" in Delft Technical University. Here he presented a paper on the hydrodynamics of compressible media, an important subject in meteorology. Undoubtedly he must at that time also have visited his former Leningrad teacher Paul Ehrenfest in Leiden University (only 14 miles from Delft), but I could not find a record of this.

Between all his duties, as an institute director, research leader and teacher, Friedman made computations about the General Theory of Relativity. In this work he discovered in 1922/1923 that if one assumes that the universe is not static but dynamic—which means that its structure changes with time—Einstein's original equations, without the Λ-term, for a universe that contains matter, allow solutions that can contract as well as expand. In contrast to Einstein and de Sitter, who had found only solutions for some special cases, Friedman had succeeded in finding the *general solutions* of Einstein's original equations. Friedman discovered that without the Λ-term for repulsive gravity, in fact all solutions can start with an expanding universe.

In such an expanding universe there are two forms of energy: the positive energy of motion, also called *kinetic energy*, of the expanding matter, and the negative potential energy of the attractive gravitational forces that all stars and galaxies exert on each other. The expansion of the universe is counteracted by these attractive gravitational forces, which gradually slow down the expansion. One of the basic laws of physics is that the *total energy* of a system—which is the sum of the kinetic energy plus the gravitational potential energy—cannot change. This is the law of the conservation of (total) energy. Because of the cosmological principle, which states that everywhere the universe, on average, looks the same, one can apply this conservation law on any arbitrarily chosen element (amount of matter) in the universe. For such a limited amount of matter the total energy is conserved during the expansion of the universe, as explained in Appendix C.

The absolute value of the potential energy of gravity (that is its value without the minus sign in front of it) is large when the matter is packed closely together, that is: when the universe has a small size, and goes to zero when the universe becomes very large, that is: when the matter particles are at very large distances from each other.

During the expansion, the slowing down of the expansion due to the attractive gravity forces will decrease the kinetic energy, while the potential energy of gravity becomes less and less negative: the growth of the potential energy eats up, as it were, the kinetic energy.

When the total energy of the universe is positive, one will in the end have the situation that the sum of kinetic and potential energy will still be positive when the universe becomes very large, that is: when the potential energy becomes zero, the matter will still have kinetic energy, so the universe will keep expanding and can reach in the end an infinite size. This Friedman solution is called an *open universe*. On the other hand, when the total energy is negative, the universe cannot expand to an infinite size: there will come a moment during the expansion at which the growing negative potential energy has completely eaten up the kinetic energy. At that moment the expansion stops, and after that the matter will start falling towards each other again: the universe begins to contract. This universe, which expands to a certain size, then stops and starts contracting again, is called a *closed universe*. The contracting solution of a closed universe is the one which Einstein had found in 1916/1917 and which induced him to introduce the Λ-term.

Exactly at the boundary between the cases of an open and a closed universe there is the universe with a total energy equal to zero. This is called a *flat universe*. In this case the expansion of the universe will come precisely to a standstill when the universe has reached an infinite size.

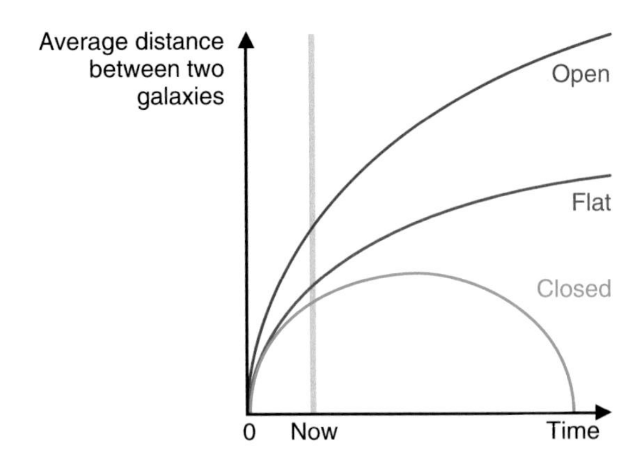

Fig. 8.4 The three different Friedman solutions for the evolution of the scale factor (the distance between two points in the universe) in the course of time. In all solutions the scale factor begins at time zero with value zero, after which its value increases. In an open universe, the scale factor keeps increasing forever, although the rate of increase decreases with time. For a closed universe, the scale factor increases up to a maximum value, after which it starts to decrease and reaches zero after a finite time (the 'Big Squeeze'). A flat universe is just at the border between these two cases. It lives forever, but after a very long time the scale factor does not increase anymore. At the time "now" indicated in the figure by the vertical green line, we observe an expanding universe for each of these three cases; the open universe expands faster than the flat one, and the flat one faster than the closed one. In order to find out which of these three cases applies to our expanding universe, one should accurately measure the density of matter and energy in our universe (see Chaps. 12 and 13)

Figure 8.4 depicts these three possible Friedman solutions for the universe. The figure shows how in these three models the distance between two points in the universe changes in the course of time. In fact, these three solutions are precisely the same as the three possibilities of what can happen to a stone that is thrown vertically upwards on Earth, as illustrated in Fig. 8.5. Due to Earth's gravitational attraction, the stone's velocity will slow down with increasing height. Whether it will be able to completely escape Earth's gravity will depend on how high the velocity was at which it was thrown upwards on Earth's surface. The escape velocity on Earth is the velocity that an object should have to be able to escape from Earth's gravitational attraction and just reach an infinite distance. That means: the sum of its kinetic energy and its (negative) gravitational potential energy at the moment it leaves the surface of Earth should be exactly zero; it then has just enough kinetic energy to be able to completely overcome the potential energy of Earth's gravitational attraction, and reach an infinite distance, where its velocity becomes zero. It turns out that the escape velocity from the surface of Earth is 11.2 km/s. If a stone is thrown upwards with a velocity lower than the escape velocity, it will reach a maximum height where its velocity becomes zero. After that it will fall down, as depicted in Fig. 8.5. On the other hand, if it is launched with a velocity larger that the escape velocity, it has so much kinetic energy that even when it reached an infinite distance, it still has kinetic energy left, so it still has a velocity larger than zero. These three possibilities for a

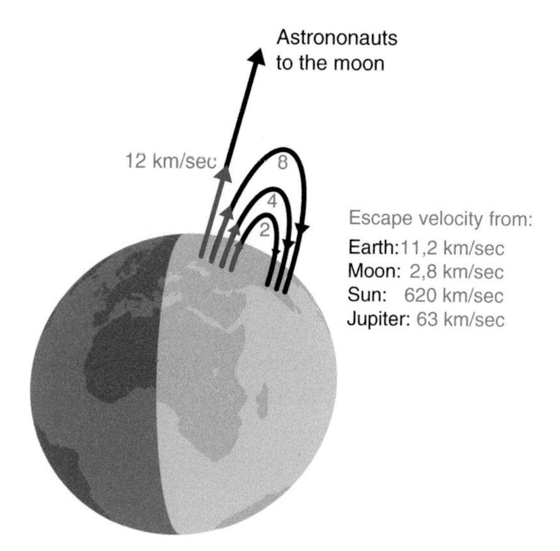

Fig. 8.5 The three Friedman solutions depicted in Fig. 8.4 are, in fact, completely similar to the three possibilities for a stone that is thrown upwards with different velocities. If the stone is thrown upwards with a velocity larger than the escape velocity (on Earth: 11.2 km/s) it will never come back and will, even at an infinite distance, still have some speed left ('open universe'). If it is thrown upwards with a speed less than the escape velocity, it will reach a highest point and then fall back ('closed universe'). If it is thrown upwards with exactly the escape velocity, it will go on to an infinite distance, where its velocity becomes exactly zero ('flat universe')

stone that is thrown upwards, depicted in Fig. 8.5, are exactly similar to the three different Friedman solutions depicted in Fig. 8.4. The stone that has negative energy and reaches a maximum height and then falls back is similar to the *closed* universe. The stone with just the escape velocity is similar to the *flat* universe, and the stone with velocity larger than the escape velocity is similar to the *open* universe. Friedman published his first paper, about the *closed* solution in 1922, and his second paper, with *open* solutions, in 1924, both in the German *Zeitschrift für Physik*. The closed solution has a finite size, but no boundaries, just like the Einstein universe. The flat and open universes are infinite: they extend to infinite distances.

Einstein's Reactions to Friedman's Discoveries

At first Einstein thought that Friedman's 1922 article was not correct, and on September 18, 1922 he published a letter to the editor of the same journal to point this out. Einstein was at the time still of the opinion that in models of the universe based on his own equations, time should not play a role. On December 6, 1922, Friedman in a personal letter to Einstein attempted to point out that this opinion was not correct. Einstein's error was that he beforehand had made the assumption that the universe should be static and eternal, and therefore does not change with time. Friedman pointed out to him that Einstein's own equations definitely allow time to play a role. It is not clear whether at the time Einstein realized the correctness of Friedman's arguments. Just at this time he was busy with a trip to Japan. It turned out that in the final resolution of this conflict unintentionally the Netherland played an important role. In May 1923 Hendrik A. Lorentz (Fig. 7.10), whom Einstein admired as a father figure, presented his farewell lecture as a professor at Leiden University. This lecture was attended by both Einstein and Friedman's Leningrad colleague Joeri Krutkov. Both were friends of Lorentz's successor Paul Eherenfest. At this occasion Krutkov was able to convince Einstein of the correctness of Friedman's work, which led Einstein to send an article to the Zeitschrift für Physik on May 21, 1923, in which he confessed his error and confirmed the correctness of Friedman's work[1].

Einstein's "Biggest Blunder"

At first nobody outside the Soviet Union paid any attention to Friedman's work. Only in the 1930s his work became accepted after the American Robertson and the Englishman Walker had found an elegant formulation of Friedman's general solutions of Einstein's equations. After the discovery of the expansion of the universe and the acceptance of the correctness of Friedman's work, people realized that Einstein's repulsive gravitation, produced by adding the Λ-term to his equations, was not at all needed to explain the expansion of the universe. This led Einstein to his statement, in a private conversation with George Gamow during the Second World War, that the introduction of the Λ-term was his *greatest blunder*.[2]

From this time on, de Sitter's model of the universe disappeared to the background and was viewed as a somewhat strange anomaly, until almost 70 later it began a second life, as will be described later, in Chaps. 13 and 14. Also, it was discovered that the three Friedman solutions appear in exactly the same way in a universe with Einstein's gravity as in one with Newtonian gravity. And also that the three Friedman solutions are indeed exactly equivalent to the three possible fates of a stone that is on Earth is thrown upwards, as depicted in Fig. 8.5, and explained in Appendix C.

Lemaître's Hypothesis of the Big Bang

In 1927 young Belgian Roman Catholic clergyman Georges Lemaître (1894–1966, Fig. 8.6) of Leuven's Catholic University, without knowing about Friedman's solutions, rediscovered these solutions for a universe filled with matter. Apart from

[1] E.A. Tropp, V.Ya. Frenkel and A.D. Cherni, *Alexander Friedman, the Man Who Made the Universe Expand*, Cambridge University Press, 1993.

[2] George Gamow, *My World Line* (autobiography), The Viking Press, New York, 1970, p. 44.

Fig. 8.6 Georges Lemaître and Albert Einstein, during a visit to the California Institute of Technology in January 1933

matter, Lemaître included in his model also a cosmological constant (Λ-term), which he needed to make his models fit the observations. His model therefore is a kind of combination of a Friedman model with a de Sitter model. When in 1927 Lemaître met Einstein at a Solvay Conference in Brussels and showed him his solutions, Einstein told him that Friedman had found similar solutions, after which Lemaître added a reference to Friedman in the manuscript of his paper. At this occasion Einstein also told Lemaître that he was very sceptical about such models and that not all mathematically correct solutions of physical equations lead to good physics. Lemaître, however, had just made an effort to give his models a solid physical formulation, such that it would fit the astronomical observations. He had taken seriously the observed redshifts of the spectra of galaxies observed by Slipher and Hubble, and interpreted these in terms of an expanding universe—and 2 years before Hubble formulated his famous redshift versus distance law, Lemaître had already found this same law, as mentioned in Chap. 6. Lemaître's now famous article of 1927 was published in French language in the *Annales de la Société Scientifique de Bruxelles*, and was entitled "*Un Univers homogène de masse constant et de rayon croissant rendant compte de la vitesse radiale des nébuleuses extragalactiques*" ("A homogeneous Universe with a constant mass and a growing radius, that takes into account the radial velocities of the extragalactic nebulas"). As we saw, Lemaître's model also contains a cosmological constant Λ. Without this repulsive gravity, his model would have been a closed Friedman universe, that reaches a maximum size and then contracts to a smaller radius. The cosmological constant, however, makes that the "repulsive gravity" at a certain moment

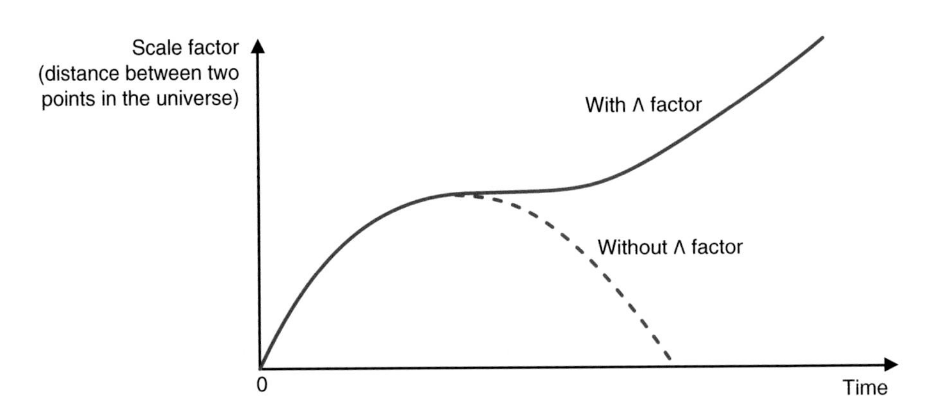

Fig. 8.7 Change of the scale factor of the universe with time for the Lemaître model. This model starts as an originally closed universe, to which Lemaître had added a cosmological constant Λ. Without this cosmological constant this universe would collapse again, as indicated by the dashed curve. The universe starts expanding, and then reaches almost a stand-still for a long time after which, due to the cosmological constant, it again resumes its expansion, and ends as an eternally expanding de Sitter model

overcomes the normal attractive gravity of thc matter, and pushes the universe further apart, such that it resumes its expansion, as depicted in Fig. 8.7. From that moment the solution approaches the expanding de Sitter universe. Lemaître had a very good observational reason for choosing just this particular model: the fact that otherwise he could not fit the age derived from the observed expansion of the universe with the age of Earth derived from the decay of radioactive elements. As mentioned in Chap. 6, both Hubble in 1929 and Lemaître already in 1927, had derived a value of the Hubble constant of about 190 km/s per million light-years. With this value, one can for either a pure de Sitter or a pure Friedman model calculate how long ago all matter of the universe would have been collected in one point: one just divides one million light-years by 190 km/s and gets then a time of *about one billion years* (we nowadays know that this Hubble constant was about ten times too large, and the age of the universe is therefore about ten times larger). Rutherford's discovery that the decay of radioactive elements can be used to determine the ages of rocks on Earth (see Chap. 2) had in the 1920s shown that Earth is at least several billion years old. As Earth cannot be older than the universe, Lemaitre had constructed his model such that, even with a Hubble constant of 190 km/s, it still was much older than one billion years. In his model the measured Hubble constant holds only for the later "*de Sitter part*" of the expansion of the universe, but the universe itself is much older: as depicted in Fig. 8.7 it started out as an expanding Friedman model, then went through a kind of a "*plateau phase*", during which its rate of expansion was close to zero, and then resumed its expansion in its later de Sitter phase. In this ingenious way, Lemaître was able to push back the time of the creation of the universe to a much earlier time than the one billion years of the *expansion age*, and make the ages of the Universe and Earth agree with each other.

After hearing about Hubble's 1929 paper, Eddington in England got very interested in Lemaître's work, and in 1930 he published a review of it in the British journal *Monthly Notices of the Royal Astronomical Society (MNRAS)*. Also, thanks to Eddington's efforts, Lemaître's 1927 article was translated in English and published in the *MNRAS*. As mentioned above, Lemaître had in his 1927 paper already derived a value of the Hubble constant that was practically the same as the one published by Hubble in 1929. But apparently Lemaître did not consider his discovery of the expansion of the universe as very important, because in the English translation of 1931 in the MNRAS he omitted the parts about this discovery, including his estimate of the Hubble constant. By doing this, he left the honour of the discovery of the expansion of the universe to Hubble, who, by the way, in his papers never referred to the work of Lemaître (See: http://hubblesite.org/newscenter/archive/releases/2011/36/full/).

Lemaître was the first who proposed that the universe started a finite time ago from an extremely dense and small situation. This was the *singularity* which was already present in de Sitter's model and which Einstein criticised in his postcard of July 31, 1917 (Fig. 8.2b). Later, in 1931, Lemaître presented, in the British scientific journal Nature, his famous idea of the *Primeval Atom*, which exploded at the beginning of the universe, creating the expanding universe and the atoms of the elements which are presently found all around us.

The Origin of the Name Big Bang

British astronomer Fred Hoyle (1915–2001), who for philosophical reasons could not accept a universe created in an event a finite time ago, as for him this smelled too much like an act of creation by a God, presented in 1946, in collaboration with Thomas Gold (1920–2004) and Hermann Bondi (1919–2005) (see Fig. 11.1) an alternative model, the so-called "*continuous creation*" model of the universe. According to this model at any point in time the universe, on average, looks the same as nowadays. In this model, due to the expansion, continuously new hydrogen atoms are created out of the vacuum in the expanding space between the galaxies. After some time new galaxies form from these new atoms, such that in the course of time the average density of galaxies (and matter) in the universe always remains the same. For this reason this model is also called the *Steady State* model of the universe.

When in 1949 Hoyle presented a popular astronomy program for the BBC radio, he mockingly said that according to Lemaître, the universe started with a *Big Bang*, a model with which he did not agree.

Since this time the word *Big Bang* has become the well-respected name of the now generally accepted model for the origin of the universe. Tragically, Hoyle till the end of his life kept discarding all additional new observational evidence for the Big Bang model and kept believing in his Steady State model.

Apart from the expansion detected by Lemaître in 1927 and by Hubble in 1929, in the past 50 years at least three other independent observational facts have been found indicating that the Big Bang has really taken place. We will explain these in the next chapter. For a long time it appeared that Lemaître's model of the universe would have been only of historical value. As described above, Lemaître had based his model on observations that produced an about ten times too large Hubble constant. When it was discovered that the real Hubble constant is about ten times smaller than he had assumed (about 20 km/s per million light years), and the universe therefore was about ten times older, his combination of a Friedman universe with a de Sitter Universe seemed no longer necessary. Until 1998 the idea was therefore that an ordinary Friedman model suffices to explain the observations. However, since 1998 the cosmological constant is back in the form of what is called "*dark energy*", a kind of vacuum energy in the universe (see Chap. 13) and the models of de Sitter and Lemaître have been revived in a new form. Still, the corrections to the Friedman model because of these new observations are quite small. For this reason, we will in what follows discuss the evolution of the universe in terms of Friedman models and only later return to the small corrections that have to be made to these models.

The Hubble Time

If one assumes that the universe has always been expanding with the present Hubble constant of about 20 km/s per million light years, it is a simple matter to calculate how long ago the entire universe was concentrated in one point: one just has to divide a distance of one million light years by the velocity of 20 km/s (this is because calculating backwards this is the time at which two points that presently have a distance of one million light years, were together in one point). The result of this division is about 15 billion years. This time is called the *Hubble time*. However, as explained in Appendix C, for Friedman models the "Hubble constant" is not constant in time, because the expansion of the universe did not always proceed at the same rate. For a flat Friedman model one finds an age of less than 2/3 of the Hubble time and for an open universe the age is between 2/3 and one Hubble time. As we will see later, the best estimate of the age of the universe, which does not precisely follow a Friedman universe, is 13.8 billion years.

Short Biography of Georges Lemaître

In 1911, at age 17, Lemaître started studying civil engineering at the Catholic University of Leuven. He interrupted his studies at the beginning of the First World war to enlist in the Belgian army, where he became an artillery officer. After the war he started studies in mathematics and physics and began also an

(continued)

(continued)
education to become a Roman catholic priest. In 1920 he received his degree of doctor in mathematics and physics and in 1923 followed his dedication as a Jesuit priest. In the meantime he had become interested in astronomy and he subsequently went to work 1 year in Cambridge in the UK with Eddington who introduced him to cosmology and stellar dynamics. Here he probable heard about the expanding Eddington-de Sitter model of 1923. After this year he left for the USA where he worked at Harvard College Observatory with Harlow Shapley, who just had discovered, in his work on globular clusters, that the centre of the Milky Way is tens of thousands of light years away from us in the direction of Sagittarius. After a year, Lemaître registered for a PhD education at the Massachusetts Institute of Technology (MIT) in the same Cambridge (USA) where Harvard University is located, where he studied and worked for another year. During his British and American years Lemaître undoubtedly heard of the work of Slipher on redshifts of galaxy spectra, and of Hubble's determination of the distance of the Andromeda Nebula in 1923, and of his subsequent work on galaxy distances, then a hot subject.

After his return to Belgium in 1925 he was appointed lecturer in astronomy at the Catholic University of Leuven. In 1927 he briefly returned to MIT to defend his PhD thesis on relativistic fluid mechanics and was awarded the American degree of PhD. Back in Belgium he was appointed that same year to full professor of astronomy in Leuven and published his famous article about the expanding universe.

The Cosmological Redshift

Galaxies at very large distances can attain redshifts larger than unity ($z = 1$; see Fig. 8.9). The ordinary Doppler formula ($z = \Delta\lambda/\lambda = v/c$) does not allow redshifts larger than unity, as a velocity (v) can never become larger than c. Here $\Delta\lambda$ is the increase in wavelength due to the Doppler effect and λ is the rest wavelength of the spectral line. In order to understand why redshifts can become larger than unity, we have to take a closer look at how the expansion of the *space* of the universe influences the wavelength of the light. Consider two points A and B in the universe which, due to the expansion of the universe move away from each other with a velocity v. The time taken by the light to travel the distance between A and B we call *the light travel time*. This time is equal to the distance between A and B divided by the velocity of light c. A light wave which leaves A will reach B after one light travel time. At its time of arrival, B has moved with respect to its original position over a distance of v times the light travel time. We call the increase in distance between A and B during the light travel time ΔR. If we divide ΔR by the distance R between A and B at the time of arrival of the light wave, the light travel time drops out and we see that in good approximation $\Delta R/R = v/c$, and this is equal to the redshift $z = \Delta\lambda/\lambda$. For very small values of the light travel time, ΔR will be very much smaller than R, so R will in fact be equal to the original distance between A and B, such that again $z = \Delta R/R$, where R now is the original distance between A and B.

The redshift therefore tells us how much the space between A and B has expanded between the time the light wave left A and the time it arrived at B. In other words, the light wave was stretched out by an amount $\Delta\lambda$ when space expanded by an amount ΔR. Thus: redshift is a measure for the expansion of space since the time the light was emitted. When the distance between A and B becomes very large, one can divide this distance into many small distances which all expand by a small amount ΔR for which the reasoning $\Delta\lambda/\lambda = \Delta R/R$ holds. The sum of all these small redshifts gives then the (large) redshift of the light waves arriving in B, which is the sum of all the small amounts of expansion of space, which add up to the total expansion of space at the time the light wave arrives at B. We thus see that the rule: redshift $z = \Delta\lambda/\lambda = \Delta R/R$ holds generally. We thus see that the redshift of a galaxy is a direct measure of how much the space has expanded since the light wave was emitted by the galaxy. One calls this the *cosmological redshift* (see Fig. 8.8). This means that redshifts can become larger than unity, because if space expanded by more than a factor of two since the light of the galaxy was emitted, the redshift of the galaxy will

Fig. 8.8 The expansion of the universe means that the entire space, including light waves in it, are stretched out. The result is that a light wave that was emitted when space was three times smaller than at present (*left figure*), now has become three times longer (*right-hand figure*). As the increase in wavelength $\Delta\lambda$ then is two times the wavelength λ, the redshift ($\Delta\lambda/\lambda$) is equal to 2. More generally holds that the light of a galaxy that we observe to have a redshift z, was emitted when the universe was $(z+1)$ times smaller than it is now. The largest redshifts that have so far been measured for galaxies are about $z = 10$

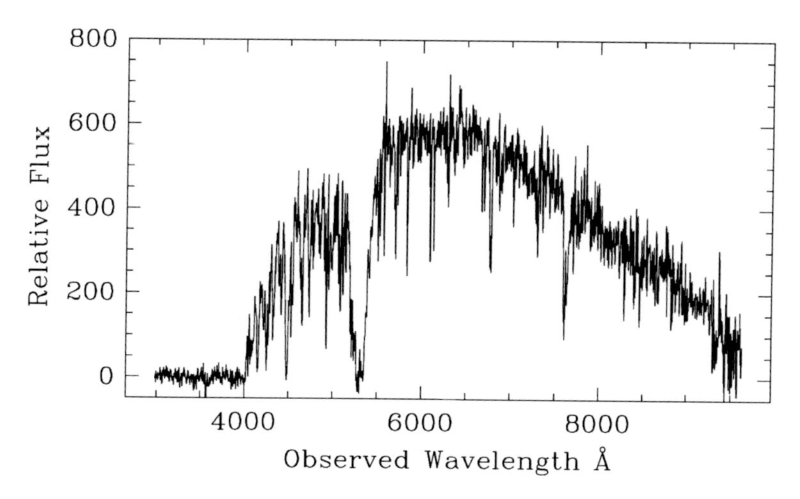

Fig. 8.9 Graph of the spectrum of the Gamma-Ray Burst of March 23, 2003, with a redshift of 3.4, as measured by our research group at the University of Amsterdam, using ESO's Very Large Telescope in Chile. The figure shows that the ultraviolet Lyman alpha line of hydrogen, with a laboratory wavelength of 121.5 nm (1215 Å), is shifted into the visible part of the spectrum, at a wavelength of 530 nm. When the Gamma Ray Burst was emitted, the universe was $(1+3.4)=4.4$ times smaller than at present, in linear measure, and its volume was 4.4^3 times smaller: about 85 times

become larger than one. For example, if since the light of the galaxy was emitted the space expanded by a factor four, the wavelength of the light we receive will have become four times the original wavelength, so the wavelength increase is three times the original wavelength, and the redshift will be three. In other words, if the redshift that we observe for a galaxy is z, the universe has increased in size by a factor $(z+1)$ since the time the light was emitted. These very large redshifts mean, in terms of the Doppler effect, that the velocity with which a galaxy flies away from us approaches the velocity of light c. In that case the simple Doppler effect formula $\Delta\lambda/\lambda = v/c$ is no longer valid. In Appendix C the correct formula for the Doppler effect is given that also holds for velocities approaching the velocity of light. Figure 8.9 shows as an example a graph of the spectrum of a Gamma Ray Burst with redshift 3.4. Gamma Ray Bursts are extremely bright stellar explosions that are observed in very distant galaxies. When the burst of Fig. 8.9 went off, the universe was 4.4 times smaller than nowadays. Since space is three-dimensional, the volume of the universe was at that time $4.4\times4.4\times4.4\sim85$ times smaller than today. The largest redshift measured so far in the universe is $z=11$, for a galaxy in the Hubble Ultra Deep Field. The light which we receive from this galaxy was emitted when the universe was less than 0.5 billion years old.

Can Galaxies Move Away from us with Velocities Larger than the Speed of Light?

If due to the expansion of the universe a galaxy that is twice as far away from us than another one moves twice as fast from us—as depicted in the raisin-bread model of the universe depicted in Fig. 6.9—then the situation could occur that a distant galaxy moves away with a velocity larger than the speed of light. At first glance this seems in conflict with the Special Theory of Relativity. But this is not the case. The reason is that it is space itself, which is expanding faster than light, and this is not in conflict with Special Relativity. That theory only forbids objects to move *with respect to each other*, and thus also, *with respect to space*, with a velocity larger than the velocity of light. If the space between two galaxies expands faster than light, a signal from one of these galaxies can never reach the other galaxy, and there is no problem with causality, because these galaxies will never be able to notice each other's existence.

How Large is the Universe?

The horizon of our observations is in theory located 13.8 billion years in the past. Light which was emitted at the time of the Big Bang would have an infinite redshift. We will see in the next chapter that we will never be able to look back that far and that our true observational horizon is at about 400,000 years after the Big Bang, when the universe became transparent. The heat radiation emitted by the gas that filled our universe at that time is nowadays still observable in the form of short-wavelength radio waves—so-called microwaves—with a redshift of about 1100. There are small ripples in the distribution of this radiation over the sky, which contain information about still much earlier phases of the Big Bang. Hopefully in the future this information can be extracted (see Chap. 14). Taking into account the expansion of the universe, the point from which we receive this microwave radiation from the Big Bang has in the meantime moved to a distance of about 50 billion light years from us. This is the largest distance to which we can look into the universe. Still the universe itself may be very much larger, and as we will see in the next chapters, there are indications it is thousands and perhaps millions or even billions of times larger than this horizon.

Colliding galaxies

Chapter 9
The Big Bang as Origin of the Universe

Just like on the battlefield, also in the field of science front posts seldom harvest fame

Jozef Eötvös (1813–1891) Hungarian physicist, writer and philosopher

Lemaître's idea of the Big Bang was that of a "Primeval Atom" of unimaginably high density and temperature. He thought of matter with a density similar to that of atomic nuclei (some 10^{14} times the density of water) and consisting of protons and electrons. However, in 1931 the knowledge of the physics of such matter (nuclear physics) was still so poor that he could not further support his model with reliable physical calculations. He speculated that his Primeval Atom would have fragmented and decayed, finally producing all known kinds of atomic nuclei which combined with the electrons and made atoms. At a later stage these atoms would have assembled into stars, planets and galaxies.

The first scientists who, shortly after the Second World War, started a thorough study of the physics of the Big Bang, supported by solid nuclear physics computations, were George Gamow (1904–1968; Fig. 9.1) and his student Ralph Alpher (1921–2007) and collaborator Robert Herman (1914–1997) (The three of them are depicted in Fig. 9.3).

Gamow, who was born in the Soviet Union, in Odessa, had Friedman as one of his teachers in Leningrad, where he obtained his PhD degree in 1929. He subsequently worked with Niels Bohr in Copenhagen and at Cambridge University

E. van den Heuvel, *The Amazing Unity of the Universe*,
Astronomers' Universe, DOI 10.1007/978-3-319-23543-1_9

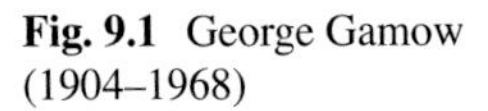

Fig. 9.1 George Gamow (1904–1968)

(UK) and made important fundamental contributions to nuclear reactions. Already before obtaining his PhD he worked some time at Göttingen University in Germany, which in these days had one of the most outstanding physics departments in the world. Here worked the later Nobel laureates Max Born and Wolfgang Pauli, who at that time were deeply involved in the development of quantum mechanics. In 1928 Gamow was one of the first in the world to apply quantum mechanics to nuclear physics. Prior to this, quantum mechanics had only been applied to light and electrons. The phenomenon of *tunnel effect*, which plays a key role in the quantum mechanics of nuclear reactions, was discovered by Gamow in the late 1920s.

Also the idea that atomic nuclei can be represented as drops of liquid with a surface tension was due to Gamow. That idea was later developed into the successful "liquid drop model" of nuclei by, among others, Niels Bohr and John A. Wheeler, who used it to explain the fission of uranium nuclei by neutron capture. This proved to be of key importance in the 1940s in the development of the atomic bomb and of nuclear reactors. Thanks to Gamow's theory of the tunnel effect, young German refugee Hans Bethe at Cornell University could in 1938 precisely calculate how the sun generates its energy by fusion of hydrogen nuclei. This earned Bethe the 1967 physics Nobel Prize. It remains a mystery why Gamow, who made so many fundamental contributions to physics, never himself was awarded this prize.

After his return to Russia in 1931 Gamow noticed that the scientific atmosphere had drastically changed. For example, the Relativity Theory of Einstein—one of Gamow's heroes—was no longer allowed to be taught and studied. Two years later, he and his wife were able to miraculously escape from the Soviet Union, after which he settled in the USA (see Gamow's autobiography "My World Line", Viking Press, New York 1970, and the biography "Gamow, van Atoomkern tot Kosmos", by Rob van den Berg, Veen Publishers, Amsterdam, 2011).

Gamow started his calculations on the Big Bang with the assumption that during the earliest phases, when the density was very high, also the temperature was very high. The reason for this assumption is simple: if one compresses a gas, its temperature increases, as anyone knows who has used a bicycle pump: the pumps get hotter due to the compression of the air. Calculating backwards in time, the matter in the universe was much more compressed and therefore must have been hotter. In his 1952 popular book "*The Creation of the Universe*" Gamow mentions that he believed the universe to be cyclical, which means: it expands to a maximum radius and then collapses again to an incredibly high density: the "Big Squeeze". During this collapse the temperature rises to an incredibly high value, and the thermal gamma radiation that is generated at this high temperature splits all atomic nuclei back to their elementary parts: protons, neutrons and electrons. Gamow proposed that when during this Big Squeeze the temperature exceeds a certain critical value, the collapse is halted and reverses into an explosion causing the entire evolution of the universe to start all over again. This would mean that the universe is presently in the phase following a new explosion and started out with an extremely high temperature and density. This model is called the *hot Big Bang*. Apart from these simple considerations, Gamow and his student Ralph Alpher had a very important other reason for proposing a very high temperature in the early phases of the Big Bang: the fact that over 70 % of the mass of the universe consists of hydrogen, the element with the simplest atomic structure. As explained above, at high temperatures very intense electromagnetic radiation ("heat radiation") is generated, with an amount of energy per unit volume (cubic feet or cubic centimetre) that increases as the fourth power of the temperature (this is Stephan-Boltzmann's law, see Appendix D). The enormous amount of photons at these high temperatures, which consist mainly of gamma rays, causes the heavier atomic nuclei to be split into the smallest elementary particles: protons, neutrons and electrons; and since the neutron is radio-active and decays in 11 min into a proton and an electron, the final product of this splitting process will be just a soup of protons and electrons. Protons are the nuclei of hydrogen atoms (see Appendix B). Had this heat radiation not existed at that time, a large part of the protons would with electrons have re-assembled into neutrons, and neutrons and protons would have assembled into heavier nuclei, such that the high percentage of hydrogen would have disappeared. This was discovered in 1948 by Alpher in his PhD research with Gamow. The fact that this large amount of heavier elements was not formed and that hydrogen remained the dominant element in the universe, can therefore only be understood in terms of a *hot Big Bang* with a very large number of photons (gamma rays) per nucleon (proton or neutron).

Blackbody Radiation and Planck Curves

Every object with a temperature above absolute zero temperature (zero degrees Kelvin = −273° C) radiates "heat" radiation of all wavelengths. When the temperature is very low, most of these waves are radio waves; when the temperature gets higher most of the emitted radiation becomes infrared radiation and at still higher temperatures most of the emitted waves become light

(continued)

(continued)

waves and waves of even shorter wavelengths. When one heats a piece of iron, it first starts to glow deep red, at higher temperatures it becomes white-hot and at still higher temperatures it glows in blue light. Around 1860 German physicist Gustav Kirchhoff discovered that the best possible radiators are objects that are completely black. When a black object is heated its starts radiating at all wavelengths: it radiates electromagnetic radiation from the shortest wavelengths (X-rays and gamma rays, ultraviolet radiation and light) to the longest possible wavelengths (infrared radiation and radio waves). Kirchhoff discovered that the most perfect "blackbody radiation" exists in a completely closed box, for example a metal box which is heated from outside. He also found that the distribution of the amount of energy radiated at different wavelengths depends *only on the temperature T* of the "black body" and is independent of the material from which the walls of the closed box is made. Figure 9.2 shows the distribution of the amount of radiation energy emitted at different wavelengths, for a number of different

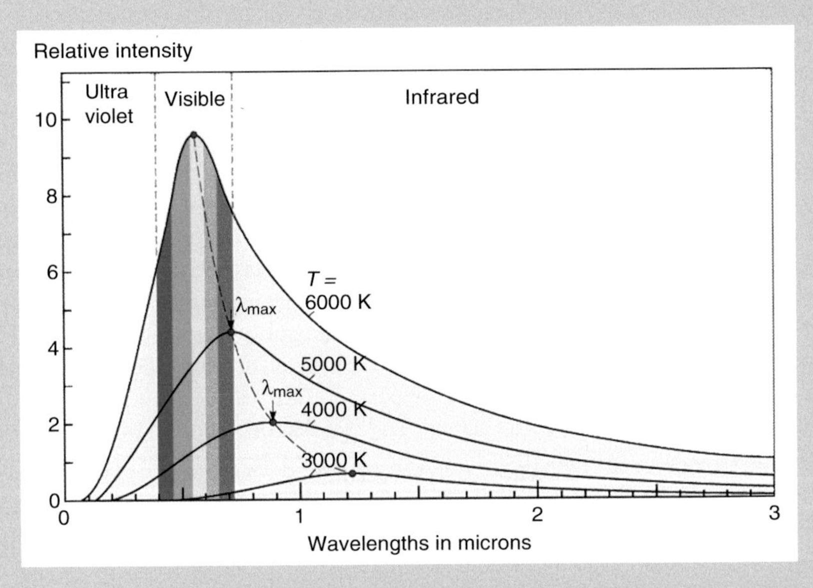

Fig. 9.2 Curves showing the amount of radiation energy emitted by an ideal black body per square metre per second, at different wavelengths. These curves, for different temperatures, are called *Planck curves*. The *total* amount of energy emitted by the black body, summarized over all wavelengths, is given by the area under the curves. This area increases as the fourth power of the temperature (*Stephan-Boltzmann's law*). One clearly sees that the wavelength where the curve reaches its maximum shifts to shorter wavelengths when the temperature increases: the wavelength of the maximum is inversely proportional to the temperature (*Wien's law*)

(continued)

(continued)
temperatures. The figure shows that with increasing temperature, the peak of the distribution of the emitted radiation shifts towards shorter wavelengths. This is called "*Wien's law*", after the German physicist which discovered this phenomenon. Wien discovered that the wavelength where most of the radiation energy is emitted is inversely proportional to the temperature. (The temperature is here: the absolute temperature, counted in Kelvins, that is: from −273 °C, which is the absolute zero temperature at which all motions of atoms and molecules have stopped. So, 10 K is −263 °C, 20 K is −253 °C, etc.). The area below the black body radiation curves is a measure for the *total amount of radiation energy* emitted by the black body, per square meter per second, summarized over all wavelengths. This area appears to increase as the *fourth power* of the absolute temperature. So, at a twice as large temperature, 16 times more energy is emitted, at a three times higher temperature, 81 times more energy is emitted, etc. This is called *Stephan-Boltzmann's law* (see Appendix D). All these laws of the behaviour of blackbody radiation were discovered by laboratory experiments in the second half of the nineteenth century. But it was only in 1900 that Max Planck succeeded in theoretically deriving these laws. As we saw in Chap. 7, for this he had to assume that electromagnetic radiation is emitted in the form of a kind of particles (photons = wave quantums = packets of waves). For this reason the blackbody radiation curves as depicted in Fig. 9.2 are also called *Planck curves*, and the radiation that a "black" object—which can also be a gas or a liquid—emits is called "*Planck radiation*".

Radiation in the Early Universe

In the hot Big Bang, at a temperature of trillions of Kelvins, the Planck blackbody radiation is incredibly intense. As mentioned above, at such temperatures, the peak of the Planck radiation curve is at gamma ray wavelengths, and this radiation will split any heavier nucleus that might form immediately back into protons, neutrons and electrons. At still higher temperatures, even the protons and neutrons are split into their still smaller constituents, the quarks. A proton consists of two up-quarks and one down-quark, the neutron consists of two down-quarks and one up-quark (see Appendix B for further information). In the very earliest phases, the universe consisted of *quark-gluon plasma*, a soup of quarks and gluons—the photon-like particles that bind quarks together in protons and neutrons. Gamow in the 1940s did not yet know about quarks and gluons, which were discovered only in the 1960s.

In a series of articles, written by Gamow, Alpher and Herman (see Fig. 9.3) between 1946 and 1949, they calculated how the different elements could have been created in the expanding early hot Big Bang universe. They assumed that the universe started out with only neutrons. The intense gamma radiation split part of

Fig. 9.3 Gamow (*middle*) with his students Ralph Alpher (*right*) and Robert Herman (*left*) who in 1949 predicted that nowadays still heat radiation from the Big Bang should be present in the universe. Alpher and Herman predicted this radiation to still have a temperature of about 5 K. Gamow had invented the name "Ylem" for the primordial material of the Big Bang, and in this picture one observes him to materialize from a bottle of Ylem

the neutrons in protons, electrons and anti-neutrinos, such that a soup of all these kinds of particles was created. During the further expansion and cooling of the universe also heavier nuclei were formed, largely by neutron capture. When a proton captures a neutron, a nucleus of *deuterium* ("heavy hydrogen") forms, which by neutron capture is transformed into ^{3}Helium, which in turn by proton capture transforms into ^{4}Helium, and so on. The intense gamma radiation splits a part of the thus-formed nuclei back into smaller particles. Gamow and Alpher found that to ensure that at the end of the "hot" period some 70 % hydrogen and some 28 % helium remains (as observed in the universe), a very strong radiation field is needed during the first few minutes of the expansion, of more than a billion photons per proton or neutron. After these first few minutes, the formation of nuclei had ended, because the gas had cooled due to the expansion. Gamow and his collaborators had hoped that their calculations would have produced nuclei heavier than helium, as Lemaître had proposed in his 1931 Big Bang model, but this proved to be extremely difficult. The problem they encountered was that in nature there do not exist stable nuclei with masses 5 or 8. If one sticks a neutron or proton to a ^{4}Helium nucleus, or a deuteron to a ^{3}Helium nucleus, a nucleus of mass 5 is produced, which in an extremely short time decays back into smaller pieces. And also, if one tries to stick two ^{4}Helium nuclei together to form a ^{8}Beryllium nucleus, that nucleus also is violently unstable and almost immediately fragments into smaller pieces. Without stable nuclei of masses 5 and 8, one cannot in the Big Bang succeed in making the nuclei with masses 6, 7, 9 and higher, which we find in nature. The problem in the early universe is that due to the rapid expansion, the duration of the period during which the temperatures were high enough to build up heavier nuclei, was very short, less than 3 min. If the phase of the Big Bang when the temperatures were high enough for element building had lasted very much longer, one could have reached a situation in which an equilibrium is established between formation and decay of the nuclei of masses 5 and 8. In such a situation there would have been a small amount of these nuclei always present in the universe, which could have been used as a stepping stone to build higher-mass nuclei by neutron capture. However, as the sufficiently hot phase lasted only a few minutes, never such an equilibrium could have been established, as Gamow and Alpher noticed in their calculations, and so no elements heavier than ^{4}Helium can have been made in the Big Bang. Later, in the 1950s, it was found By Fred Hoyle and Edwin Salpeter that the only places in the universe where a long-lasting hot and dense equilibrium state can be established, such that the barrier at nuclei with masses 5 and 8 can be overcome, is in the interiors of evolved massive stars, where for millions of years a hot dense equilibrium state can exist. Therefore these are the places where the elements heavier than helium are formed in nature.

As it later turned out, the computations of the late 1940s by Gamow and his collaborators also had not been fully correct. As mentioned above, they had assumed the matter in the universe to have started out only with neutrons, which subsequently decayed into protons, electrons and anti-neutrinos. It was well known at the time that a neutron decays, on average in about 11 min into these three other particles. However, in 1950, Japanese astronomer Chushiro Hayashi realized that in the hot early universe the conversion of neutrons into protons proceeds much faster, by collisions of neutrons with positrons (positively charged anti particles of electrons) and neutrinos. He realized that the intense gamma-radiation field in the hot early universe created gigantic numbers of pairs of electrons and positrons and of neutrinos and anti-neutrinos. The collisions of the neutrons with positrons and neutrinos enormously accelerated the conversion of neutrons into protons and electrons. In 1953, Alpher and Herman, together with James Follin, included these processes in improved Big Bang calculations and in this way carried out the first really correct calculations of the evolution of the ratio of the numbers of protons and neutrons in the universe, and the resulting speed of the formation of helium in the Big Bang. The later calculations of the physics of the Big Bang, carried out by other scientists around 1964, which will be discussed here below, basically added nothing new to the 1953 results of Alpher, Herman and Follin. After 1953, Gamow and his collaborators did not further continue this work, presumably because they realized that they could not succeed in explaining the formation of the heavier elements in terms of the Big Bang model. Their work, however, led to an enormously important prediction which, 15 years after it was made, was beautifully confirmed by a totally unexpected observational discovery.

The Prediction of the Cosmic Microwave Background Radiation

As already mentioned, the computations by Gamow, Alpher and Herman showed that due to the rapid decrease of the temperature in the expanding universe the phase of nuclear fusion in the early universe would be over after only a few minutes, and produced a few tens of per cents of helium. When the temperature dropped below 100 million Kelvin, fusion of hydrogen to helium was no longer possible, but the expanding universe was still very hot. Due to this high temperature the atomic nuclei and electrons had such high speeds that their collisions were extremely violent and kept the matter in the universe fully ionized: a neutral atom, if ever it formed, would be immediately ionized by the high energy collisions with other particles. The matter was therefore a mixture of the nuclei of hydrogen and helium and of electrons and neutrinos and anti-neutrinos. Apart from this, there were enormous numbers of photons: about one billion per proton. Free electrons have the property to scatter photons ("light rays") in all directions. A light ray could therefore in these early times travel only a very small distance in this dense soup of electrons and nuclei before meeting an electron and being scattered into another direction. As a result the universe was in these times completely opaque to light and other electromagnetic radiations.

Only when the temperature dropped very much further, this situation could change: at temperatures below 3000 K, the collisions of electrons and neutral hydrogen atoms are no longer powerful enough to ionize these atoms. So, when the temperature dropped below this value, the free electrons were captured by the protons and formed neutral hydrogen atoms (neutral helium had already formed at slightly higher temperatures), and the soup of electrons and ions was converted into electrically neutral hydrogen and helium atoms, as depicted in Fig. 9.4. As the free electrons had disappeared, the result of this "recombination" was that the universe had suddenly become fully transparent, and the light rays of the heat radiation of a temperature of 3000 K that still were present at that time could now traverse the entire universe without ever being absorbed or scattered. The calculations of Gamow and his collaborators in 1949 showed that the temperature of 3000 K was reached about 400,000 years after the beginning of the Big Bang, and that the heat radiation that was present at that time should still be present in the universe today. Due to the expansion of the universe, the wavelengths of the photons of this radiation were since very much stretched out. According to the calculations of Alpher and Herman in 1949 the universe had expanded by a factor 600 since it became transparent, such that all the wavelengths had increased by this factor, including the peak-wavelength of the Planck-distribution of the heat radiation. According to Wien's law (see Appendix D) this implies that the distribution of this radiation nowadays should correspond to a Planck curve for a temperature of 5 K (=3000/600). (It was later found that the universe has expanded by a factor about 1100 since it became transparent, such that the present temperature will be close to 3 K). Blackbody radiation of a temperature of 5 K peaks at a wavelength of 0.6 mm (see for example

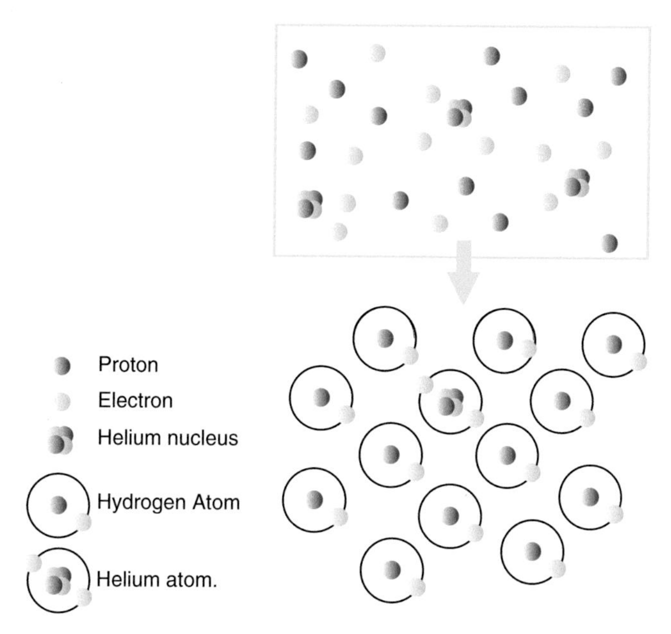

Fig. 9.4 *Top*: at temperatures above 3000 K the matter of the expanding Big Bang was still completely ionized. It was a soup of mostly hydrogen nuclei and free electrons (plus some helium ions). Free electrons strongly scatter in all directions the heat radiation that was present in the universe, and therefore made the universe completely opaque. *Below*: after the temperature dropped below 3000 K, the electrons disappeared as they were captured by the hydrogen nuclei and helium ions to form neutral hydrogen and helium atoms. Due to this 'recombination', which took place 380,000 years after the Big Bang, the universe suddenly became completely transparent: photons from that time could traverse the entire universe without ever being absorbed again or scattered. Alpher and Herman predicted in 1949 that the light waves of the heat radiation from these times—*reddish* light waves corresponding to a temperature of 3000 K—were stretched out due to the subsequent expansion of the universe by about a factor of 600, to become radio waves with wavelengths of millimetres to centimetres, corresponding to a blackbody radiation curve for a temperature of about 5 K (nowadays we know that the stretching was even larger, about a factor 1100)

Fig. 9.2). At this wavelength one calls the radiation *microwaves*, which are short-wavelength radio waves, the same type of radio waves that are used in a microwave oven in the kitchen.

When in 1949 Gamow's collaborators Alpher and Herman predicted that these microwaves from the Big Bang should be present in the universe, radio astronomy was still in a very primitive state. It would take until 1952 before the first radio spectral line was detected: the 21-cm wavelength line of neutral hydrogen, and the radio telescopes at that time were still very insensitive. Although Alpher and Herman informed radio astronomers of their prediction, these were sceptical both about the Big Bang model and about the possibility that these waves could be detected (see: R.A. Alpher and R. Herman in *Cosmology, Fusion and Other Matters*, F. Reines, *editor*, Colorado Associated University Press, 1972, pp. 1–14). It would therefore take until 1964 before this direct radio signal from about 400,000 years after the Big Bang was detected, and the way this happened was purely by chance and completely unintended, as many astronomical discoveries.

The Discovery of the Cosmic Microwave Background Radiation from the Big Bang

This discovery was made by two employees of the Bell Telephone Research Laboratories in New Jersey. These laboratories had in the early 1960s developed a special type of radio telescope—a so-called *horn antenna*—with an opening (*aperture*) of 6 m (Fig. 9.5). This instrument was originally built for testing satellite communication by means of microwaves, by beaming microwaves off the Echo satellite and receiving the reflected signal on the ground. Echo was a large gas-filled balloon of 100 ft diameter, covered with a thin layer of gold. It was in the early 1960s visible as a bright star slowly moving along the night sky. Bell Laboratory's technicians soon found out that reflection is not a very efficient way of satellite communication and the project was abandoned. Bell scientists Arno Penzias and Robert Wilson (Fig. 9.5) obtained permission to use the horn antenna for astronomical

Fig. 9.5 Arno Penzias (*right*) and Robert Wilson in front of the Holmdel radio telescope with which they in 1964 discovered the microwave radiation from the Big Bang

observations, for which is was very well suited because of its extremely low noise level. Penzias and Wilson wanted to use it for carrying out high-precision measurements of the absolute strength[1] of the radio emission of supernova remnants, such as Cassiopeia A—radio emission that is produced by electrons with relativistic speeds that are moving through the magnetic field of the supernova shell. A problem in radio astronomy is that astronomical radio sources are almost always very weak, and that apart from detecting the emissions from the source itself, there are all kinds of interfering background signals which are also recorded by the radio telescope. These may be due to nearby radio stations or to the Earth's atmosphere. That Penzias and Wilson succeeded to record with this telescope radio waves from the Big Bang was due to the very great effort they made for tracing all possible sources of background noise. To this end they improved the electronics and developed highly accurate ways to precisely calibrate their instruments. When a radio telescope is pointed at the sky and one subtracts the own noise of the receivers from the recorded signal, the signal that is left will consist of the noise produced by the radio telescope itself plus noise from the atmosphere and the noise from the astronomical radio sources that may be present in the sky.

In order to find out the noise produced by the telescope and the atmosphere Penzias and Wilson pointed the telescope at places on the sky where no astronomical short-wave radio sources are known, far outside the plane of the Milky Way. They carried out their observations at a wavelength of 7.35 cm, where absolutely no source signal was expected. At this wavelength our atmosphere is a weak radio source, of which the intensity varies with the height above the horizon: near the horizon the telescope looks through a much thicker layer of air than at higher altitudes, which produces much more heat radiation. If this altitude-dependent atmospheric radiation—which is well known—is subtracted from the received signal, one would expect that all that is left is the radio noise produced by the telescope itself. Penzias and Wilson had expected this noise to be negligibly small, as it was known that the horn antenna had an extremely low noise level. In the spring of 1964, however, they discovered that after subtraction of all known noise sources at 7.35 cm, still a considerable amount of background noise remained, with a strength that was the same in all directions on the sky and did not vary in the course of the day or night, or in the course of the year. They were very puzzled about what could be the cause of this noise and at first thought that something in the telescope could be the noise source. They discovered that pigeons had built a nest in the telescope horn and had left white droppings. They caught the pigeons and released them at a large distance, but they were soon back, after which Penzias and Wilson "took more drastic measures with them". After complete cleaning of the telescope the noise still remained.

[1]Measuring the *absolute strength* of a radio source is much more difficult than measuring the *difference in strength* of two different sources. For example, by subtracting the strength of the one source from that of the other, the contributions of background sources are eliminated, and one obtains the real difference in source strengths.

Assuming that there was nothing wrong with the measurements, the fact that the intensity of the radiation did not show any variation in time or direction indicated that this radiation, with a "temperature" of between 2.5 and 4.5° K, must originate in the distant universe. Penzias and Wilson were completely unaware of the fact that Alpher and Herman, with the guidance of Gamow, had predicted in 1949 the existence of microwave radiation from the Big Bang with a temperature of about 5 K—very close to the measured temperature of their background noise. They also did not know that just at the time when they made their discovery, at three places in the world scientists had resumed making computations of the physics of the Big Bang, and that also these computations predicted the presence in the universe of microwave radiation from the Big Bang.

One of the groups making such calculations was in Princeton, not far from the Bell Laboratories in New Jersey, and at the time when Penzias and Wilson made their discovery, this group was just building instruments for detecting this radiation. The idea for this research had come from Robert Dicke, professor of experimental physics in Princeton who, just like Gamow, believed in a pulsating universe that prior to the Big Bang had collapsed and become extremely hot. Under his guidance, young Canadian theoretical physicist Jim Peebles had carried out computations about the physics of the hot Big Bang and made the prediction—just like Alpher and Herman 15 years earlier—that microwave radiation from the Big Bang must still be present in the universe and observable today. According to his calculations the temperature of this radiation should nowadays be close to 10° K. At Dicke's initiative, Peter Rol and David Wilkinson built a small microwave antenna which they installed on the roof of Princeton's Palmer Laboratory. But before Dicke and his collaborators could start observing, they heard from M.I.T.'s radio astronomer Bernard Burke about the background noise that Bell Laboratory's astronomers had discovered with their horn antenna. After getting into contact with Penzias and Wilson in 1965 they immediately realized that they had been scooped and that these scientists had discovered the heat radiation from the Big Bang. It was then decided that the two groups would each publish an article in the *Astrophysical Journal*: the first one by Penzias and Wilson with their discovery of the microwave background radiation and the second one by Dicke, Peebles, Rol and Wilkinson, with the cosmological interpretation of this discovery. Already earlier, Peebles had written an extensive article about his Big Bang computations which he had submitted to the journal *Physical Review*. In this original article there were no references to the earlier work of Gamow, Alpher and Herman. Editors of scientific journals always send submitted manuscripts to referees, experts in the same field, who are asked to judge whether the article is of sufficient quality to merit to be published in the journal. In this case the referees pointed out that Peebles had, in fact, rediscovered the results of Gamow and his collaborators of 15 years earlier, and that he should refer to that work. (Many years later it came out that that the referees had been Alpher and Herman, who in 1949 themselves had predicted the microwave background radiation from the Big Bang). It turned out that the referees still were not happy with the revised version of the paper, in which these references had been included, and the paper was never published. In the *Astrophysical Journal* paper of Dicke, Peebles, Rol and Wilkinson, references to the work of Gamow and collaborators were included, but not in the paper of Penzias and Wilson (see Chap. 18 of this book and Fig. 18.1).

In 1963 also Russian theoretician Jakov Zeldovich—who at an earlier stage played a key role in the development of the Soviet atomic and hydrogen bombs—without knowing about the work of the Princeton group of Dicke, had started making computations about the physics of the Big Bang, together with his collaborators Adrej Doroschkevich and Igor Novikov. In a paper by the last-mentioned two in a Russian journal they concluded in 1963 that microwave radiation from the Big Bang could still be observable in the universe today, and they suggested that the horn antenna of Bell Laboratories in New Jersey would be an excellent instrument to observe this radiation. Apparently, in the Sovjet Union it was exactly known what instruments were available in this laboratory! At the moment when Penzias and Wilson made their discovery, in May 1964, nobody outside Russia had read this article.

Also British astronomer Fred Hoyle and his collaborator Roger Taylor had in 1964, without knowing about the works of the others, carried out computations about the hot Big Bang. Their goal was to see if in this way the high fraction of helium in the universe, about 25 % of all mass, could be explained. They knew that this large amount of helium could not have been made by nuclear fusion of hydrogen inside stars: this fusion can explain at most the presence of only a few per cent of helium in the universe. The only other possibility is that this large amount of helium was produced in an early phase of the Big Bang when the universe was very hot. Their calculations showed that indeed the hot Big Bang easily explains the large helium fraction in the universe, and we nowadays know that the Big Bang is indeed the true explanation for the large helium fraction. It should be noticed, however, that before as well as after reaching this result, Hoyle always kept maintaining that he did not believe in the Big Bang, and kept believing in his *Steady State/Continuous Creation* model of the universe (see also Chap. 11).

Zeldovich and his collaborators, as well as Hoyle and Taylor were well aware of the work of Gamow's group and used this as a starting point for their work. Hoyle and Taylor assumed that the helium production started at a temperature of five billion Kelvin, which results in 36 % helium. This early temperature produces a temperature of the present microwave background radiation of about 10 K, similar to that of Peebles' work. Although this is too high, the basic idea behind this work is correct. If a helium fraction of 25–30 % is assumed, the temperature for helium formation is lower and the temperature predicted for the present-day cosmic microwave radiation is close to the observed about 3 K.

Apart from Hubble's discovery of the expansion of the universe, the discovery of the cosmic microwave background radiation is arguably the most important cosmological discovery ever made. It constitutes the proof that the universe originated in a hot Big Bang phase. In 1978 Penzias and Wilson were awarded the physics Nobel Prize for this discovery.

Further Proofs of the Big Bang

Apart from the expansion of the universe and the cosmic microwave background radiation, several other proofs have been discovered for the origin of the universe from a big bang event that took place a finite time ago. We will discuss here three of them.

The Abundances of Light Isotopes

Apart from helium there are three more isotopes of light elements that are expected to be produced in the hot Big Bang, in small quantities. These are the light isotope ^{3}He of helium, the heavy isotope ^{2}D (deuterium) of hydrogen, and the isotope ^{7}Li of lithium. The mass fractions of ^{3}He and ^{7}Li found in stars and the interstellar gas are probably unchanged from the fractions produced in the Big Bang, but at relatively low temperatures in the interiors of stars the deuterium undergoes nuclear fusion, and is converted in other elements. Since much of the present interstellar gas in galaxies has been expelled during the late evolutionary phases of earlier generations of stars, a large part of the interstellar gas consists of matter that has been recycled through stellar interiors. As a result, the present deuterium content of this gas presumably is not a good measure for the quantity of this isotope that has been produced in the Big Bang. The observed mass fractions of ^{4}He, ^{3}He and ^{7}Li are in good agreement with the predictions from the hot Big Bang model, in which it is assumed that there are about one billion times more photons than nucleons (protons and neutrons), and also the observed deuterium abundance does not disagree with the predictions from this model. The observed ratio between the number of cosmic microwave background photons and the number of nucleons is about one billion, and therefore is in very good agreement with the ratio required by the hot Big Bang model for predicting the fractions of the light isotopes.

The Evolution of Galaxies

Comparison of the structure and shapes of high-redshift galaxies with those of galaxies in our neighbourhood shows clear signs of galaxy evolution. At large redshifts we look far back into the past. At redshift $z = 1$ we observe the universe as it was nine billion years ago, and in the Hubble Ultra Deep Field (Fig. 6.9) we even see galaxies with redshifts greater than 4, corresponding to ages of 12 billion years or more. Already from redshift 0.5 onwards one observes the distribution of galaxy types to be different from that in our neighbourhood (which represents the present-day universe). At redshift 4 the distribution is totally different: here the galaxy population is dominated by small blue galaxies, resembling the Magellanic Clouds;

spirals and elliptical galaxies of the types that we observe in our neighbourhood, are rare. Galaxies with redshifts larger than 1.5 often are irregularly shaped (see Fig. 9.6a for some examples). They often are colliding galaxies. In a collision the gas in the galaxies is compressed in a number of places, which leads to rapid formation of large numbers of new massive stars in a so-called "starburst". Because of the irregular distribution of these clumps of new stars resulting from the collision, these shaggy-shaped galaxies also are also called "train wrecks". At large redshifts, that is: in the distant past, we encounter many more of these "train wrecks" than in our neighbourhood: nowadays they are rare. Nearby examples are the starburst galaxy M82 (Fig. 9.7) and the Antenna galaxies (Fig. 9.8).

The collisions of small galaxies that often occurred in the past have led to the formation of larger galaxies, and the present view is that also our Milky Way galaxy, during the first billions of years of its existence has swallowed many small galaxies. Groningen University astronomer Amina Helmi, using stellar proper motions measured with ESA's Hipparcos satellite, discovered traces of such swallowed galaxies in the form of large groups of stars with similar velocities, that deviate from the mean velocities of the bulk of the stars in their part of the Milky Way.

Apart from the "train-wreck" galaxies one finds at redshifts between 2 and 3 also 100–1000 times more *quasars* per unit volume in the universe than in our neighbourhood. *Quasars* are relatively rare objects. They are super-bright nuclei of large galaxies of a type that often is found in the centres of large galaxy clusters. The word quasar is an abbreviation of *quasi-stellar radio source*, which means: a radio source that on the sky coincides with a visible object that looks like a star. At first, when these objects were discovered in the early 1960s, it was indeed thought that they were *stars* in our Galaxy. However, in 1963 Maarten Schmidt—a Dutch astronomer working at Caltech (Fig. 9.9)—discovered that the spectrum of these "stars" has a large redshift, and that therefore the quasars must be distant extragalactic objects. This has since been fully confirmed and we now know that at close scrutiny quasars are surrounded by a faint glare produced by a giant galaxy whose light is outshone by a large factor by the light of its extremely bright nucleus: the quasar.

Quasars emit colossal amounts of energy in the form of X-rays and radio radiation, and their very small sizes (only a few light days) and gigantic energy emission indicate that we are dealing here with supermassive black holes, often with masses of several billion times that of the sun. Their enormous energy production is due to the inflow of matter towards the black hole. Before the matter disappears into the black hole (behind its horizon) the gravitational attraction of the hole has accelerated it to velocities approaching the velocity of light. The atoms of the gas spiral inwards through an accretion disk—since all matter always has some angular momentum (see Fig. 9.10)—and collide with one another with velocities close to the velocity of light when they are approaching the horizon of the hole. These violent collisions heat the inner part of the disk to temperatures of ten to hundred million Kelvin, causing it to become a strong source of X- and Gamma rays. The black hole often cannot swallow all of the inflowing disk matter—it is a "sloppy eater"—and ejects the excess of the inflowing matter in jets perpendicular to the disk, with relativistic velocities (Fig. 9.10). The relativistic particles moving in the

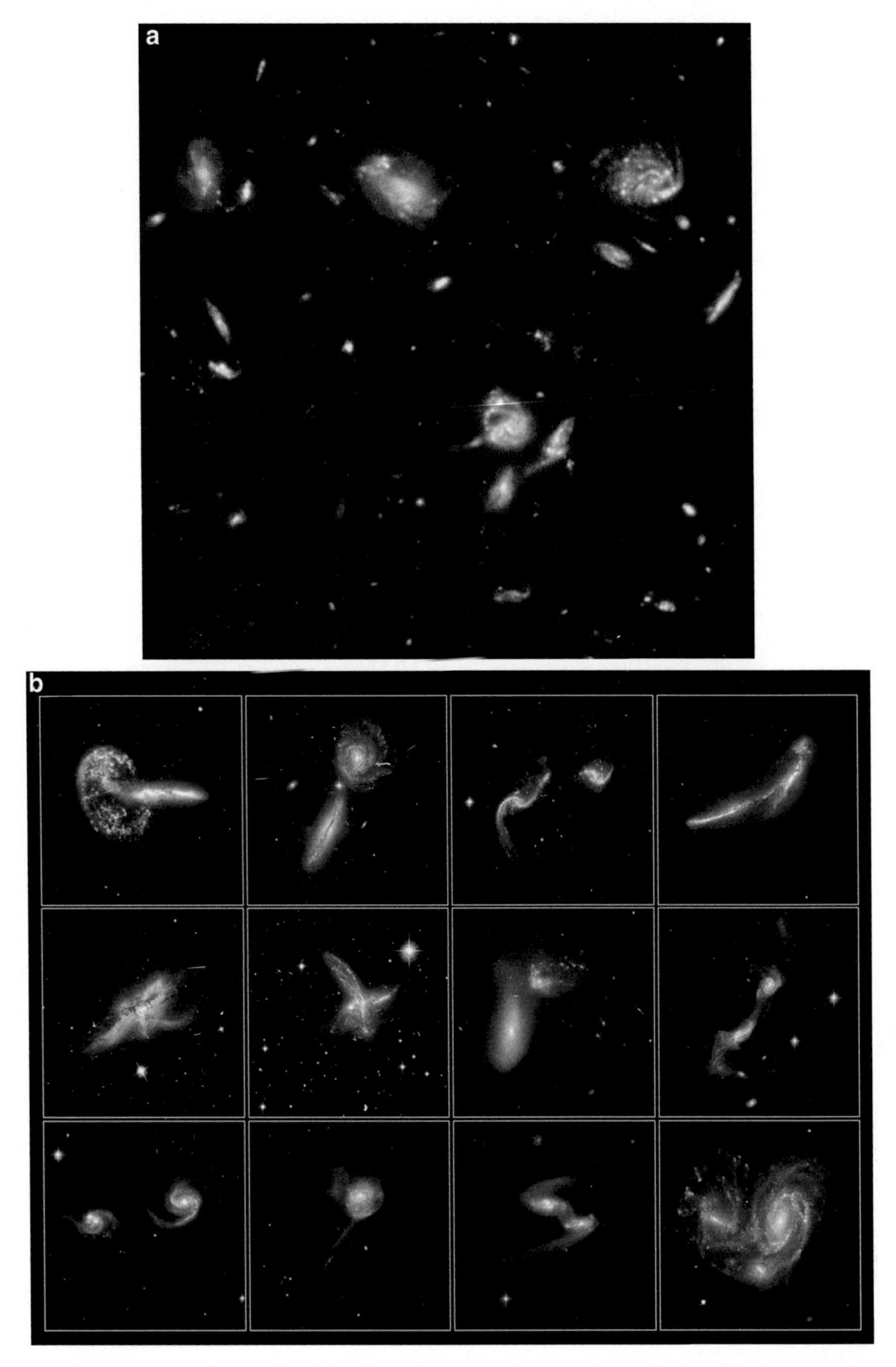

Fig. 9.6 (**a**) A small piece of the Hubble Ultra Deep Field with galaxies in the early universe, many of which are strongly disturbed by collisions. In the early times galaxy collisions were very common. (**b**) Pictures of 12 pairs of colliding galaxies taken by the Hubble Space Telescope

Fig. 9.7 The 'starburst' galaxy M82 is strongly disturbed by its recent encounter with the spiral galaxy M81. This gravitational encounter led to a compression of the interstellar gas that gave rise to a burst of formation of short-lived massive stars. The strong stellar winds and the supernova explosions with which these stars end their lives produced a large high-velocity outflow of gas and relativistic electrons that is visible as the ragged red and blue (blue = X-ray emitting relativistic electrons) material perpendicular to the disk of the underlying galaxy

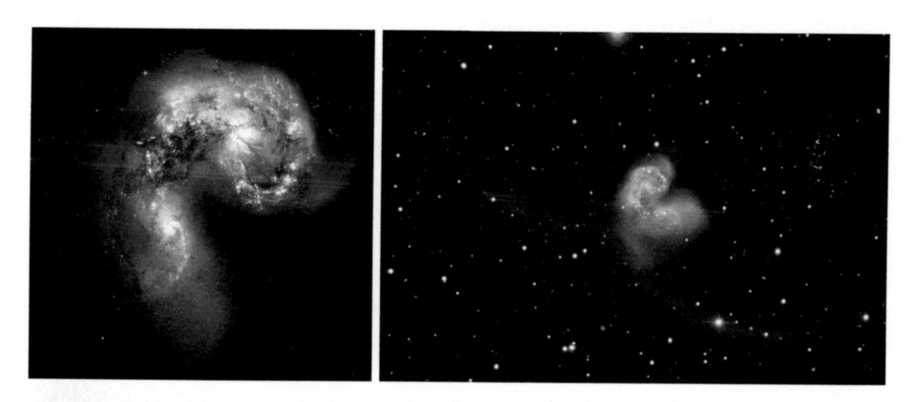

Fig. 9.8 The 'Antennae galaxies' (*right*) are the result of a collision between two spiral galaxies. They presently experience a large wave of star formation, a so-called 'starburst'. *Left*: close-up of these galaxies

Fig. 9.9 Because of his discovery of the large redshifts of the quasars, Time magazine in March 1966 put Maarten Schmidt's picture on its cover

jets, with their magnetic fields, emit large amounts of radio waves, optical light and X-rays as illustrated in Fig. 9.11. (This type of radiation is called *synchrotron radiation.*) In the 1970s, Schmidt discovered that the largest numbers of quasars are found at redshifts between 2 and 3, and that towards smaller and larger redshifts their numbers per unit volume in the universe decrease. This "quasar birth peak" is probably due to the fact that at that time there were still many small galaxies in the universe, which were merging with larger galaxies, such that the gas of these merging galaxies could feed the supermassive black hole of the quasar. This black hole had already formed at an earlier time, at redshifts beyond 3. Indeed, there are quasars with redshifts as large as 6–7, indicating that these massive black holes formed already at very early times. At redshifts between 2 and 3 the supermassive black holes in the giant galaxies could be copiously fed with the gas of small galaxies swallowed by the giant galaxies. But when more and more small galaxies had merged into larger galaxies, this "food source" of the quasars dried up and the

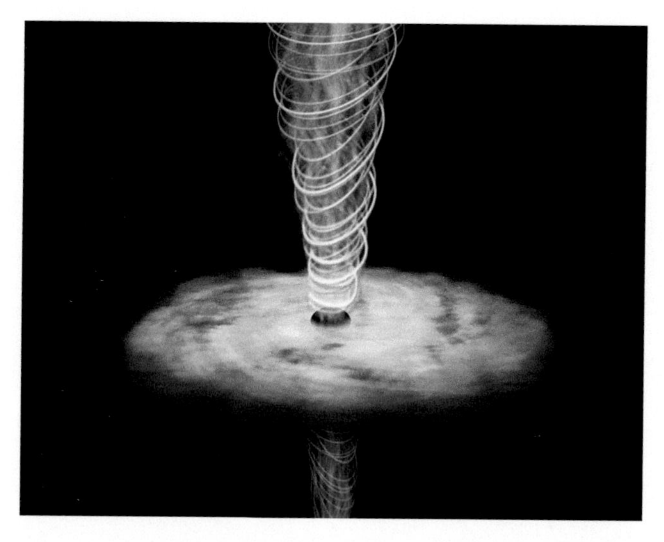

Fig. 9.10 A quasar is powered by a *black* hole with a mass of billions of solar masses. Inflowing gas has angular momentum ('rotation') which causes it to spiral inwards to the *black* hole through a disk. When the gas approaches the horizon of the *black* hole (the distance from the *black* hole centre where the escape velocity equals the velocity of light) the orbital velocity of the gas in the disk approaches the velocity of light. Violent collisions of particles with these velocities cause the disk gas to be heated to tens of millions Kelvins, making it a strong source of X-rays and optical light. The excess inflowing gas that the black hole cannot swallow is ejected perpendicular to the disk, in the form of two relativistic jets on opposite sides of the disk, with velocities approaching the velocity of light

Fig. 9.11 *Left*: in visible light, the quasar 3C273 looks like a star. However, the strange "jet" pointing diagonally downwards makes this 'star' peculiar. *Middle*: the quasar and its jet pictured in X-rays by NASA's Chandra satellite. *Right*: the jet at radio wavelengths. It is now known that quasars are the nuclei of large galaxies, resembling M87 (see Fig. 6.5)

quasar activity of the supermassive black holes died out. At present—that is: in our Galaxy's neighbourhood—only one out of 1000 supermassive black holes is an active radio and X-ray source (see Fig. 9.12a), the rest of them is sleeping.

The fact that already at redshift 6 to 7 some quasars are found shows that within the first billion years after the Big Bang some very large galaxies with a

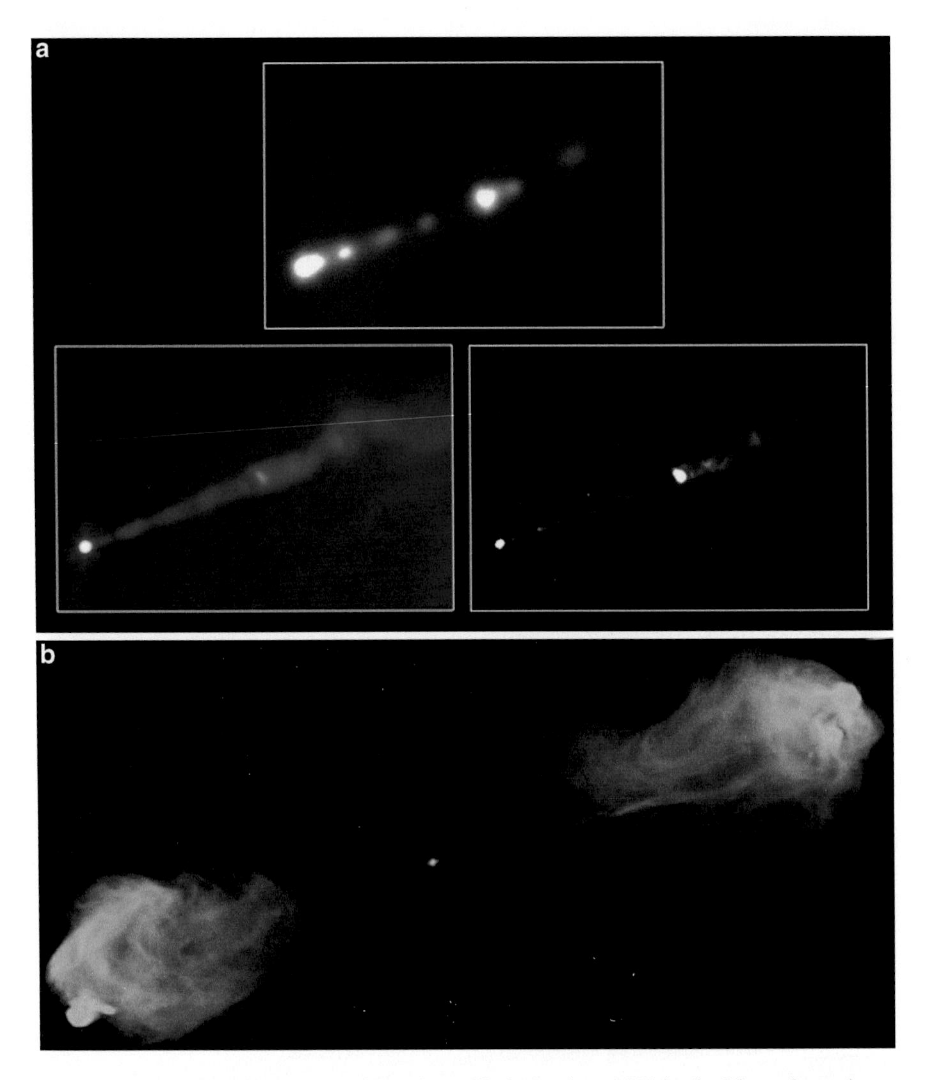

Fig. 9.12 (**a**) *Above*: the nucleus of the giant elliptical galaxy M87 in the Virgo cluster has a visible relativistic jet that also is observable in radio waves and in X-rays. The 'radio picture' (*lower left*) was made with the Very Large Array (VLA) in New Mexico, the X-ray picture (*top*) is by the Chandra X-ray satellite and the optical picture (*lower right*) was made by the Hubble Space Telescope. (**b**) *Middle*: the two radio jets of the strong radio source Cygnus A extend over millions of light years and consist of material that is being ejected by the nucleus of the elliptical galaxy in its centre. (**c**) *Bottom picture*: The same is the case with the nearest Active Galaxy Nucleus (AGN), the radio source Centaurus A, which has ejected a two-sided radio jet, of which the different radio intensities are indicated here by different colours. Here the central giant elliptical galaxy has collided with a spiral galaxy, of which we observe the dark dust band. The gas and dust of the spiral galaxy is now being swallowed by the elliptical galaxy, and is feeding the gigantic central black hole of this elliptical system, which has led to the production of the two opposite relativistic jets

Fig. 9.12 (continued)

supermassive black hole in the centre had formed. Such galaxies are almost always found in the centres of galaxy clusters, and therefore also the formation of galaxy clusters must have started at a very early stage.

In galaxy clusters many collisions take place between galaxies. In a collision much of the gas of the colliding galaxies is compressed and converted into new stars, and also much of the gas is blown out of the galaxies. The result is that nowadays the majority of galaxies in rich clusters are so-called *elliptical galaxies*, which are practically free of gas, dust and young stars. A good example of such a cluster rich in elliptical galaxies is the Coma Cluster (see first page of Chap. 13). The biggest galaxies in the clusters underwent the largest numbers of collisions and in this process have swallowed ("cannibalized") many small galaxies. Every time a small galaxy is swallowed by a big one, the supermassive black hole of the big one is fed with gas and stars, making it active as a quasar or another type of Active Galaxy Nucleus (AGN) for several hundreds of millions of years. The above-mentioned changes as a function of redshift of the properties of galaxies and quasars shows that the universe is not at all in a *Steady State* (as had been suggested by Hoyle, Gold and Bondi in the 1940s and 1950s) but shows many signs of evolution. That fits well with a universe that started from a beginning, in which it consisted only of gas and radiation, from which at a later stage the stars and galaxies formed—starting with a very large number of small galaxies and a few large ones, and in which most of the smaller galaxies gradually, in a few billion years, coalesced into larger galaxies.

Olbers' Paradox: Why Is It Dark At Night?

Physician-pharmacist and amateur astronomer Wilhelm Olbers of Bremen (see also Chap. 3) realized in 1823 that the fact that sky is dark at night gives us important information about the state and structure of the universe. If the universe has an infinite size and an infinite age, and is filled with stars like in our Milky Way, then the stars should overlap each other on the sky. And since the surface of a star is as bright as the surface of the sun, the night sky must then be as bright as the surface of the sun, and should blind us all. The fact that this is clearly not the case, is called "*Olbers' paradox*". (Later it was found that already in the early seventeenth century Johannes Kepler had put forward the same problem.)

The solution of this problem is found in the finite age of our universe in combination with the finite velocity of light. Since the universe has an age of 13.8 billion years, we cannot observe stars or galaxies of which the light after travelling for 13.8 billion years still has not reached us. Taking into account the expansion of the universe, this means that we cannot look further into the universe than a distance of about 50 billion light years. As this is not infinitely far away, the surfaces of the stars cannot overlap one another on the sky, and it is dark at night. Hence, the simple fact that it is dark at night in combination with the finite velocity of light shows us that the universe cannot be infinitely old! Olbers himself had thought that one could solve the paradox by assuming that there is so much dust present between the stars that distant stars are invisible. This solution does, however, not take into account that because of the conservation of energy, the dust would absorb so much energy from the starlight that it would become just as hot as the surface of a star, and the night sky would still glow as hot as the solar surface.

Cosmology as a Real Science

In the meantime we know that the temperature of the cosmic microwave background radiation is 2.725 K (see Fig. 9.13). This temperature has been very accurately measured by the FIRAS instrument in NASA's Cosmic Microwave Background Explorer (COBE) satellite, launched in 1989. This instrument (its name stands for Far Infrared Absolute Spectrophotometer) was built by the team of John Mather, and has produced the most precise Planck curve ever measured in physics, depicted in Fig. 9.13. This temperature means that since the universe became transparent at a temperature of 3000 K, it has cooled by a factor of about 1100 (3000/2.725) and

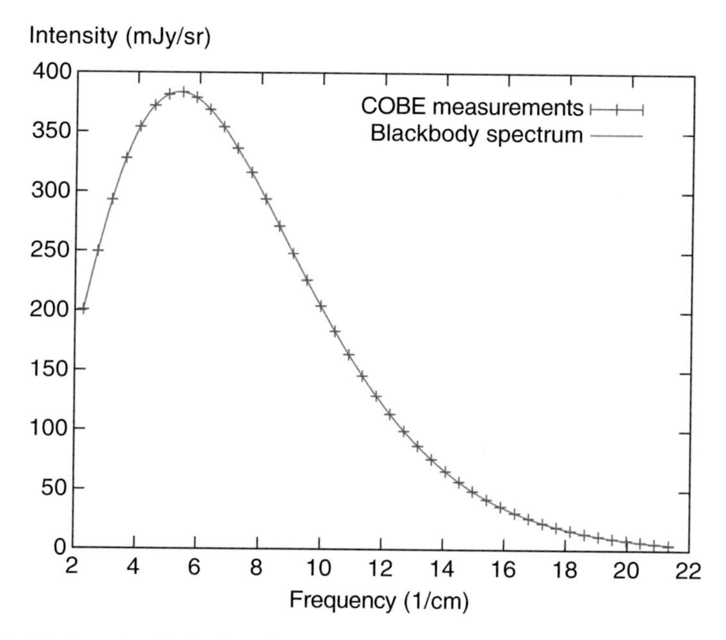

Fig. 9.13 The intensity distribution of the cosmic microwave background radiation, as measured by the Cosmic Background Explorer satellite (COBE). The sizes of the crosses indicate the measurement uncertainties. The curve perfectly fits a Planck curve of 2.725 K. This is the most perfect Planck curve ('black-body radiation curve') ever measured in physics

because of Wien's law (Appendix D) this means that the wavelengths of the radiation increased by the same factor, which implies that the linear size of the universe increased by the same factor of about 1100 since it became transparent, about 380,000 years after the Big Bang.

The discovery of the cosmic microwave background radiation has transformed cosmology from a speculative theoretical science into a quantitative observational science, based on solid observational facts. Prior to this, physicists often mockingly repeated the statement "Cosmologists are often in error but never in doubt" of Russian theoretical physicist Lev Landau. The discovery of the cosmic microwave background radiation has completely changed this situation. In the first place, the existence of this radiation shows that the universe really originated in a hot Big Bang. In the second place, the precise measurement of the temperature and intensity of the radiation allows one to reconstruct—by simple computation—the entire history of the universe back to one billionth of a second after the Big Bang, by just using presently well-known physics. One of the most amazing characteristics of the cosmic microwave background radiation is its strength. Already with a modest radio telescope it can be measured. For people that still receive their TV broadcasts with an antenna on the roof (that is: not through a cable), about 1 % of the "snow" on their TV screens is due to the photons of the cosmic microwave background radiation. So, they can directly see the Big Bang on their TV screen! The number of these photons in the universe is extremely large: about one billion per nuclear particle (proton and neutron), just as Gamow and his collaborators Alpher and Herman had predicted in 1949. The equivalent energy of these one billion photons is today much smaller than the equivalent energy of the mass of a proton or neutron: about one million electron-volts against 938 million electron-volts. But this was not always the case. When the universe was twice as small as today, the wavelengths of the cosmic microwave photons were twice as short, so the energies of their photons was twice as large as today. This means that when the universe was 938 times smaller than today, the energy of one billion cosmic microwave photons was just as large as the rest-mass energy of a proton or neutron. So, before that time, there was more energy in the cosmic background radiation than in the matter of the atoms and molecules in the universe. Surprisingly, the time at which the energy of the background photons became similar to that of the nucleons, differs very little from the time when the universe became transparent—which was when it was 1100 times smaller than at present. It is not clear whether this is just a coincidence or whether it has a deeper physical meaning. In any case, it is quite amazing. A further key characteristic of the microwave background radiation is its very high degree of isotropy: the distribution of its intensity over the sky is—apart from a Doppler effect of a few hundred km/s, due to the motion of the solar system relative to the mean background of the universe (see Fig. 9.14 and also Chap. 14)—extremely smooth. The local deviations from the mean intensity are less than one hundredth of a per cent (less than 1 in 10,000). This means that the cosmological models that were developed in the 1920s and 1930s, starting with those of Friedman, which were based on the *cosmological principle* of homogeneity and isotropy, are an excellent approximation of the real situation in the universe!

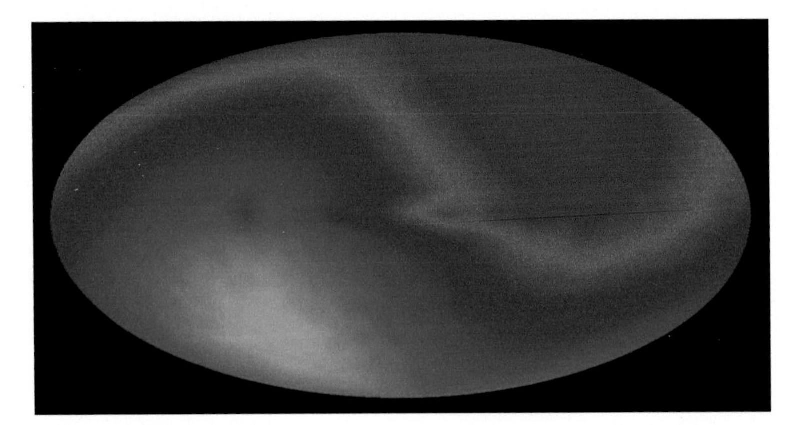

Fig. 9.14 The sky distribution of the cosmic microwave background radiation shows a systematic *blue-shift* on one side of the sky and a *red-shift* on the other half of the sky. This is a Doppler effect produced by the motion of our solar system with respect to the background radiation, with a velocity of 390 km/s. This velocity is a combination of the motion of the sun around the centre of our Milky Way Galaxy and the velocity of our Galaxy relative to the background radiation

Quantum foam

Chapter 10
The Origin of the Matter in the Universe

Almost everything originates from almost nothing

Henri-Frédéric Amiel (1821–1881), Swiss philosopher

Where did the protons, neutrons and electrons in the hot Big Bang come from? Gamow and his collaborators had, without giving an explanation for it, assumed that everything started with neutrons and a very large amount of radiation energy. In 1953 Alpher, Herman and Follin had, just as the later workers mentioned in the last chapter, assumed a mixture was present of neutrons, protons, electrons, neutrinos and anti-neutrinos, plus very much heat radiation. But where did these particles come from? The answer is: they originated from the photons of this heat radiation of the Big Bang.

In the early phases of the hot Big Bang the energy per unit volume (for example: per cubic metre) in the form of electromagnetic radiation was enormously much larger than in the form of matter. In the last chapter we saw that in the first 380,000 years following the Big Bang the energy in the form of radiation exceeded that in the form of particles. The further one goes back in time, the larger the dominance of the radiation energy becomes: at every reduction of the linear size of the universe by

E. van den Heuvel, *The Amazing Unity of the Universe*,
Astronomers' Universe, DOI 10.1007/978-3-319-23543-1_10

a factor of two, the energy in the form of radiation doubles, while the energy in the form of particles remains the same. For this reason, during the first seconds of the Big Bang the radiation energy was colossal and overwhelmed everything else in the universe.

In 1930 British theoretical physicist Paul Dirac, next to Einstein one of the greatest physicists of the last century, predicted that every elementary particle has an antiparticle. The antiparticle of the electron is the positron: an electron with a positive electric charge. This particle was discovered in the 1930s in the secondary cosmic rays in the Earth's atmosphere.[1]

Also the proton has an antiparticle, with a negative electric charge, called the antiproton. In the same way there are antineutrons and antineutrinos. All these particles have been discovered in particle accelerators or near nuclear reactors.

When a particle collides with its antiparticle they completely annihilate one another: they are completely converted into electromagnetic radiation, in the form of two photons, each with an energy which according to Einstein's formula $E=mc^2$ is equivalent with the mass of one destroyed particle (they both have the same mass). According to this formula the mass of an electron is equivalent to an energy of 0.511 million eV (1 eV is the kinetic energy obtained by an electron when it is accelerated by an electric voltage difference of 1 V). Photons in this energy range are gamma rays. Thus the annihilation of an electron-positron pair produces two gamma-ray photons of 0.511 million eV.

It appears that also the reverse of this annihilation process is possible: when two gamma-ray photons with energies larger than 0.511 million eV meet each other, they can spontaneously produce an electron-positron pair. This sudden creation of an electron-positron pair out of the vacuum is called *pair-creation*. When the photon energy is some 1840 times larger, this energy is high enough to create a proton-antiproton pair or a neutron-antineutron pair out of the vacuum.

The fact that the Big Bang created particles means therefore that in the beginning it must have been very hot, such that the gamma rays of the heat radiation were energetic enough to create pairs of particles and their antiparticles. Every 0.00008617 eV energy more requires a temperature of 1 K higher. Therefore, to reach an energy

---

[1] *Primary cosmic rays* are atomic nuclei with extremely high energies which, coming from outer space, continuously bombard the Earth's atmosphere. These primary particles are mostly protons (nuclei of hydrogen atoms), but also some heavier nuclei are present, as well as electrons and positrons. The nuclei can have energies up to 10^{20} eV, ten million times higher than the energies that the largest particle accelerators on Earth can achieve. The origin of these primary cosmic rays is only partly known. A part of them originates in supernova explosions, another part is accelerated near the supermassive black holes in the nuclei of galaxies. Low-energy cosmic rays can also originate in eruptions on the sun, so-called solar flares, and their analogues on other solar-like stars. When primary cosmic ray particles collide with the nuclei of oxygen or nitrogen atoms in the Earth's atmosphere, their energy is converted into thousands of new elementary particles, the so-called *secondary cosmic rays*. These are electrons and positrons, and also various types of mesons, such as *pions* and *muons*, which later also decay into electrons and positrons. The avalanche of these secondary particles produced by one primary particle at several tens of kilometres height in the atmosphere, is called and *air shower*. It can on the Earth's surface be hundreds of meters wide. It was in such an air shower that the first positron was discovered.

of 0.511 million eV requires a temperature of six billion Kelvin. And to form proton-antiproton pairs requires a temperature still 1840 times higher: of order 12 trillion Kelvin. One thus sees that early in the Big Bang, matter is spontaneously created out of the heat radiation. According to quantum mechanics indeed the vacuum (empty space) is not at all empty: due to quantum fluctuations in the vacuum, continuously pairs of particles and their antiparticles are formed, which immediately annihilate one another again after a tiny fraction of a second. This is a consequence of the *uncertainty principle* of Heisenberg, which states that the product of the *uncertainty in velocity* of a particle times its *uncertainty in place* of a particle equals $\hbar/2 = h/4\pi$, where h is Planck's constant (see also below). One can write Heisenberg's principle also as *uncertainty in lifetime* times *uncertainty in energy* of the particle equals $\hbar/2$. Therefore, for a very short time the uncertainty in energy can become very large, such that a large energy can be created out of nothing, which then causes a pair to be created out of the vacuum, which then after the same very short time—of order one billionth of a billionth of a second—annihilates again to form two photons. Such photons, which appear for a very short time in the vacuum are called "*virtual photons*" and the pairs that appear and disappear in the vacuum are called "*virtual pairs*". The vacuum therefore teems with pairs of particles and antiparticles that continuously form and annihilate. This continuous formation and destruction of pairs from the vacuum is schematically depicted in Fig. 10.1. If further nothing happens, these virtual pairs and virtual photons cannot escape from the vacuum. However, if one brings in real photons with sufficient energy to create a pair, or if one exerts a very strong force (by means of a strong electromagnetic field or gravitational field) then spontaneously a virtual pair may take up the real energy of the passing photons or of the force field, and become a real pair, that escapes out of the vacuum. In the early universe, photons of sufficient energy for pair creation were abundantly present and continuously created real pairs of particles and their antiparticles.

Why Is There Matter in the Universe?

The above scenario for the creation of the matter in the universe does, however, lead to a problem as in this way an exactly equal number of particles and their antiparticles is created. When the universe expands and cools, the temperature of the radiation field decreases, and with it the energy of the photons decreases. When the temperature has dropped below 12 trillion Kelvin, the photons no longer have sufficient energy to create protons and anti-protons and neutrons and anti-neutrons. On the other hand, the protons and anti-protons that at that moment still are present in the universe, will at some time collide with one another and thus annihilate each other, and be converted into radiation, while the reverse process is no longer possible. The same is true for the neutrons and anti-neutrons. So these particles will disappear, and since the particles and their anti-particles were created in equal numbers, one would expect that no protons or neutrons will be left in the universe. And the

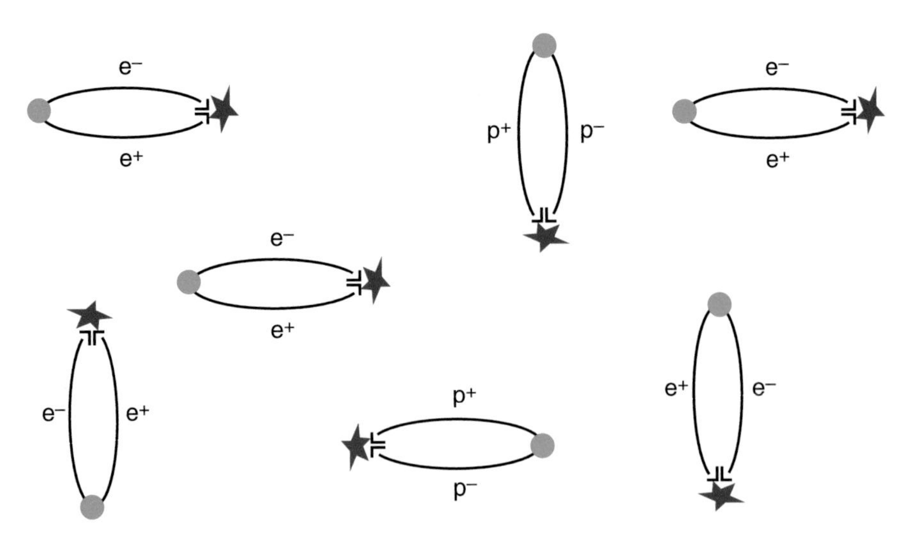

Fig. 10.1 Due to quantum fluctuations, in vacuum continuously pairs of particles and their antiparticles are formed, which almost immediately again annihilate each other. These pairs are called *virtual pairs* (electron-positron pairs, proton-antiproton pairs, etc.). They owe their energy to the Heisenberg uncertainty principle, but only if they are able to extract energy from a 'real' energy source, such as a strong electromagnetic field or a strong gravitational field (e.g. near a black hole) can they emerge from the vacuum and become a real particle-antiparticle pair. The latter process is called pair creation

same will happen to the electrons and positrons when the temperature has dropped below six billion Kelvin.

In this way, no matter is expected to be left in the universe, and one would expect the universe nowadays to be filled with only photons, that is: with electromagnetic radiation. The discovery of the microwave background radiation from the Big Bang does indeed show that there is very much radiation in the universe: about one billion photons per nucleon (proton or neutron) and per electron, but not an infinite number of photons per matter particle, as would have been expected if all particles had been annihilated by their antiparticles. Apparently, for some reason, during the creation of pairs of particles and antiparticles and their later annihilation, something has happened which caused that for every billion pairs of particles and their antiparticles that finally were annihilated, one matter particle (proton, neutron, electron) was left. This implies that for some reason the process of creation and annihilation was *not completely symmetric*, and was biased to produce one more matter particle for every billion pairs of particle plus its antiparticle. Physicists still have not found the explanation of why this asymmetry has occurred. It is, however, known that some processes in particle physics are not fully symmetric between matter and antimatter, that is: antimatter is not a completely exact mirror image of matter. A well-known example of a not completely symmetric process is the decay of the k-meson (also called kaon) and of its antiparticle. It thus seems that the presence of matter in the universe, and thus of stars, planets and humans, is due to a tiny imbalance in the basic laws of physics, which appear to slightly favour matter over antimatter. The present idea is that this surplus of matter over antimatter must have arisen in a very early phase, at the time of the *baryon creation*, which took place between about 10^{-36} and 10^{-33} s after the beginning of the Big Bang. This period, which is characterized by an extremely rapid expansion of the universe, is called the *era of inflation*. We will return to this in Chap. 12.

Brief Summary of the History of the Universe

The fact that the observed cosmic microwave background radiation and the observed abundances of the light elements (helium, lithium) agree so well with the predictions of the hot Big Bang model gives us confidence that the history of the universe can be safely based on this model. Experiments with the largest particle accelerators reach particle energies comparable with the photon energies that occurred in the universe about 10^{-11} s after the beginning of the Big Bang. As there is no reason to assume that the laws of physics have changed since then, our presently known (and experimentally tested) physics therefore allows us, starting from the observed intensity of the microwave background radiation, to compute the history of the universe backwards with confidence towards this moment. Since at still larger energies the laws of physics are not known from experiments, we cannot know with certainty what happened prior to 10^{-11} s. About this we only have some indirect evidence from the likely occurrence of inflation, between 10^{-35} and 10^{-33} s after the beginning, which will be discussed in Chap. 12. What happened between 10^{-33} and 10^{-11} s (22 powers of 10!) is unknown. This period is called *the desert*. Also about the period between 10^{-43} s (the *Planck time*, see below) and 10^{-36} s after the beginning (7 powers of 10) we have no information. Altogether there is thus a gap of 29 powers of 10 in our knowledge of the history of the universe. This gap is about the same size as the time range that we are able to know, using the presently known laws of physics: from 10^{-11} s till the present: 10^{17} s after the beginning of the Big Bang: 28 powers of 10.

Planck Time, Planck Length and Planck Mass

The various phases of the history of the universe discussed above are summarized in Table 10.1. In this table is also mentioned the Planck time: 10^{-43} s. We encountered the name Planck already earlier in this book. As already mentioned in Chap. 7, Max Planck in 1900 introduced the concept of the light quantum (photon) to explain the shape of the energy distribution curve of blackbody radiation (e.g., depicted in Fig. 9.2). He found the energy of a light quantum to be $E = h\nu$ where ν is he frequency of the light, which is equal to the light velocity c divided by the wavelength λ of the light, and h is Planck's constant, which has a value of 6.626×10^{-34} kg m^2/s.

Table 10.1 The cosmological timescale

0 s	Big Bang: start of the expansion of the universe
10^{-43} s	Planck Time: gravity force separates from the other forces
10^{-36} s	Strong nuclear force separates from the electroweak force
10^{-35} s	Start of inflation
10^{-33} s	End of inflation. From heat energy released at this termination, the quarks are formed
10^{-10} s	Weak nuclear force separates from the electromagnetic force: from now on the four known fundamental physical forces are separated from each other
10^{-5} s	Annihilation of pairs of protons and antiprotons, and of neutrons and antineutrons; per billion annihilated proton-antiproton pairs, one proton is left; for one billion neutron-antineutron pairs one neutron is left
1 s	Annihilation of pairs of electrons and positrons; per billion annihilated pairs, one electron is left over. Start of nuclear fusion
3 min	End of nuclear fusion. The universe now contains hydrogen and helium roughly in a mass ratio of 4:1, plus a small traces of deuterium, ^{3}Helium, ^{7}Lithium, Boron and Beryllium
380,000 years	Neutral hydrogen atoms form, and matter energy begins to dominate radiation energy. Universe becomes transparent. Last scattering of the cosmic microwave background radiation, at a redshift of about 1100
300 million years	First stars and proto-galaxies form. Re-ionization of the universe due to the powerful ultraviolet radiation of the first generations of massive stars. These short-lived stars explode as supernovae after a few million years, and inject the heavier elements generated in their interiors into the hydrogen and helium clouds of the universe With this the enrichment of interstellar matter with heavier elements begins, at redshifts between 20 and 10
One billion years	Birth of our Milky Way galaxy and other larger galaxies, at redshifts about 5–7
Nine billion years	Expansion of the universe begins to accelerate, due to the presence of 'dark energy' (see Chap. 13)
9.15 billion years	Birth of the solar system (4.65 billion years ago)
13.8 billion years	Today

In Planck's time it was already known that the velocity of light c has a value of about 3×10^{8} m/s, and the constant of gravity G in Newton's law is 6.7×10^{-11} m^{3}/kg s^{2}. Planck realized that the values of all these constants depend on three chosen units: of length, mass and time. The units of these quantities (metres, kilograms and seconds) have been chosen arbitrarily by us humans, to fit more or less our own lengths, masses and times (a second is about the time of our heartbeat). If one had chosen other units, for example, centimetres, grams and hours, the values of the natural constants h, c and G would have been different. For example, the velocity of light would then be $c=1.08\times10^{14}$ cm/h. And if one uses for the unit of time 1 year, and for the unit of length 1 light-year, the velocity of light c becomes $c=1$ light-year/year.

Planck saw that you can choose the units of length, mass and time such that $\hbar$ ($=h/2\pi$), G and c each become equal to one. It turns out that in this case the *natural unit* of length is about 10^{-33} cm (*Planck length*), the *natural unit* of mass is about 10^{-5} g (*Planck mass*), and the *natural unit* of time is about 10^{-43} s (*Planck time*). Planck, who in 1947 died at age 89,[2] did not know what these units could mean physically. Only later it became clear that they must play a role in a quantum theory of gravity. Unfortunately, we do not yet have such a theory, but theoretical physicists are working very hard on such theories. Many of them think that *superstring theory* (often abbreviated to just *string theory*) is a very promising candidate to lead to such a theory, but the problem is still far from being solved. Even though such a quantum theory of gravity is still missing, one can make some global statements about quantum gravity. The *Planck mass* is the mass of the smallest black hole that can exist. Such a black hole has a "size" (size of its horizon) equal to the *Planck length*, and cannot exist longer than the *Planck time*: after this it has evaporated by quantum effects. In this sense, the Planck mass can be considered as the 'elementary particle of gravitation'. This particle is extremely heavy, if compared to all other known elementary particles: it has 10^{19} times the mass of a proton: it has the mass of a dust particle visible to the naked eye. Its equivalent energy (mc^2) is 550 kW h, which is the energy in a full gasoline tank of a car. It is completely out of question that such a particle could ever be created in a particle accelerator. Even the 27 km accelerator ring of CERN in Geneva falls short for this in energy by a factor of 10^{16}. The *Planck length* tells us about the structure of space. If one would be able to look at empty space with a giant microscope, one would see that space is not entirely smooth. Due to the quantum behaviour of gravity, space is granular: it exhibits small wrinkles. The great American physicist John Wheeler has given this granular structure of space the name *quantum foam* (Fig. 10.2). The size of the wrinkles is of the order of the Planck length, which is 10^{20} times smaller than the size of a proton, and 10^{25} times smaller than the size of an atom. We have absolutely no way to ever observes this granular structure of space: for all practical purposes, for us space is completely smooth. Nevertheless, according to quantum mechanics, on the scale of the Planck length, continuously small black holes of a Planck mass are created which exist for a Planck time and then decay again. Viewed through a super microscope, space therefore is continuously in a boiling motion.

What further about the Planck time? Here Heisenberg's uncertainty principle, which we already encountered earlier in this chapter, comes in the picture again. We have seen that this relation can be written as: '*uncertainty of time*' times '*uncertainty of energy*' equals $\hbar/2$. For times shorter than the Planck time, the uncertainty in energy becomes very large, implying that large masses could be created out of nothing. It could be that on such timescales the universe could have been created out of a quantum fluctuation in space-time. This idea was suggested, for example, by American physicist Edward Tryon ad British astrophysicist Stephen Hawking (see Chap. 13).

[2] Planck suffered many disasters in his personal life. His two daughters both died during childbirth, one of his sons was killed in the First World War, and the second one was executed in 1944 by the Nazis for his participation in the Stauffenberg assault on Hitler.

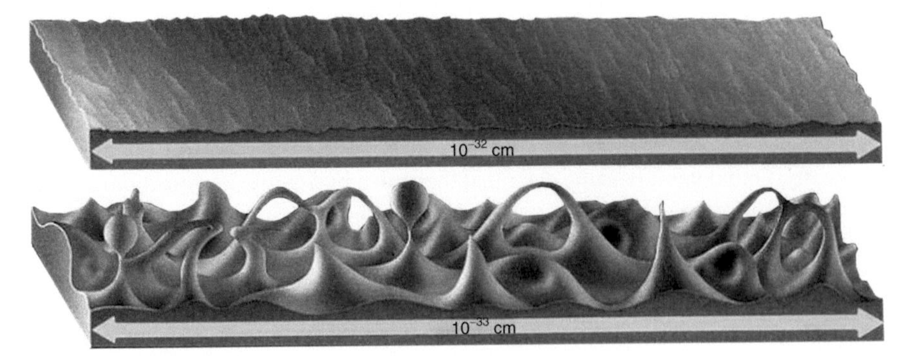

Fig. 10.2 On the scale of the Planck length space is no longer smooth, but becomes rough because of quantum effects. This structure of space is called 'quantum foam'. As the Planck length is twenty orders of magnitude smaller than the diameter of a proton, it is very questionable whether we will ever be able to directly observe this granular structure of space

The Unification of All Forces of Nature

Physics knows four basic forces of nature, with corresponding particles, which together can explain all what we observe in the universe. The two forces known to everyone in daily life are the gravity force and the electromagnetic force (electricity and magnetism). The other two are the *weak nuclear force*, which binds a proton and an electron together in a neutron, and the *strong nuclear force*, which binds protons and neutrons together in atomic nuclei. Protons and neutrons, in turn, consist of other elementary particles, called *quarks*, which also are held together with the strong nuclear force (see Appendix B for a brief explanation).

According to the Standard Model of particle physics, all forces originate from the exchange of special particles called *bosons* (see Appendix B). For the electromagnetic force these bosons are the *photons*, for the weak nuclear force they are the very energetic (high-mass) W^+, W^- and Z^0 bosons, which were discovered some 35 years ago at the CERN particle accelerator in Geneva, Switzerland, earning Italian Carlo Rubbia and Dutch physicist Simon van der Meer the 1984 physics Nobel Prize. Already during the 1960s, Steven Weinberg, Abdus Salam and Sheldon Glashow had realized that the electromagnetic force and the weak nuclear force have certain properties in common, and they put forward a theory that unified these two forces into one, the so-called *electroweak force*. According to this theory, at very high energies, corresponding to temperatures higher than 10^{15} K, these two forces would start behaving exactly the same. The electroweak theory predicts that at energies corresponding to temperatures below 10^{15} K, which occurred in the universe after 10^{-10} s, the electromagnetic and the weak nuclear force would split off from each other and each go its own way, with its own bosons: the photons for the electromagnetic force and the W and Z bosons for the weak force. This is why we today perceive these two forces as different ones.

In nature, the distance over which a force acts is smaller the more massive the force-carrying particle is. Hence, the weak force works only over very short distances, whereas the electromagnetic force can, in principle, work over infinite distances (see Table B.2 of Appendix B). The electroweak theory was considered so convincing by the physics community that already before the massive W and Z bosons predicted by this theory were discovered in experiments at CERN in 1983, Weinberg, Salam and Glashow in 1979 were awarded the physics Nobel Prize, even though they had not been able to give the rigorous final mathematical proofs of their theory. The important missing proofs of the theory were provided by—at the time—graduate student Gerard't Hooft and his supervisor Martinus Veltman, of Utrecht University, which earned these two the 1999 physics Nobel Prize.

Now that it had been proven that the electromagnetic and the weak force are, in fact, two aspects of the same force, it became obvious to examine whether, at still much higher energies, the *electroweak* force could be unified with the *strong nuclear force* and ultimately, also with the *force of gravity*. A variety of possible theories has been proposed by theoretical physicists for achieving this. Such theories are called *Grand Unified Theories* (GUTs). These theories are, however, still far from complete, and their predictions have so far not been verified by experiments. In any case, the energies at which the electroweak force and the strong force are to be unified are extremely high, and can have occurred in the universe only at times shorter than about 10^{-35} s after the start of the Big Bang. Only in the period between the Planck time of 10^{-43} s and about 10^{-35} s, these two forces are expected to have been one and the same force, and at the Planck time and earlier, this force is expected to have been unified also with the force of gravity. Sadly, the temperatures at which the unification of the electroweak force and the strong force took place are so high (above 10^{39} K) that it will forever be impossible to reach them with particle accelerators on Earth. The only way to study the predictions of a GUT is to examine whether according to this GUT the unification at the early times in the universe, before 10^{-35} s, may still have left traces that are observable in the present-day universe. Thanks to the work of American theoretical physicist Alan Guth and Russian physicists Alexei Starobinsky and Andrei Linde, we now know that such traces indeed exist. These theorists have argued that prior to about 10^{-32} s an extremely fast expansion of the universe took place, called "*inflation*", which solved two fundamental cosmological problems: the *horizon problem* and the *flatness problem*, which will be explained and discussed in Chap. 12. A second trace that a GUT must have left in the present universe is the already mentioned slight excess of normal matter over anti-matter that occurs in the process of pair creation: one particle of normal matter extra being created in the creation of one billion pairs of matter and anti-matter particles. This excess must also have been created in the inflation period prior to 10^{-32} s.

The great dream of theoretical physicists is to unify gravity with the other three forces. The idea is that at and prior to the Planck time of 10^{-43} s such a total unification must have been the case, as depicted in Fig. 10.3. The dream and hope is to find

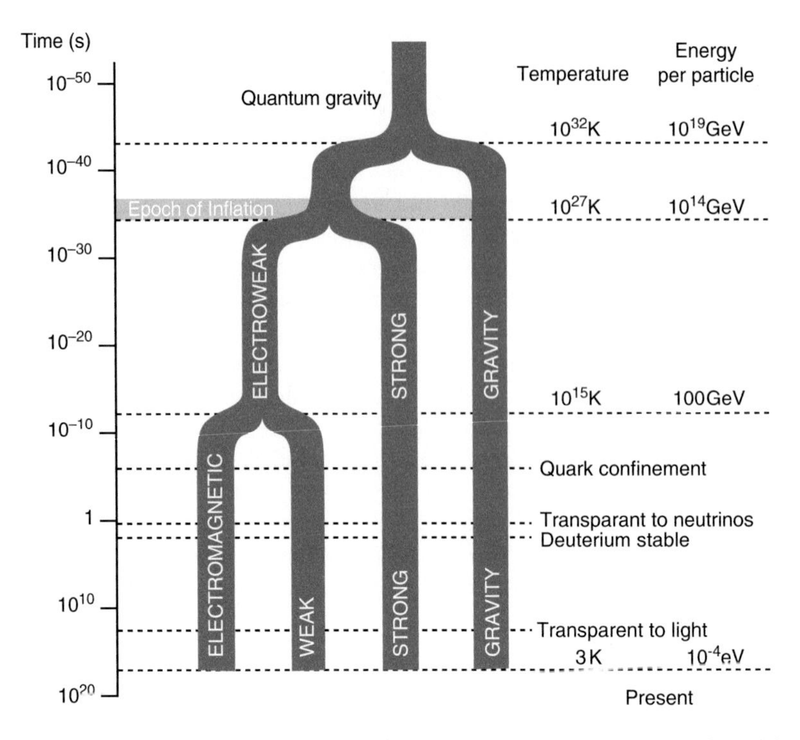

Fig. 10.3 Unification of the four fundamental forces of physics. The present view of physicists is that these forces are different aspects of one single unified force, and that at the moment of the Big Bang all four were one force and had the same strength. The thought is that after the Big Bang each subsequently split off from the others and went its own way, as depicted here. The so-far unknown over-arching theory that describes this unification is called the 'theory of everything'. A promising candidate for such a theory is the super-string theory, that attempts to unify gravity with the other three fundamental forces. One Planck time (10^{-43} s) after the beginning gravity was the first to separate from the other three, which at that time were still unified through a 'Grand Unified Theory' (GUT). At the beginning of the epoch of Inflation (10^{-35} s) the strong nuclear force separated from the electroweak force, and about 10^{-10} s after the beginning the latter one separated into the weak nuclear force and the electromagnetic force. As a result we nowadays know four fundamental forces. Physicists are actively searching for a 'Grand Unified Theory' and for a 'theory of everything', but no generally accepted solution for either of them is presently known. This in contrast to the unified theory for the electroweak interaction, which is well-known and has been experimentally tested, because the energy required for this unification is within reach of the largest particle accelerators on Earth. The fact that a unified theory for these two interactions exists, gives confidence that also a GUT should exist, as well as a 'theory of everything'

an ultimate *Theory of Everything,*[3] which can, in principle, explain the entire universe and all of nature. Many theorists focus their hopes on the earlier mentioned string theory—a mathematically highly elegant and tempting theory—which is, however, still far from complete. According to this theory all elementary particles can be represented by vibrations of extremely small strings in a space of 11 (or more) dimensions. We briefly return to this in Chap. 16.

[3] See Steven Weinberg: *Dreams of a Final Theory*, Hutchinson Radius, London 1993.

Globular cluster: very old and poor in elements heavier than helium

Chapter 11
We Are Made of Stardust; Timescales of the Universe and of Life

Also with errors one can be ahead of one's time

Hans Krailsheimer (1888–1958), German writer

In Chap. 9 we saw that shortly after the World War II, British scientists Fred Hoyle, Hermann Bondi and Tommy Gold (Fig. 11.1) proposed the *Steady State Theory* of the universe, because on philosophical grounds they could not believe in a universe that started a finite time ago with a Big Bang. To ensure that, despite its expansion, the universe would in the course of time keep looking always the same, they proposed that, as mentioned in Chap. 9, out of the vacuum between the galaxies spontaneously new hydrogen atoms gradually are created, which in the course of time concentrate in new stars and galaxies. The Steady State Theory is therefore also

E. van den Heuvel, *The Amazing Unity of the Universe*,
Astronomers' Universe, DOI 10.1007/978-3-319-23543-1_11

Fig. 11.1 The three founders of the 'steady state' theory of the evolution of the universe. According to this theory the universe will in the course of time always on average look the same in every direction and at every distance. From *left* to *right*: Thomas Gold (1920–2004), Herman Bondi (1919–2005) and Fred Hoyle (1915–2001)

called the *Continuous Creation Theory*. Since there was no Big Bang in this theory, and since in the 1940s, due to the work of Gamow, the idea still was that all elements were created in the Big Bang, Hoyle had to look for other ways to make the elements in the universe. He proposed therefore in 1946 that the elements heavier than hydrogen were all made in the interiors of stars. He was not the only one that had this idea. In the Netherlands Bruno van Albada, who later held the chair of astronomy at the University of Amsterdam as my predecessor, also in 1946 put forward this model for the formation of the elements.

The most interesting fact here is that while Hoyle came to this theory for the formation of the heavier elements *for a wrong reason*, namely that there has not been a Big Bang, this theory has turned out to be the *correct theory* for the origin of the elements heavier than helium in the universe, and is now generally accepted.

Stellar Populations in Galaxies

Hoyle at the time had realized that there are important observational facts indicating that the heavier elements are made inside stars. That came about when during the war he worked for the British defence research, and in this job had to visit industries in California. When during this visit he had a few days off, he visited German-born astronomer Walter Baade at the Mount Wilson Observatory, not far from Los Angeles. Baade's American-born colleagues had left the observatory to work in defence research, but as a German he himself was not allowed to do such work. Therefore he had the 100 in. telescope, then the largest telescope of the world, all for himself. In addition there was the for an astronomer lucky fact that, because of the war with Japan, the lights of Los Angeles were blacked out, which caused the sky over Mount Wilson to be pitch dark. This allowed Baade to take long-exposure pictures of the Andromeda Nebula in different colours. In this work he discovered to his great surprise that the disk of the Andromeda galaxy is embedded in a faint halo of stars that, on average, are much redder than the stars in the disk of this galaxy. Also the globular clusters of the Andromeda galaxy—twice as many as in our Galaxy—appear to belong to this halo. The much bluer stars in the spiral arms of the Andromeda galaxy apparently belong to a different *stellar population* than the halo. In this way Baade had discovered that galaxies consist of different stellar populations, and also in our Galaxy these different populations were subsequently recognized. In the rapidly rotating flattened disk of our Galaxy, where also the gas and dust of our stellar system are concentrated, we find hot, blue and very luminous stars, that have masses from several solar masses to over 30 solar masses, some even as high as 100 solar masses. These are the so-called O- and B-stars, that radiate between one hundred and several hundred-thousand times more energy per second than the sun, largely in the form of ultraviolet radiation. The reflection of their light lightens up the gas and dust clouds that surround them, turning them into *reflection nebulae*, which are observed in many parts of the Milky Way (e.g. see Figs. 3.7 and 4.4). With their ultraviolet radiation the O- and B-stars also ionize the hydrogen clouds that surround them, and by recombination of protons and electrons into neutral hydrogen atoms, these clouds radiate mostly in the light of the red hydrogen-alpha line, making them so-called *emission nebulae*. It is the combined light of these bright reflection and emission nebulae that surround massive stars, that produces the light that we observe as the spiral arms of our Galaxy and other galaxies (see Figs. 3.7, 4.6, 6.3, 6.4 and 11.2).

The spherical halo of our Galaxy, which rotates much slower than the disk, consists of redder lower-mass stars of much lower luminosities than the O- and B-stars in the disk. Typically these stars have luminosities smaller than that of our sun. The *disk population* of stars is therefore very different from the *halo population*. The disk population is called *Population I*, and the halo population is called *Population II*. Study of the spectra of the stars of these populations showed that the Population I stars in the Galactic disk have about the same percentage of elements heavier than helium as the sun: between 1 and 2 % in most of the disk, up to 3 % in the central parts of the disk. On the other hand, the spectra of the Population II stars in the Galactic halo show a lower percentage of heavy elements than the sun, sometimes even as much as a hundred to a thousand times lower. Such stars are called *super metal-poor* (astronomers have the strange habit of calling all elements heavier than helium *metals*). Particularly, the stars in globular clusters are poorer in heavier elements than the sun and the other stars in the Galactic disk. For example,

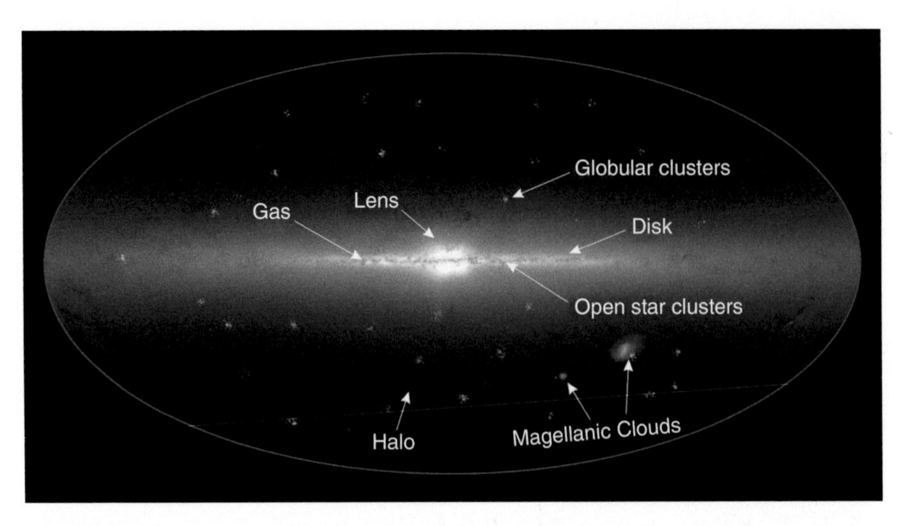

Fig. 11.2 Spiral galaxies like our Milky Way have two distinctly different stellar populations: the population of the disk (population I) and that of the spherical halo that surrounds the system (population II). In the disk one finds the gas and dust from which recently the massive and luminous *blue* O and B stars formed, and stars of all masses are still forming. Also one finds in the disk thousands of open star clusters. The stars in the disk, like our sun and the interstellar gas, consist for some 1 to 2 % of elements heavier than helium. On the other hand, the halo stars, including the globular star clusters, form a redder stellar population of stars with masses like that of our sun and smaller, and their heavy element content is considerably lower than that of the stars in the disk, and ranges from one thousandth of a percent to about half a percent

the fraction of heavier elements in the globular clusters M3, M5, M15 and M92 (see Fig. 4.8) range between 20 and 200 times lower than in the sun. These large differences in the abundances of the heavier elements led Hoyle to conclude that the elements heavier than helium have not been made in a Big Bang, but must have formed later in our Milky Way galaxy itself. The general idea, which Hoyle already realized at the time, is that the stars in the halo of the Galaxy were the first stars that formed out of the gas cloud from which our Galaxy condensed, at the time that this cloud was still slowly rotating and spherical, while the disk only formed later when the remaining gas of the cloud had contracted further, and as a result of this contraction began to spin faster, causing it to flatten into a disk (see Figs. 4.16, 11.2 and 11.3). The fact that the rate of rotation increases when an object contracts is a consequence of a law of physics called the *law of conservation of angular momentum*. Angular momentum of an object is a measure for its "amount of rotation", which is the product of *mass* times *size* times *rotation speed*. This is a quantity that for an isolated system—like a gas cloud—cannot change. Ice dancers use this law to obtain fast spin by first turning around at a moderate rate with their arms spread widely, and then suddenly pulling their arms in. By reducing their "size" (the spread of the arms), their rotation speed increases to a very fast spin.

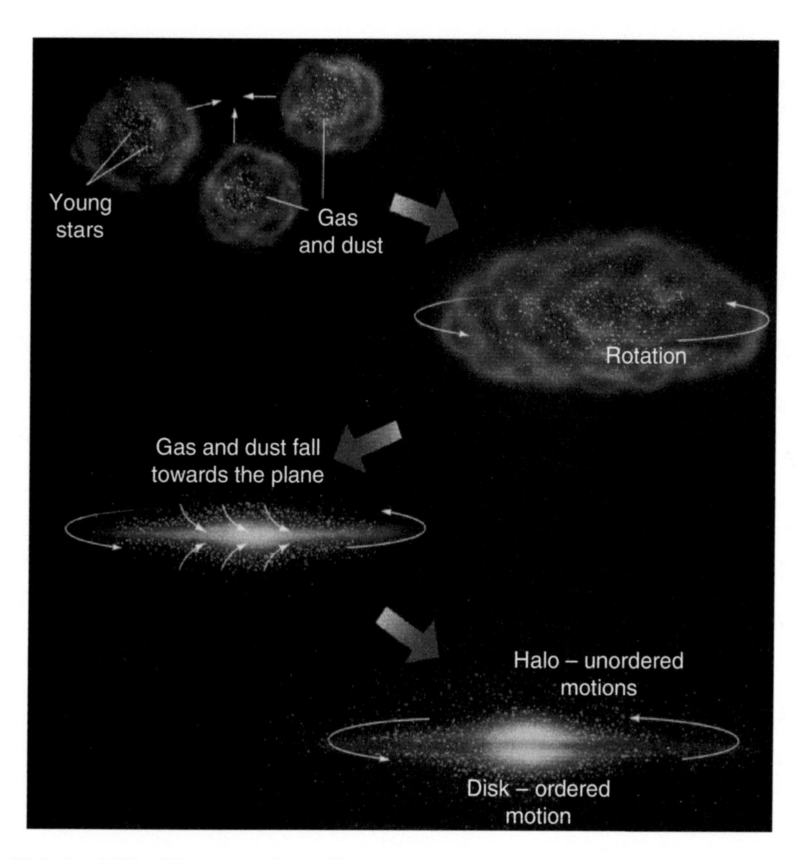

Fig. 11.3 Our Milky Way system formed by the merging of a number of more or less spherical gas clouds of almost pure hydrogen and helium, in which already the first stars were forming. In this way a slowly rotating almost spherical system of gas and stars was formed. These first stars, which hardly contained elements heavier than helium, formed a nearly spherical system: the halo. The short-lived massive stars in the halo exploded as supernovae and enriched the remaining gas with elements heavier than helium, such that stars which later formed in the halo were already somewhat enriched with heavier elements. The remaining enriched gas and dust contracted further, causing it to spin faster, due to the law of conservation of angular momentum ('conservation of rotation'); this faster spin resulted in the flattened disk of gas and dust of the Milky Way in which the later generations of stars formed. In the disk the stars orbit the galactic centre much faster than in the halo. Due to the presence of gas and dust, star formation in the disk still continues, and new massive short-lived stars keep forming, which at the end of their lives explode and continue to enrich the disk gas with heavy elements

Following Hoyle, the present idea is therefore that the stars that were born when the gas cloud from which our Galaxy formed was still large, spherical and slowly spinning, nowadays form the halo, and these stars are the oldest ones in the Galaxy. The stars which we find in the disk formed later when the gas had contracted to a faster spinning disk of gas. The observation that the oldest stars in the Galaxy have a thousand times lower percentages of heavy elements than the stars in the disk proves that our Galaxy formed from a gas cloud that was almost pure hydrogen and helium. Hoyle therefore concluded that the heavier elements must have formed later in our Galaxy itself. The same holds for other galaxies, such as the Andromeda Nebula, where Baade had discovered the existence of the different stellar populations.

How did this formation of the heavier elements take place? It was already known in 1946 that massive stars "burn" their hydrogen fuel much faster than lower mass stars and therefore live very much shorter. For example, while the sun needs 10 billion years to exhaust its supply of hydrogen fuel, a star of about 30 solar masses consumes every second some 60,000 times more hydrogen fuel than the sun (it shines 60,000 times brighter than the sun) and therefore consumes all its hydrogen fuel in only 5 million years: it lives therefore 2000 times shorter than our sun. Likewise a star of 16 solar masses lives only 10 million years, one of 6 solar masses 80 million years, one of 2 solar masses: 1 billion years, etc. Hoyle proposed that the elements that are produced by nuclear fusion processes in the interiors of the short-lived massive stars, were ejected at the end of the life of such a star in a so-called *supernova explosion*. In this way the supernovae of massive stars would be enriching the interstellar gas in the galaxy with heavier elements produced in the interiors of these stars, and by doing this for billions of years would have built up the present about 2 % of heavy elements in gas of the Galactic disk from which the youngest stars are forming. This is indeed the nowadays generally accepted picture of element formation in galaxies.

In the *box* a brief explanation is given of how the various kinds of elements are formed in stellar interiors.

Star Cluster Ages and the Cycle of Enrichment of the Gas of Galaxies with Heavier Elements

Astronomers have found ways to determine the ages of clusters of stars. When a gas cloud collapses and forms a star cluster, stars of all kinds of masses, up to the most massive ones, will be formed. The first stars of the cluster that will end their lives are the most massive ones. When all stars more massive than 30 solar masses are gone, and stars of 30 solar masses are still present but are just on the verge of exploding as a supernova, that cluster will have an age of 5 million years. If all stars more massive than 16 solar masses are already gone, but the most massive remaining star is just 16 solar masses, the cluster will be 10 million years old. If the most massive remaining star is just 6 solar masses, but all more massive ones are gone, the cluster will be 80 million years old. An example of the latter is the Pleiades cluster of Fig. 3.7, in which the most massive star is 6 solar masses. In the Ursa Major moving cluster of Fig. 3.6, the most massive stars still present are 2 solar masses, so that cluster is about 1 billion years old. One finds that in globular clusters even stars as massive as the sun are already gone, so these clusters must be older

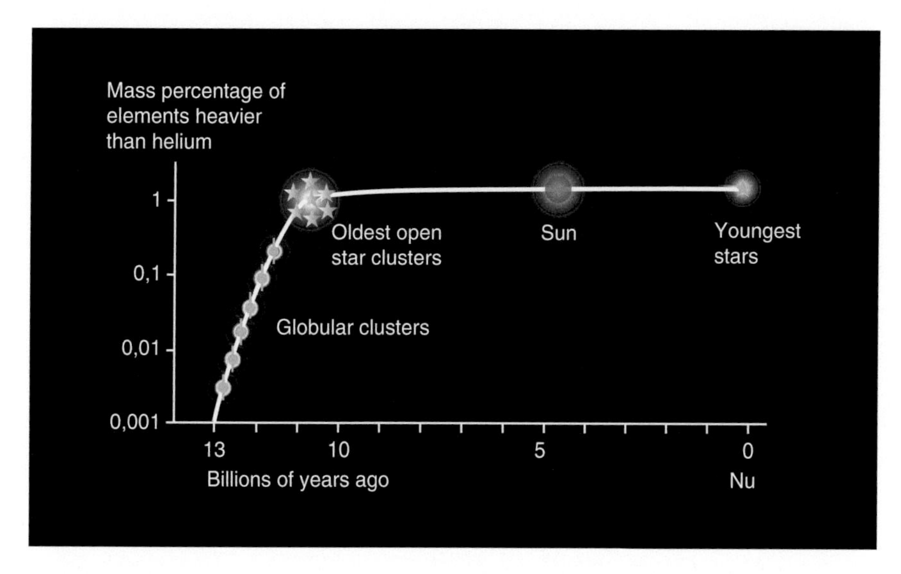

Fig. 11.4 The increase with time of the percentage of elements heavier than helium ('metals') in the Milky Way system, as derived from the spectra of stars in globular and open star clusters with well-determined ages. During the first few billion years the heavier element content increased very rapidly to the present amount of between 1 and 2 %. Between 11 billion years ago and the present the heavy element content further slightly increased, but at a very slow pace. Apparently, during the first few billion years after the formation of our Galaxy the rate of formation of massive stars—which after their short lives exploded as supernovae—was very much larger than at present

than 10 billion years. The most massive stars still remaining in them are about 0.9 solar masses. Such stars (with low heavy element contents) live for about 12–13 billion years, so we conclude that the ages of the *globular clusters* are of order 12–13 billion years. While the globular clusters all have considerably lower heavy element contents than the stars in the galactic disk, one amazingly finds that the oldest *open star clusters* in the galactic disk have an age of 11 billion years and already contain almost as much heavier elements as the sun. Plotting the heavy element contents of star clusters against their ages, one obtains the graph depicted in Fig. 11.4. One sees here that, apparently, between 13 and 11 billion years the heavy element content of the gas from which the stars in our Galaxy formed increased rapidly from almost zero to over one per cent, and that in the subsequent 11 billion years it only very slowly further increased. It thus appears that in the first 2 billion years of the life of our Galaxy, there must have been an enormous number of massive stars formed which exploded and rapidly increased the heavy element content of the interstellar gas. Apparently, our Galaxy went through a huge burst of star formation in these early first 2 billion years, in which very large amounts of gas were converted into stars in a very short time. The first generation of exploding massive stars enriched the gas, from which a second generation of stars formed, of which the most massive ones again rapidly exploded, further enriching the gas, from

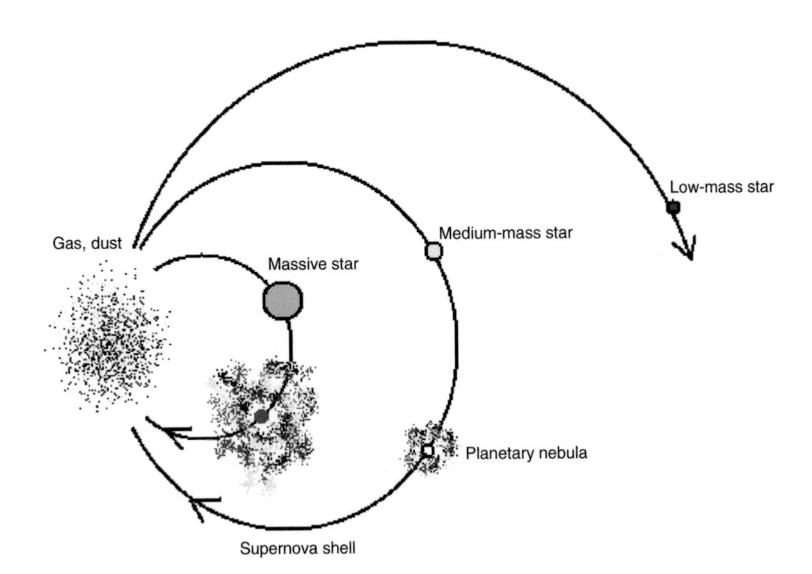

Fig. 11.5 The cycle of matter in the Milky Way system: Stars more massive than eight times the sun live less than a few tens of millions of years and at the end of life inject heavier elements into the interstellar medium. Stars of medium mass (1–8 times the sun) take longer (20 million to 10 billion years), and also return heavier elements—mainly nitrogen, carbon and oxygen plus elements formed by slow neutron capture—into the interstellar medium. The lightest stars, less massive than the sun, live so long that they not yet had time to return anything back into the interstellar medium. The sizes of the circles symbolize the lifetimes of the stars (not to scale)

which a new third generation of stars formed, and so on. This recycling of matter through stars and the enrichment of interstellar matter with heavier elements, is schematically depicted in Fig. 11.5, and still goes on today, though as Fig. 11.4 shows, nowadays at a much slower pace than in the first few billion years of the life of our Galaxy.

The matter cycle depicted in Fig. 11.5 produced the enrichment of the interstellar gas clouds from which the solar system formed 4.65 billion years ago. All atoms heavier than helium in our bodies, such as those of carbon, nitrogen, oxygen, phosphor, iron, calcium, etc. were produced sometime between 13 and 4.65 billion years ago in the interiors of many different massive stars and mixed through the hydrogen and helium clouds in the Galactic disk by the supernova explosions with which those stars terminated their lives. We can therefore literally say that *we are made of stardust*. The exception is the element hydrogen in our bodies, which originated in the first 3 minutes of the Big Bang.

The Formation of Elements in Stars

We nowadays know that the production of the heavier elements mainly takes place in the interiors of stars more massive than about eight times the mass of the sun. Such stars live shorter than 20 million years, which astronomically speaking is a very short time. In their interiors they produce by nuclear fusion all elements up to iron-56. This production goes in a number of steps. First, stars go through a phase in which hydrogen nuclei are fused into nuclei of helium, a process that produces the energy that stars radiate in the form of light. This phase occupies the main part of the life of the star, and also our sun is in this phase, in which it will remain for another 5.3 billion years. Some 90 % of all stars that we see in the sky are in this phase of hydrogen fusion. After the hydrogen in the core of a star is exhausted, it starts to fuse helium into nuclei of carbon-12 and oxygen-16. Also this process still produces quite a lot of energy, and in this phase the star appears at the outside as a red giant star. When the helium is exhausted, and the star has a mass above 8 solar masses, it begins to fuse nuclei of Carbon into Neon-20 and Magnesium-24, and somewhat later to fuse nuclei of Oxygen into Silicon-28, followed by Neon- and Magnesium fusion to elements between Silicon and Iron, for example calcium. Finally, it starts fusion of Silicon into Iron-56. These different fusion stages take place at higher and higher temperatures. Helium fusion at 100 million Kelvin, Carbon and Oxygen fusion at temperatures close a billion Kelvin, and Silicon fusion at several billions Kelvin. At the end, when the iron core forms, this core is surrounded by layers in which, going outwards, the temperature decreases. Around the Iron core there is a layer of Silicon-fusion, then a layer of Neon- and Magnesium fusion, then a layer of Carbon- and Oxygen fusion, a layer of Helium fusion and a layer of Hydrogen fusion. The star with this "onion structure" is then close to the end of its life. When the mass of the Iron core that forms inside the massive star exceeds the so-called *Chandrasekhar limit* of about 1.4 solar masses, it collapses to a *neutron star*—an extremely compact star with a diameter of only 20 km (15 miles), not larger than Manhattan, but containing some 420,000 times the mass of the Earth. The amount of gravitational energy released in this collapse is enormous: as much energy as our sun would emit in 1000 billion years. This gigantic amount of energy is released partly as heat: enough heat to blow up the star in a supernova explosion, in which all the layers of the star outside the collapsing iron core are blown into space with a speed of many thousands of kilometres per second. In the most massive stars the mass of the collapsing iron core is larger than 3 solar masses, which causes it to collapse to a *black hole*: a star from which even light cannot escape (the escape velocity from the surface of a neutron star is already half the velocity of light. If one fuses three neutron stars together, the result will also be a black hole). Also when the stellar core collapses to a black hole, most of the remaining part of the star blows up in a supernova explosion. During the

(continued)

(continued)
supernova explosion also a lot of free neutrons are produced. When these bombard the about 0.07 solar masses of Iron atoms ejected in the supernova, a small fraction of the Iron nuclei is converted into nuclei of heavier elements, up to Uranium. As neutrons have no electric charge, they can easily penetrate other nuclei and in this way build up even the heaviest nuclei known in nature.

The theory of the formation of the heavier elements in stars, including the production of the elements beyond Iron by neutron captures, was developed in the 1950s by Fred Hoyle, together with William Fowler of the California Institute of Technology (Caltech) and the British astronomer couple Margaret and Geoffrey Burbidge. This theory was summarized in a famous, now classical, publication in 1957 in the journal *Reviews of Modern Physics*. In the astronomical literature the names of the four authors are often abbreviated as B^2FH. The model was so successful that Fowler—who was leading the nuclear physics laboratory of Caltech where the rates of astrophysically important nuclear reactions were measured—was awarded the physics Nobel Prize in 1983. Fowler was very upset that Fred Hoyle, who had set him on the track of the research of the synthesis of elements in stars, was not included in this prize (see further in Chap. 16). Fowler shared the 1983 Nobel Prize with Chandrasekhar, who had 53 years earlier—at age 19 years—discovered the limiting mass for burned-out stellar cores: the above-mentioned *Chandrasekhar limit*.

Once More the Cosmic Timescale

We saw earlier that according to the best current estimates, 13.8 billion years ago the universe was born and started to expand from a phase in which all matter of the universe was concentrated in an extremely small volume—not larger than an orange for the matter of the universe that can presently be observed out to our cosmic horizon. We also know that our Galaxy was born less than a billion years later, and we saw in Chap. 2 that the solar system and Earth are much younger: "only" 4.65 billion years old. The universe existed already for over 9 billion years when our Earth and the other planets in the solar system were born. To make the solar system with planets, heavier elements are needed. At the birth of our Galaxy these were not yet available. It is therefore not surprising that our solar system is younger than our Galaxy. Figure 11.6 shows the timescale of the universe, in which we zoomed in on the timescale of our solar system and of life on Earth.

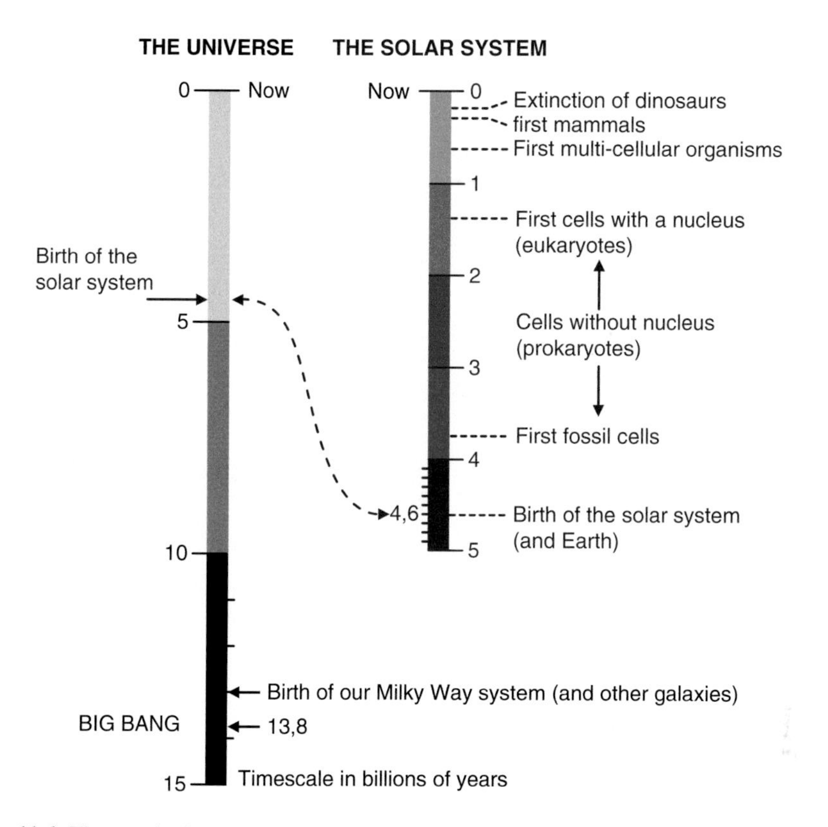

Fig. 11.6 The cosmic timescale

The Development of Life on Earth

In Chap. 2 we saw that, while the solar system and Earth have an age of 4.65 billion years, in Earth's crust no rocks older than 4.3 billion years have been found. The absence of older rocks is thought to be due to the huge bombardment with asteroids of the inner planets during the first half billion years after the formation of the solar system. This bombardment was, in fact, part of the formation process of the inner planets. Due to this bombardment much of the outer parts of Earth remained hot and molten during the first half billion years, although the presence of some rocks with an age of 4.3 billion years in Canada shows that locally solid crust had here and there started to form by that time. In these times, oceans did not yet exist, and Earth was a glowing inferno. The many icy asteroids and comets coming from the outer parts of the solar system that hit Earth in these times are thought to have deposited water, carbon and carbon-dioxide. Due to the high temperatures this water presumably was present only in the form of steam, and a considerable part of it may have escaped again into space. Only after the termination of the large asteroid

Fig. 11.7 Almost immediately after the termination of the large asteroid bombardment, 3.8 billion years ago, the first life on Earth started; 3.2 billion years later this would lead to the multi-cellular organisms like us

bombardment around 4.0 billion years ago, could oceans of liquid water begin to form. The fact that in Greenland and Australia fossil bacteria have been found in rocks with an age of 3.8 billion years shows that very soon after liquid water became available, primitive life appeared on Earth. Also, one finds in Greenland a 3.7 billion year old layer of sediments—with a thickness of 150 ft—that contains abundant graphite particles: pure carbon. In this carbon the ratio of the numbers of atoms of the isotopes ^{12}C and ^{13}C has a value characteristic for living organisms, which is higher than in carbon in inorganic rocks (in living creatures, the ^{12}C atoms are taken up more easily than the heavier ^{13}C atoms, such that the isotope ratio becomes different than the "natural" ratio in inorganic matter). This layer therefore suggests that 3.7 billion years ago there were already abundant primitive organisms in what is now Greenland. It thus appears that almost immediately after the termination of the large asteroid bombardment, liquid water was present and primitive life emerged (see Fig. 11.7).

We humans often have the tendency to think that the evolution of life concerns only organisms that are visible to the naked eye, such as plants and animals. These organism evolved from a primitive state which appeared some 600 million (0.6 billion) years ago in the oceans, into the plants and animals with which we are familiar today. The picture, however, that all evolution took place only during the past 600 million years, is a completely distorted one. Multi-cellular organisms that are visible to the naked eye are only a relatively recent phenomenon. They started to appear within a timespan of a few tens of millions of years at the beginning of the Cambrian epoch, about 600 million years ago. In this so-called *Cambrian Explosion*

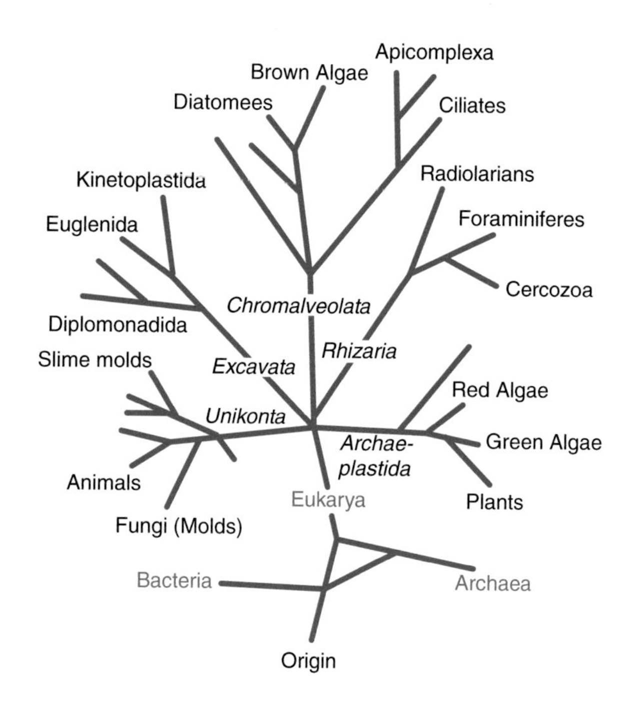

Fig. 11.8 The tree of life. The cells with nuclei—Eukaryotes—probably originated as a branch of the Archaea, a group to which belong the primitive bacteria found near deep-ocean volcanic chimneys and in the volcanic pools in Yellowstone Park, living at temperatures above the boiling point of water (100 °C)

suddenly multicellular organisms appeared with hard internal or external skeletons and of sufficient size, such that they left fossils clearly recognizable with the naked eye. All living species that are presently found on Earth are descendants of these new early Cambrian species. But these new species did not come out of nowhere: they, of course, had predecessors that were much smaller, and softer-bodied and harder to recognize.

Already hundreds of millions of years before the Cambrian explosion there were simple microscopic multi-cellular organisms, without hard skeletons, which did not leave easily recognizable fossils. And already 1.2 billion years ago there were microscopic little plants, just short strings of not more than a few dozen cells, closely resembling present-day red algae. But during the immensely long preceding 2.6 billion years, life on Earth was just single-celled. In this by far the longest period of the evolution of life on Earth, evolution was the evolution of single-celled organisms, from extremely primitive ones to highly complex ones, resembling the cells of multi-cellular organisms like ourselves. Figure 11.8 depicts the evolutionary "tree" from the simplest organisms to the most complex ones. The simplest single-celled organisms that existed 3.7–3.8 billion years ago and left fossils in rocks in Greenland and Australia were very primitive bacteria that did not differ much from some types of bacteria that still exist today (Fig. 2.16). As a group, these bacteria are called the *prokaryotes*—cells without a nucleus. In the cells of bacteria the hereditary information in the DNA genetic blueprint is not stored in chromosomes inside a cellular nucleus, but it is just a free-floating string of DNA in the cell. Already very early, probably about 3 billion years ago, one type of bacteria, the so-called cyano-bacteria (blue algae) invented the process of photosynthesis, which enabled them to utilize sunlight as a source of energy. They represent seven-eighth of the history of life on Earth. Even today these bacteria are still found at many places on Earth where they form large colonies, called stromatolites (see Fig. 2.16). Fossils of stromatolites, with ages up to 3 billion years, have been found on locations as diverse as Greenland, Spitsbergen and the high Andes in Bolivia. Another branch of bacteria that already existed at least 2.7 billion years ago are the so-called Archaea. They do not use sunlight or oxygen as energy sources (till 2.4 billion years ago Earth's atmosphere did not contain oxygen). The Archaea obtain their energy from chemical reactions, for example, between hydrogen-sulfide (H_2S) and ions of iron. They can survive even in boiling volcanic water, like in Yellowstone Park's geysers, or near the volcanic "chimneys" that are found near mid-ocean ridges, some 10,000 ft below the ocean surface. Thanks to the work of the cyanobacteria, 2.4 billion years ago some oxygen started to appear in Earth's atmosphere. From this moment on, cells could start using the energetic-rich mechanism of oxidation ("burning"). The appearance of atmospheric oxygen is visible in the occurrence of the so-called *banded-iron formations* on the ocean bottoms, which at this epoch reached their peak. How these iron-rich sediments precisely did form is not yet fully understood, but it is clear that they were produced by reactions between atmospheric oxygen and iron ions dissolved in the ocean water. These reactions produced insoluble iron oxides that sank to the ocean bottoms.

The Origin of the Eukaryotes

The prokaryotes that for some 2 billion years reigned Earth gradually evolved into more complex types of cells. Between 2 and 1.5 billion years ago, after the appearance of oxygen in the atmosphere, a new type of single-celled organism appeared: cells in which the genetic DNA blueprints are stored in a cellular nucleus, in the form of chromosomes, surrounded by a membrane. Organisms—single-celled as well as multi-cellular—with cells of this type cells are called *eukaryotes*. Eukaryotes have a much more complex cellular architecture than the prokaryotes. Practically all multi-cellular organisms belong to the eukaryotes: plants, fungi, animals. Particularly noticeable constituents of the eukaryotic cells, apart from the nucleus, are the *mitochondria*—structures in the cell which resemble bacteria, and which are the energy-factories of the eukaryotic cells. It is nowadays generally accepted that—as proposed in 1965 by American biologist Lynn Margulis—the mitochondria originally were independent single-celled organisms that penetrated the cells of other single-celled organisms and, instead of killing them or being killed by them, began a symbiosis with these host cells, from which both the host and the penetrator profited. Apparently, their survival strategy was to give the host something it needed: energy, while the host in return provided the mitochondria with food. In this way they both profited from each other and increased their chances for survival. No doubt this combination was the result of pure random chance: one successful experiment of nature among millions of failed ones, where cells fought and killed each other. In this way a much more advanced and complex type of single-celled organism arose. (It has been suggested that also the cell nucleus itself originated from a similar type of symbiosis). With this new cell type the way was opened for evolution to more complex lifeforms. Indeed, some 1.2 billion years ago simple organisms consisting of more than one cell began to appear. The cells of these simple multi-cellular organisms, which finally evolved in the plants and animals of today, already have the same basic building plan with a nucleus with chromosomes and mitochondria for the energy production.

The evolutionary advantage of bringing a large number of cells together in one individual proved to be great, and opened the way to an enormous number of different new possibilities. From the moment, 600 million years ago, that the first organisms with a strong skeleton began to appear, the speed of evolution became much faster than in the preceding 3 billion years. But all the basic inventions at the level of the cells that were required for starting this *multi-cellular revolution* had already been made at the level of microscopic single-celled organisms of types that we still find everywhere on Earth today. We still carry in our bodies these basic inventions made in the 3 billion years preceding the "Cambrian explosion" of life. A striking example of such an invention is the DNA repair mechanism. This mechanism ensures that the dozens of mistakes made during cell division, when copies are made of the millions of DNA molecules of our chromosomes, are nicely repaired before the new copy of the cell begins to function. Without DNA-repair, multicellular organisms would never be able to survive.

The Timescales of Earth and of the Universe

The timescales depicted in Fig. 11.6 are incredibly long compared to the on average 70–80 years lifetime of human beings. In order to get a better feeling for the cosmic timescales, I often compare the history of the universe to a series of books, as depicted in Fig. 11.9. The entire 13.8 billion years lifetime of the universe is depicted here as a series of 24 books, each of 285 pages, with on every page forty lines of text, each with on average seventy letters. One page is then 2 million years, one line is 50,000 years and one letter is 700 years. The Big Bang started at the top of the first page of the first volume. The formation of the galaxies, including our Milky Way system, started at about page 200 of the first volume. Our solar system, including Earth, appeared at the beginning of volume 17, and near the end of that volume life began to appear on Earth: the prokaryotes. The eukaryotes made their appearance around page 90 of volume 22, and microscopic multi-cellular organisms began to appear at the beginning of volume 23. Multi-cellular organisms with hard skeletons that were big enough to be visible with the naked eye appeared at the beginning of volume 24, the dinosaurs became extinct at page 252 of volume 24, our apelike ancestors in Africa, that began to make primitive stone tools appeared at the top of the last page of volume 24 (see Fig. 11.9). Neanderthals appeared two

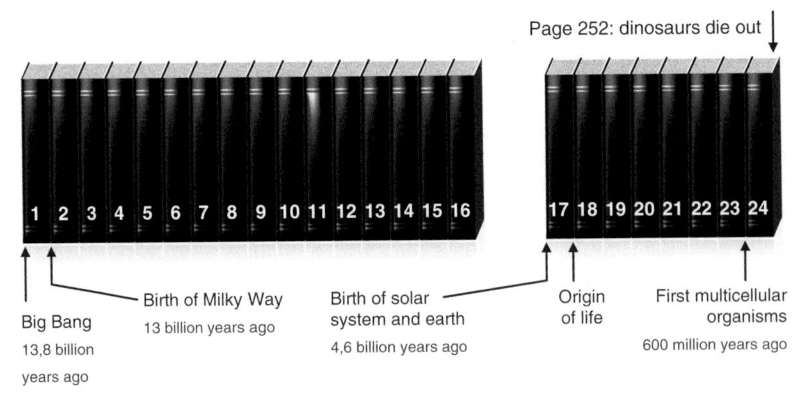

Each book has 285 pages of 40 lines of 70 letters

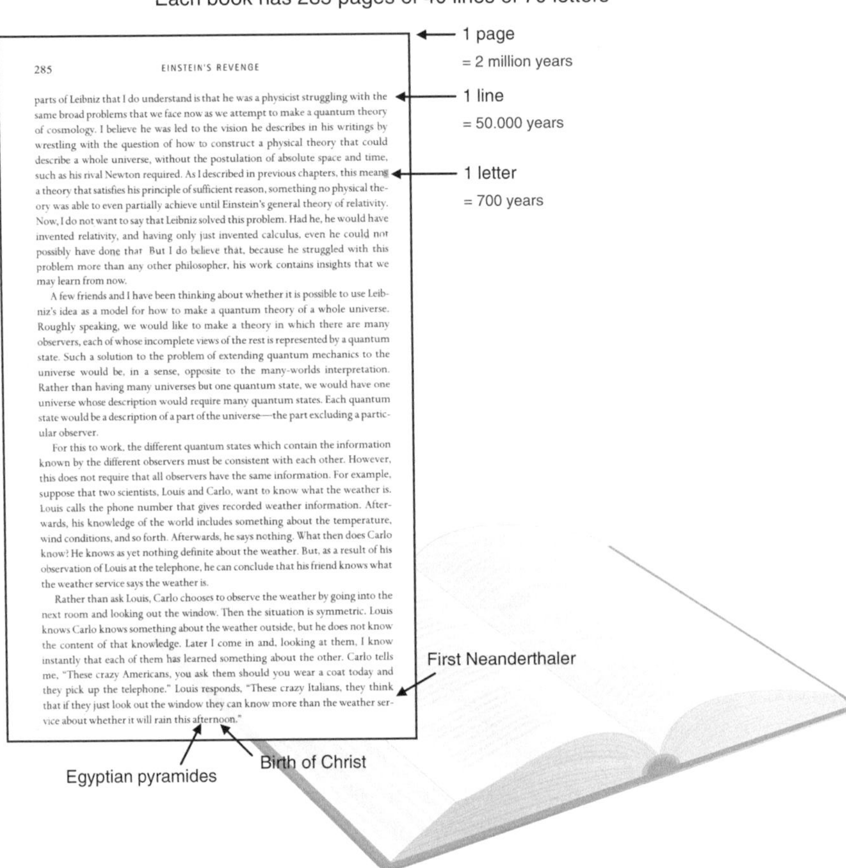

285 EINSTEIN'S REVENGE

parts of Leibniz that I do understand is that he was a physicist struggling with the same broad problems that we face now as we attempt to make a quantum theory of cosmology. I believe he was led to the vision he describes in his writings by wrestling with the question of how to construct a physical theory that could describe a whole universe, without the postulation of absolute space and time, such as his rival Newton required. As I described in previous chapters, this means a theory that satisfies his principle of sufficient reason, something no physical theory was able to even partially achieve until Einstein's general theory of relativity. Now, I do not want to say that Leibniz solved this problem. Had he, he would have invented relativity, and having only just invented calculus, even he could not possibly have done that. But I do believe that, because he struggled with this problem more than any other philosopher, his work contains insights that we may learn from now.

A few friends and I have been thinking about whether it is possible to use Leibniz's idea as a model for how to make a quantum theory of a whole universe. Roughly speaking, we would like to make a theory in which there are many observers, each of whose incomplete views of the rest is represented by a quantum state. Such a solution to the problem of extending quantum mechanics to the universe would be, in a sense, opposite to the many-worlds interpretation. Rather than having many universes but one quantum state, we would have one universe whose description would require many quantum states. Each quantum state would be a description of a part of the universe—the part excluding a particular observer.

For this to work, the different quantum states which contain the information known by the different observers must be consistent with each other. However, this does not require that all observers have the same information. For example, suppose that two scientists, Louis and Carlo, want to know what the weather is. Louis calls the phone number that gives recorded weather information. Afterwards, his knowledge of the world includes something about the temperature, wind conditions, and so forth. Afterwards, he says nothing. What then does Carlo know? He knows as yet nothing definite about the weather. But, as a result of his observation of Louis at the telephone, he can conclude that his friend knows what the weather service says the weather is.

Rather than ask Louis, Carlo chooses to observe the weather by going into the next room and looking out the window. Then the situation is symmetric. Louis knows Carlo knows something about the weather outside, but he does not know the content of that knowledge. Later I come in and, looking at them, I know instantly that each of them has learned something about the other. Carlo tells me, "These crazy Americans, you ask them should you wear a coat today and they pick up the telephone." Louis responds, "These crazy Italians, they think that if they just look out the window they can know more than the weather service about whether it will rain this afternoon."

Fig. 11.9 The history of the universe and of Earth, in the form of a series of 24 books (after Francis Crick)

lines before the end of the last page of the last volume, the oldest Egyptian pyramids were built seven letters from the end of the last page, and Jesus Christ was born three letters before the end of this page.

This comparison of the timescale of the evolution of the universe with a series of books, which I owe to Sir Francis Crick (in his book "Life Itself"), who together with James Watson discovered the double-helix structure of the DNA molecule, shows the incredible length of the timescale of the development of the universe and of the evolution of life on Earth. Also, it shows how incredibly slowly both the evolution of the universe and of life proceeded. In the course of a human lifetime, and even on a timescale of thousands of years, the structure and looks of plants and animals exhibit no noticeable changes. Only after many thousands of years some small changes may become noticeable. The bodies and faces of present-day people are in no way different from those of our ancestors in the times of the Romans, Greeks and Egyptians, and even not from those of the Cro-Magnon people that between 40,000 and 20,000 years ago made their beautiful paintings of animals on the walls of the caves in Southern France and Northern Spain.

Most people are not aware of the enormous timescales of the evolution of life and the slowness of this evolution. Many people are unable to imagine these timescales, which makes it often hard for them to believe in Darwin's theory of the evolution of life, the cornerstone and basis of biology as a science. This inability to conceive these enormous timescales makes many people vulnerable to alternative pseudo-scientific ideas such as "intelligent design".

It is important to notice both in Figs. 11.6 and 11.9 that the timescale of cosmic evolution and of the evolution of life on Earth are not very different from each other: they only differ by a factor of three. The universe is about three times older than life on Earth. This is very remarkable. Apparently, the development of life into higher forms, such as humans, requires of order of 4 billion years. Of course, the universe must be older to be able to produce us. But why is it not a hundred or a thousand or a million times older than those 4 billion years, but only three times? Why do we already exist so shortly after the beginning of the universe? We will return to this question in Chap. 16.

Group of distant galaxies

Chapter 12
Is the Universe Open, Closed or Flat? The Horizon Problem, the Flatness Problem and Inflation

The "silly" question is the first intimation of some totally new development

Alfred North Whitehead (1861–1947), British-American philosopher

We would very much like to know if our universe will keep expanding forever, as in the case of an open or flat universe, or whether the expansion will after some time halt and thereafter will reverse sign and become contraction, leading to a final "*big crunch*". Crucially for answering this question is knowledge of how large the present mean density of matter and energy in the universe is (energy has an equivalent mass, given by Einstein's formula $E = mc^2$). As depicted in Fig. 8.4, open as well as flat and closed universes all start expanding from a very small and compact beginning. If for a closed universe the time when it reaches its maximum size is still very far in the future, it will at present be very difficult to distinguish the expanding Friedmann solution for this state from a flat or open one. The curves that depict the increase with time of the distances between galaxies for these three solutions are then at the present time very close to each other. To determine the precise shape of the curve that represents the increase of the distance between galaxies with time requires very accurate measurements of galaxy distances and redshifts, particularly for very distant galaxies, where the shapes of the curves for the three different Friedmann solutions begin to deviate from each other.

E. van den Heuvel, *The Amazing Unity of the Universe*, Astronomers' Universe, DOI 10.1007/978-3-319-23543-1_12

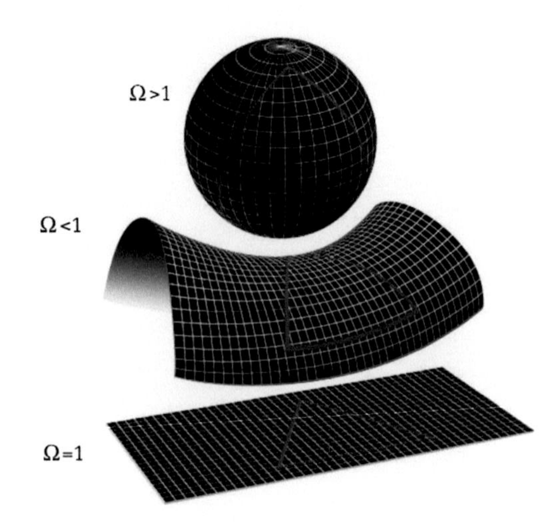

Fig. 12.1 For a closed universe the value of Ω, the mean density divided by the critical density, is larger than unity. For an open universe it is smaller than unity, and for a flat universe it is exactly unity

The present rate of expansion of the universe is given by the *Hubble constant H*. If we know the mass density of the universe (in which the energy density of all kinds of gravitating masses and of radiation are included), we can in principle calculate the kinetic energy of a given amount of mass in the universe, as well as its negative gravitational potential energy. If the sum of the kinetic and potential energy is negative (smaller than zero), the universe will be closed. If it is zero the universe is flat, and if it is larger than zero, the universe is open. In the case of a flat universe, where the total energy is zero, the mass density is called *the critical density*. The value of the real mass density in the universe, divided by the critical mass density, is indicated by astronomers with the capital Greek letter Ω (omega). In a closed universe, the mass density is larger than the critical density, so Ω is larger than one ($\Omega > 1$); in a flat universe Ω is exactly equal to one ($\Omega = 1$), and in an open universe, the density is smaller than the critical one ($\Omega < 1$). These three cases are schematically depicted in Fig. 12.1. In an open universe, space-time is *negatively curved*, in a closed universe it is *positively curved* and in a flat universe it is *flat*.

The Horizon Problem

The largest speed possible for the propagation of a signal through space is the velocity of light *c*. The finite value of c, combined with the finite age of the universe implies that there is a limit to how far we can see in the universe. If the universe would be static (non-expanding) and 13.8 billion years old, signals from distances beyond 13.8 billion light-years cannot have reached us. Our "*observing horizon*" is then located at 13.8 billion light-years. For an expanding universe, this horizon distance is larger, but also in that case there is an observational limiting distance from beyond which no signal or information can have reached us. If we look backwards in time, the observing horizon shrinks faster than the size of the universe (with "size" we mean here: the scale factor, which is the distance between two points in the universe). This is due to the fact that *forward in time* the horizon expands with the velocity of light, which is faster than the expansion of the size of the universe. Exactly how this shrinking of the horizon backwards in time, or expanding forward in time, proceeds is explained in Appendix C. Briefly summarized the result is: in a matter-dominated universe the size (scale factor) increases with the power 2/3 of time, while in a radiation-dominated universe, the size increases proportional to the square root of time. On the other hand, the size of the horizon increases directly proportional to time. For this reason, when going backwards in time, one will notice less and less of the curvature of the universe: in very good approximation, the early universe looked flat, as depicted in Fig. 12.2. For this reason, in calculations about the physics of the early universe it is allowed to ignore the curvature of the universe.

The existence of a horizon creates great problems for fundamental physics. The first problem is that in the part of the universe that we can observe within a horizon that stretches to a few tens of billions of light-years the same laws of physics are valid. Out to the largest distances, the spectra of stars and galaxies show absorption and emission lines of the same 90 chemical elements that we know on Earth, in the sun and in the Milky Way. It thus appears that the same laws of fundamental physics (of particle physics, quantum mechanics, electromagnetism, etc.) are valid everywhere inside our horizon. *This demonstrates the great unity of the universe*[1], the explanation of which has been one of the greatest challenges of cosmology, for the following reason. This unity implies that in some way the different parts of the universe—also those that billions of years ago were beyond our horizon and therefore had no way to communicate with each other—must have "known" that they should have the same laws of physics that we have. The same holds for the new parts of the universe that at present are entering our horizon, which all the time is expanding with the velocity of light: every second, our horizon extends 300,000 km further in every direction. How is it possible that these new regions entering our horizon have the same laws of physics while earlier in the history of the universe these parts of the universe have never been able to communicate with our part? How can they have communicated to each other that they should have the same laws of physics?

A closely related aspect of this *"horizon problem"* is the question of the extreme smoothness of the distribution of the intensity of the Cosmic Microwave Background radiation over the sky. Nowhere this intensity (and what basically is the same: its temperature) deviates more than a few hundred-thousandth from the intensity averaged over the sky (after the correction for our motion with respect to this radiation, see Figs. 9.14, 14.1 and 14.2). This means that the locations in the universe from which this radiation was emitted when the universe had an age of 380,000 years and became transparent, had basically the same temperature everywhere (within a few hundred-thousandth of a Kelvin). This is very strange, because one can calculate that two locations that nowadays are located diametrically opposite to one another

[1] E.g. see "*The Unity of the Universe*" by Dennis Sciama, Faber and Faber, London, 1959.

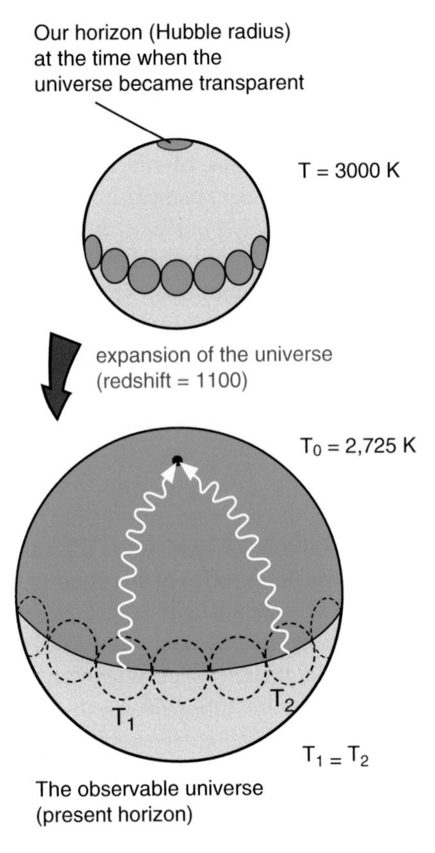

Fig. 12.2 Going back in time the area of the region inside one's horizon (the *blue* region) gets smaller and smaller, and the curvature of space becomes very difficult to notice: space seems flat. In the *upper* picture the horizon is drawn for the moment that the universe became transparent, 380,000 years after the Big Bang. The *lower* picture shows the present situation, where the universe should clearly show an observable curvature. Also is indicated that the regions from which we nowadays receive the cosmic microwave background radiation (the *dashed circles*) were not in contact with each other at the time when this radiation was emitted. Nevertheless, they emitted radiation with exactly the same temperature: $T_1 = T_2$. How could these regions, which were not in physical contact with each other, know that they should emit radiation of exactly the same temperature?

on the sky, and from which we now receive the Cosmic Microwave Background radiation, were 380,000 years ago located 90 million light-years away from each other. Therefore, 380,000 years after the Big Bang, they were located very far outside each other's horizons. How then could they have communicated that they should have exactly the same temperature 380,000 years after the Big Bang?

The only conceivable solution for this *horizon prob*lem is that the different parts of the universe have, at some time in the past, in a very early phase of the Big Bang, been in contact with each other, that is: inside a common horizon, and therefore were able to communicate the same laws of physics to each other, and reach the same temperature.

Inflation

To make the existence of such an early state fit with the presently observed expanding universe, one has to assume that in its very early history, the universe experienced a phase in which it expanded with a speed far higher than the speed of light. This idea does not violate the theory of relativity, which states that *with respect to space* nothing can move faster than light. There is no problem with space itself expanding much faster than light, even with a speed billions of times faster than the speed of light. So, it is conceivable that at a very early instant in time the matter that now fills the universe to far beyond our horizon, has been packed so closely together that all parts could reach the same temperature and communicate the same laws of physics to each other. If now this super-compact state was followed by a very brief period of expansion at a speed very much larger than the speed of light, a situation will have resulted in which the expansion of the horizon could not keep up with the expansion of the universe. In this way an initial situation in which all parts of the universe were inside each other's horizon could have been transformed in a situation in which the different parts of the universe ended up far outside each other's horizons. The early phase of extremely fast expansion of the universe required for achieving this solution of the horizon problem, is called *inflation*.

Amazingly, the idea of inflation was proposed around 1980 for entirely different reasons. The young American particle physicist Alan Guth was at that time searching for the solution of the problem that in the universe no *magnetic monopoles* exist. Every magnet has two poles: a North pole and a South pole, which are inseparably tied together in a magnetic dipole: they cannot be split into separate North or South poles. This in contrast with electric charges, where we have separate positive and negative charges: electric monopoles exist, but magnetic monopoles don't exist. However, particle theories that aim at unifying all forces of nature, the so-called Grand Unified Theories (GUTs), mentioned in Chap. 10, predict that in the very early universe lots of magnetic monopoles must have been created. Alan Guth discovered that an early phase of extremely rapid expansion of the universe can explain the absence of magnetic monopoles in the present universe. Without knowing about this work, in Russia physicists Andrei Linde and Alexei Starobinski around the same time proposed the same idea, and also predicted that the very early universe went through a phase of extremely rapid expansion which nowadays is called *inflation*.

According to this model, between 10^{-35} and 10^{-33} s after the beginning, the universe expanded by at least a factor 10^{30} (see Fig. 12.3). During this epoch the universe was, in fact, a de Sitter universe with a gigantic cosmological constant Λ with a value of order 10^{72}. A de Sitter universe always expands exponentially as a function of time, such that its dimensions can grow by a gigantic factor within a very short period of time. Only after he had done his calculations, Guth discovered to his surprise that this inflation also solves the horizon problem, plus several more problems that had bothered cosmologists for a long time. One of these problems is the *flatness problem*.

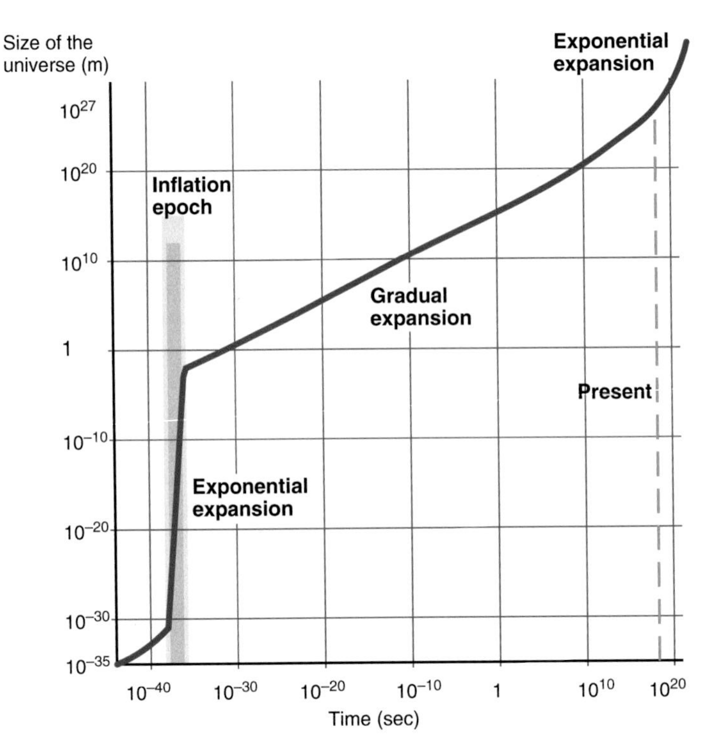

Fig. 12.3 During the period of inflation, between 10^{-35} and 10^{-33} s, the scale factor of the universe ('size' of the universe) increased by a factor of at least 10^{30}

The Flatness Problem

In the 1970s Princeton physicist Robert Dicke, whom we already encountered in Chap. 9 in connection with the renewed 1964 prediction of the Cosmic Microwave Background radiation, drew attention to a peculiar property of the universe that had not been realized before. This is the fact that it is amazing to see that the mean density of the universe is quite close to the critical density: it has a value that is between about 0.1 and 2 times the critical density. This means, as Dicke pointed out, that the universe must have started out with a matter density that was extremely close to the critical density. This is because even the slightest deviation from the critical density in the very beginning will during the later expansion of the universe grow to a gigantic deviation from the critical density, as illustrated in Fig. 12.4. In order to make sure that in the present universe the density has a value between 0.1 and 2 times the critical density, the density 1 s after the start of the Big Bang must have differed less than one part in 10^{14} from the critical density, that is: Ω should have differed from exactly $\Omega=1$ by less than 10^{-14}. In order to obtain the present universe, 1 s after the

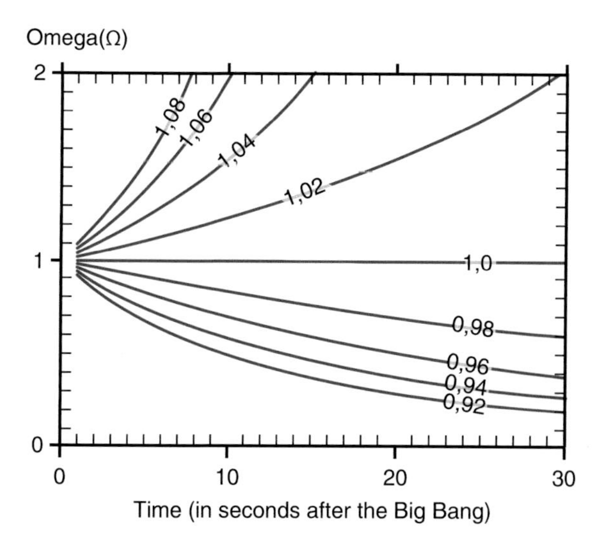

Fig. 12.4 Change of the value of Ω with time, for cases where its starting value was close to Ω = 1 (the starting values are indicated at the curves). The figure shows that even if the starting value of Ω in the Big Bang was very close to unity, its value will already within a few tens of seconds after the Big Bang deviate strongly from unity. The fact that the present value of Ω, after 13.8 billion years, is still somewhere between 0.1 and 2 then means that the starting value of Ω must have been incredibly close to unity. Its present value so close to unity is explained by inflation (for explanation: see the main text)

Big Bang, Ω should thus have had a value between 0.999,999,999,999,99 and 1.000,000,000,000,01. This means that shortly after the Big Bang the universe must have been extremely flat. This seems like an incredible coincidence. However, Guth discovered that also this problem is immediately solved by inflation (see Fig. 12.5). This is, because he showed that during inflation just the opposite happens of what happened during the later expansion: during inflation the flatness of the universe *increases* very much, which means: the deviation from $\Omega = 1$ becomes smaller and smaller, so the universe gets flatter and flatter. If at the beginning of the Big Bang Ω had a value different from $\Omega = 1$, inflation will have erased this difference and made Ω reach a value extremely close to $\Omega = 1$. The reason for this is that during inflation the size of the universe is blown up by a gigantic factor (Fig. 12.5), such that any curvature that was present prior to inflation is no longer observable. It is just like blowing up a balloon to a very large size. As long as the balloon is small, any curvature it has is easily observable. However, if you blow it up to a size of thousands of kilometres, a small piece of the balloon surface, that is comparable to the size of our universe, will look completely flat. This then immediately implies that the universe is gigantically much larger than the part that we are able to observe within our horizon. It thus appears that the fact that Ω is so close to one nowadays implies that *the universe is absurdly large*!

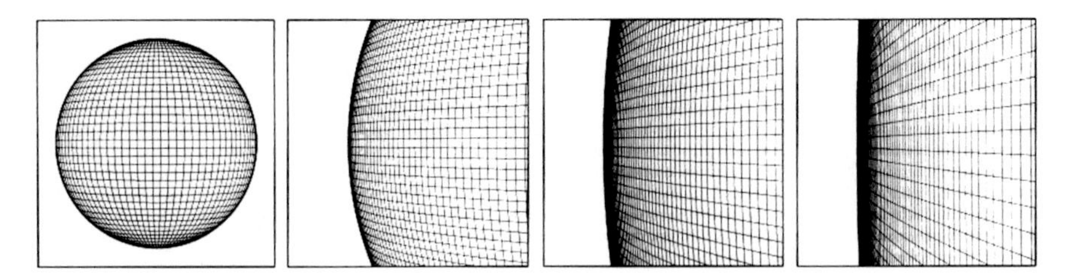

Fig. 12.5 The fact that the present value of Ω of the universe is very close to unity means that in very good approximation the universe is flat. This is nicely explained by inflation, because in the epoch of inflation the universe expanded by a gigantic factor. If a balloon is expanded to a gigantic size (from the *left-hand* picture to the *right-hand* one) and one can observe only a very small piece of it, this piece will in very good approximation be flat. This explains the fact that the part of our universe that is inside our observing horizon is flat. This does imply, however, that the universe is enormously much larger than the part of it that we can see inside our horizon

We thus see that the problem of the flatness of the universe can be solved by assuming that shortly after the beginning of the Big Bang the universe underwent a short but very powerful *super-expansion*. This, however, will mean that nowadays the density of the universe not only is between 0.1 and 2 times critical, but must be extremely close to the critical one. Until 1998 the observations seemed absolutely *not to support* a density close to the critical one, which would mean $\Omega = 1$. The best and most precise measurements of the density of normal matter, consisting of atoms and molecules, gave a value Ω_{matter} of order 0.02–0.04. Even if one added to this the amount of mysterious *Dark Matter* (matter which does not emit detectable radiations but only betrays its presence by its gravitational attraction in galaxies and galaxy cluster (see next chapter)), one would not obtain a value of Ω larger than 0.30.

However, in 1998 a discovery was made which suddenly brought the value of Ω very close to one. This was the discovery of *Dark Energy*, which causes the expansion of the universe to accelerate. The matter-equivalent of this energy produces a contribution to Ω of order 0.70–0.75, which suddenly made the universe completely flat, exactly as predicted by Dicke's argument, and expected on the basis of inflation. *Dark Matter* and *Dark Energy* are the subject of the next chapter.

Coma cluster of galaxies

Chapter 13
Dark Matter and Dark Energy: Our Strange Universe

Man cannot live without seeking to describe and explain the universe

Sir Isaiah Berlin, British social and political scientist and philosopher (1909–1997)

In 1933 Swiss-American astronomer Fritz Zwicky (1898–1974) measured the velocities of galaxies in the Coma Cluster (see Fig. 13.1). He discovered that the velocity differences between the galaxies are larger than you would expect on the basis of the gravitational attraction that the galaxies exert on each other. This attraction he estimated from the amount of visible matter, in the form of stars in the galaxies. The amount of light that a galaxy emits is determined by how many stars there are in the galaxy, so from the light he could estimate the number of stars and thus the mass of the galaxy. Zwicky measured the average velocity differences between the galaxies in the cluster to be about 700 km/s, while the escape velocity from the cluster, which he calculated from the estimated masses of the galaxies to be only a few hundred km/s. One would therefore expect the galaxies to escape from the cluster within a few billion years. However, galaxy clusters were already formed

E. van den Heuvel, *The Amazing Unity of the Universe*,
Astronomers' Universe, DOI 10.1007/978-3-319-23543-1_13

Fig. 13.1 Swiss astronomer Fritz Zwicky (1898–1974)—born in Varna, Bulgaria—worked as a professor of physics at the California Institute of Technology. He made many astronomical discoveries and in 1933 he was the first to discover that clusters of galaxies contain a large amount of unseen matter

early in the universe, over 13 billion years ago, and the galaxies of the Coma cluster have not escaped during these 13 billion years. The only way to explain why the galaxies in the cluster have stayed together and have not escaped is: that the escape velocity from the cluster is larger than 700 km/s. This is possible only if there is more matter with gravity in the cluster than we see in the form of the stars of their galaxies. There is, of course, also gas in clusters of galaxies, and this was discovered later with radio telescopes and particularly with X-ray telescopes (as a large part of the gas is very hot and emits X-rays). However, the contribution of this gas is far too small to explain an escape velocity larger than 700 km/s. There is therefore a large amount of matter with gravity in the cluster that is not observable as "normal" matter, consisting of atoms and molecules that make up the stars and the gas in the cluster. The same effect of *missing mass* is found in all other galaxy clusters and also in the galaxies themselves. In the 1970s it was found by American astronomer Vera Rubin that the velocities with which the stars in the Andromeda galaxy M31 describe their orbits around the galaxy centre hardly decrease with increasing distance from the centre, as would be expected from Kepler's laws. The rotation curve of this galaxy is *flat*. The same phenomenon of *flat rotation curves* was also found in the early 1970s in other spiral galaxies by Groningen radio astronomer Albert Bosma, by measuring the velocities of hydrogen clouds, using the Westerbork Synthesis Radio Telescope (see Fig. 4.15). These findings were very puzzling, as the amount of starlight in these galaxies decreases rapidly going outwards from their centres. If the amount of starlight would be indicative of the amount of mass present in the galactic disk, one would expect the orbital velocities of stars and hydrogen clouds to decrease quite rapidly outwards, just like the velocities of the planets around the sun in the solar system. The findings of Vera Rubin and Albert Bosma that the rotation curves of these galaxies are almost flat can only be understood if the disks of spiral galaxies contain a lot of invisible matter that exerts a force of gravity.

Another indication that such matter must be present was found in the 1970s by Princeton astronomer Jerry Ostriker, who showed that the disks of spiral galaxies are unstable if the galaxy consists only of normal matter of the visible stars. He calculated that the edge of a galaxy disk would start "flapping" up and down like a flag in the wind. Real galaxies do not show this flapping, and Ostriker showed that the disk can be stabilized by a *halo* of matter that exerts gravity but is not visible in any other way. Combined with the findings of Zwicky about galaxy clusters, and of Rubin and Bosma on rotation in disks of galaxies, Ostriker's work fitted very well into the picture that galaxies and galaxy clusters contain a large amount of invisible matter that exerts gravity.

This *missing matter*, in galaxies and galaxy clusters, which exerts gravity but is undetectable in any other way, is nowadays called *Dark Matter*. The amount of this mysterious *Dark Matter* is about 20 times the mass in the form of stars, and some 5–7 times the total amount of "normal" matter (stars, gas, etc.) consisting of atoms and molecules, present in the galaxies and galaxy clusters.

Gravitational Lenses

In the past decades astronomers succeeded in measuring the amounts of dark matter in clusters of galaxies by making use of the effect of gravitational lensing. Using the Hubble Space Telescope, astronomers discovered arcs and rings of light around many galaxy clusters; some examples are depicted in Fig. 13.2. When they measured the redshifts of the spectra of these arcs and rings, they found them always to be much larger than the redshift of the galaxy cluster itself. The light of the arc or ring is therefore due to a galaxy that is much farther away than its galaxy cluster. Amazingly, all parts or the ring or all arcs around a cluster turn out to have exactly the same spectrum and redshift. It therefore became clear that they all are *images* of the same galaxy that is far behind the cluster and whose shape has been distorted by the lensing effect of the gravity of the matter of the cluster. According to Einstein's General Theory of Relativity, the gravity of this matter deflects the light rays emitted by the galaxy that is behind the cluster in such a way that these rays are bent around the cluster and, as sketched in Fig. 13.3, produce a number of arc-like images of this galaxy around the cluster. In the most ideal case of a galaxy placed exactly behind the centre of the cluster, the image will be a complete ring, as so-called *Einstein ring*. The size of the Einstein ring, together with the redshift of the ring and of its "lensing" galaxy cluster allow one to exactly calculate how much gravitating mass is present in the cluster. Again one finds that the cluster's gravitating mass is some 20 times larger than one would expect on the basis of only the combined starlight of the cluster, and some 5–7 times the total amount of "normal" matter in the cluster.

What Is Dark Matter?

Presently nobody knows the physical nature of the dark matter. It is clear that it cannot be normal matter consisting of atoms and molecules, as this would already have been detected by ground-based optical and radio telescopes or space-based

Fig. 13.2 Hubble pictures of clusters of galaxies often show sections of 'Einstein rings'. These arcs or rings have much larger redshifts than the clusters themselves. They are images of galaxies that are located far behind the clusters, distorted and magnified by the gravitational lensing effect of the cluster, as explained in Fig. 13.3. The *left* picture is of cluster Abell 2218 and the *right* one of Abell 1689. Frcm the strength of the lensing effect one calculates that the clusters contain some 5–10 times more invisible matter than 'normal' matter in the form of stars and gas

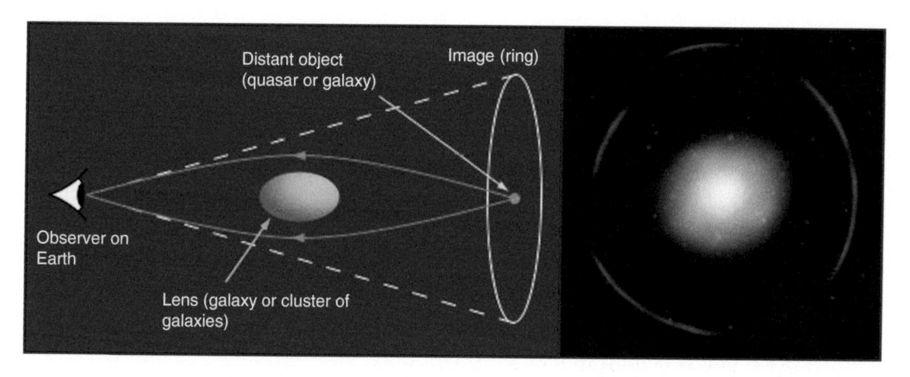

Fig. 13.3 Schematic picture of how an 'Einstein ring' is formed by the gravitational lensing effect of a cluster of galaxies. The gravitational attraction of the cluster bends the light rays of the galaxy or quasar that is located behind the cluster, such that for a distant observer a ring around the cluster is formed

X-ray, Infrared, ultraviolet or gamma-ray telescopes. Also, if all dark matter would consist of normal atoms and molecules, it would have been impossible for the universe to have produced in the Big Bang the observed fraction of helium in the universe (between 23 and 30 %), nor the amounts of the other light isotopes and Lithium. The observed amounts of these light elements and isotopes can only be produced in the Big Bang if "normal" matter does not constitute more than about 5 % of the total mass density of the universe.

Most probably, dark matter consists of a presently unknown kind of elementary particle (or perhaps of a combination of several types of unknown elementary particles) that has mass and therefore attracts other masses gravitationally, but has no other interactions with "normal" matter. We do know already for decades that there are particles that have a very small mass, but hardly interact with other particles. These are the neutrinos, of which at least three species exist, each with its anti-particle: the electron-neutrino, the muon-neutrino and he tau-neutrino (see Table B1 in Appendix B). Neutrinos fly without any problem through a several-light-year thick layer of lead. Each second, 7×10^{10} neutrinos produced by the nuclear fusion in the centre of the sun fly through every square centimetre (the size of a finger nail) of Earth and of our bodies. Although they take with them 4 % of the total energy produced by in sun, we notice on Earth nothing of this gigantic flow of neutrinos passing every second through our bodies. They just pass through Earth and through us and disappear into space. Unfortunately, the combined mass of all neutrinos in the universe is far too small to be able to account for the observed amounts of dark matter. One should therefore find a new type of elementary particle—or a group of particles—with a combined mass large enough to account for the observations.

That there would still exist unknown elementary particles in nature would not at all be surprising. It would be very strange if now, by the year 2016, humans would know all kinds of elementary particles that exist in the universe. In the course of the last century time and again new elementary particles have been discovered, such as the tau-particle that was accidentally discovered in 1974 in experiments with two different particle accelerators. It turned out that it beautifully fitted into the generally accepted "standard model" of elementary particles (see Appendix B). Particle physics theories, particularly those that attempt to unify all fundamental forces (see Chap. 10) predict a variety of still undiscovered particles. These theories are, however, still speculative and there are so many possibilities of such theories that it is hard to take all predictions very seriously. One of the theories, called *supersymmetry*, is viewed by many physicists as promising. This theory predicts particles with names such as neutralinos, photinos, gravitinos and axions, which possibly could account for dark matter. Also, some physicists favour the possibility of a so-called *sterile neutrino*, that may still be allowed by the standard model. But it is also very well possible that the dark matter exists of some kind of particles that so far nobody had thought of or predicted. The uncertainty of the properties of the dark matter particles (what mass, what spin, etc.), makes it difficult to build detectors for finding them. The dark matter detectors that are built nowadays by groups in, for example, Italy, Japan and the USA, aim for the above-mentioned particles with more or less predicted properties. But it is very well possible that other particles exist for which completely different instruments would be needed to find them. We must admire all the groups of physicists that are building instruments and carry out heroic efforts to detect the possible dark matter candidates! The search goes on.

Does Dark Matter Really Exist?

Some physicists believe that dark matter does not exist, and that the flat rotation curves of spiral galaxies and the large gravitational lensing effects of the clusters of galaxies are due to the fact that the normal Newtonian or Einsteinian gravity law (the $1/r^2$ dependence of the strength of the gravitational force) is no longer true for very large values of the distance r. Proponents of this idea have put forward a theory called Modified Newtonian Dynamics (MOND). According to Newton's law the gravitational force is strictly inversely proportional to the square of the distance. This gravity law has been tested only out to the distance of the planet Neptune in the solar system, and proponents of the MOND theory, such as Israelian physicists Jacob Bekenstein and Mordechai Milgrom of the Weizmann Institute in Rehovot, have proposed that on distances as large as the size of a galaxy or a galaxy cluster, there is a small deviation from this law which can very well explain the flat rotation curves of spiral galaxies and the large escape velocities from clusters of galaxies. Thus according to the MOND theory dark matter does not exist.

Fig. 13.4 Bullet cluster, the result of the collision of two galaxy clusters, proofs the real existence of dark matter. This picture is based on a combination of data obtained with the Hubble Space Telescope and NASA's Chandra X-ray observatory. In *blue* the locations of the dark matter of the clusters are indicated, as obtained from gravitational lensing studies of Hubble images of the clusters. *Red* indicates the locations of hot X-ray emitting gas with temperatures of order millions of Kelvins, obtained with Chandra. Before the collision, the cluster at right was at left and the cluster at left was at right. When colliding, the dark matter particles of the clusters—which constitute most of the cluster masses—just passed along each other, without actual particle collisions, and they dragged the cluster galaxies along with them; as there is much distance between galaxies in clusters, there also were no real physical collisions between galaxies. On the other hand, the hot gas masses present in each of the clusters (with a mass larger than that of the galaxies), did physically collide and therefore have stayed behind with respect to the galaxies and dark matter of the clusters, as is seen in the two *red* (X-ray emitting) blobs which are still located between the two clusters, while normally the hot cluster gas coincides with clusters. The gravitational attraction of the dark matter now makes the hot gas fall back towards the two clusters as can be seen from the shapes of the hot gas clouds. This clearly shows that dark matter with its gravity is real

In recent years, however, quite convincing evidence for the reality of the existence of dark matter has been found that makes it difficult to maintain alternative explanations like MOND. This evidence comes from colliding clusters of galaxies, called *bullet clusters*. Figure 13.4 shows an example of such a collision between two clusters. If we assume that dark matter is real, clusters consist of three ingredients: (i) dark matter, (ii) galaxies which consist of stars and gas, and (iii) hot intra-cluster gas in which the galaxies are embedded. The latter part is very important: it contains more mass than the cluster galaxies combined and we know it to be very hot and emitting X-rays. When two clusters collide, the dark matter of one cluster, which has no other interaction with normal matter than gravity, will just shoot through the other cluster, as will the galaxies, as there is enough room between galaxies to avoid collisions with the galaxies of the other cluster. Thus, one expects that if cluster nr. 1 starts at left and nr. 2 at right, that after the collision, the galaxies and dark matter of nr. 1, after passing through cluster nr. 2, will now be at right and the galaxies and dark matter of nr. 2 be at left. The gas of the two clusters, however, will really collide physically and produce a shock between the two clusters. The gas will, of course, be gravitationally attracted back towards its cluster, but the gas of nr. 1 will have lagged behind with respect to nr. 1, and the same is true for the gas of nr. 2 with respect to nr. 2. This is the situation that we now see in the picture of the colliding clusters in Fig. 13.4. The collision has already happened and we are seeing the situation now after the collision. The blue-coloured dark matter of the clusters (the amounts of which were measured with gravitational lensing), together with their galaxies have already passed through each other and separated, but the colliding cluster gas has lagged behind in the form of the two pink patches of hot X-ray emitting gas that are now falling towards the respective clusters. Normally in isolated clusters of galaxies, this gas is symmetrically distributed throughout the cluster, like the galaxies and the dark matter, but here we see that after the collision, due to its lagging behind, it is now highly asymmetrically distributed for each of the two clusters. The situation of the picture is just a temporary one: within a few hundred million years the hot gas will have fallen into their respective clusters and will have become symmetrically distributed through their clusters again. Calculations show that this requires the gravitational pull of real dark matter in the clusters, and that the gravitational pulls provided by the MOND theory would not be sufficient to drag the gas back towards the clusters again, while the picture of Fig. 13.4 shows that in reality this pulling back is taking place. There are several of these bullet clusters now known and they all provide evidence that dark matter really exists.

Dark Energy: From Greatest Blunder to a Genial Stroke

Even stranger than dark matter is *dark energy*. While dark matter has been known for over 80 years, the existence of dark energy was discovered only in 1998. This discovery was made when studying the light of supernovae, exploding stars at distances of billions of light-years.

Supernovae come in different types. Most supernovae mark the end of the life of massive stars, that started out life with masses higher than about 8 solar masses. The evolution of these stars was briefly described in the box in Chap. 11. As described there, at the end of the life of these stars, their burned-out cores collapse to a neutron star or black hole, while the energy liberated in this collapse causes the rest of the star to be ejected into space with velocities of thousands of kilometres per second in a so-called *core-collapse supernova*.

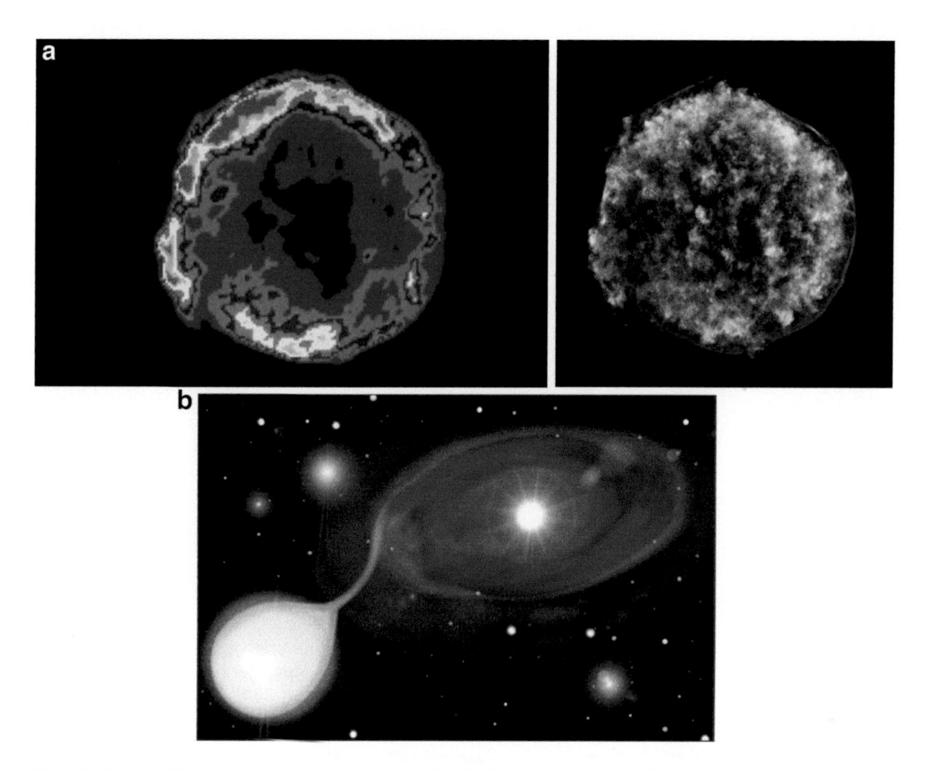

Fig. 13.5 (**a**) (*Above*) X-ray (*right*) and radio (*left*) picture of the shell of Tycho's supernova. This supernova was seen in 1572 as a very bright star in the constellation Cassiopeia, and was carefully observed for over 1 year by Tycho Brahe (see Fig. 3.1). This was a supernova of Type Ia; its shell expands with a speed of about 10,000 km/s and now has a diameter of over 10 light years. (**b**) (*Below*) Supernovae of Type Ia are exploding carbon-oxygen white dwarf stars in binary systems, that are triggered to explode due to the feeding of mass by their binary companion star, as sketched here

Apart from these supernovae, there is a second type of supernova which is powered by nuclear fusion of carbon and oxygen to iron—a colossal "nuclear fusion bomb"—which rips apart an entire star. The exploding star in this case is a carbon-oxygen *white dwarf* star, and the supernova is called a *Type Ia supernova*. These supernovae have no hydrogen or helium in their spectra, but only lines of heavier elements such as silicon, carbon, oxygen and nickel. Their spectra and light curves suggest that these explosions produce between 0.2 and 1.4 solar masses of ^{56}Nickel, which is radioactive and decays to ^{56}Iron. The Type Ia supernovae are the iron factories of the universe! The characteristics of their spectra and light curves fit exactly with what one expects from a thermonuclear explosion of a carbon-oxygen white dwarf star. Figure 13.5a shows the remnant of a Type Ia supernova that exploded in our Milky Way galaxy in the year 1572.

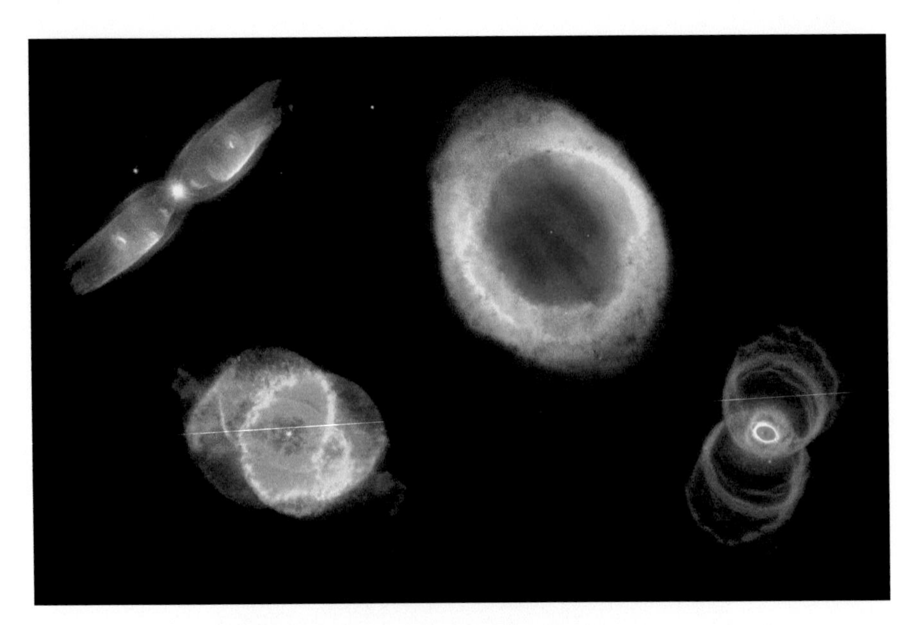

Fig. 13.6 Hubble images of four planetary nebulae. These nebulae are produced when a moderate—mass star (mass between 1 and 8 solar masses) has reached the end of its life. During the later phases of life such stars become *red* giants, which have a *white-dwarf*-like core, and gradually blow away their *red*-giant envelope. The planetary nebula is the last phase of this ejection process, and the burned-out nucleus of the star remains behind as a hot *white dwarf* star, visible in the centre of the nebula. This star has a size similar to that of Earth and a mass of less than 1.4 times that of the sun, and it quite rapidly cools down to a temperature of a few tens of thousands of Kelvins. The nebula flows out with a velocity of some 10 km/s and remains visible during about 10,000 years, after which it becomes so diluted that it is no longer visible. The often strange shapes of these nebulae are thought to be due to the presence of a companion star or a planetary system

White dwarfs are the end products of the life of stars that started out life with masses smaller than about eight times the sun. In these stars, the fusion of hydrogen and helium leaves behind a core consisting of carbon and oxygen with a mass smaller than about 1.4 solar masses: the *Chandrasekhar limit* mentioned in the box in Chap. 11. This core shrinks to a size similar to that of Earth, but has a mass similar to that of the sun: some 300,000 times that of Earth. At the same time, during the shrinking of the stellar core, the hydrogen-rich cooler outer layers of the star swell up enormously and on the outside the star has the appearance of a *red giant.* In this giant phase it slowly blows away the hydrogen-rich envelope and after a few million years only the burned-out core is left as a white dwarf, with a size similar to that of Earth. The blowing away of the outer layers produces a slowly-expanding nebula, often ring-shaped, with an expansion velocity of some 10 km/s. These nebulae are called *Planetary Nebulae*, and remain observable for tens of thousands of years. One finds thousands of these nebulae in our Galaxy. A few examples are shown in Fig. 13.6. In the centre of the nebula one finds a hot young white dwarf.

White dwarfs consisting of carbon and oxygen still contain a large amount of nuclear fuel: if this fuel would in a short time fuse to iron (through a number of intermediate steps), sufficient energy is produced to power a bright supernova explosion. White dwarfs have the nasty property that once nuclear fusion in their interior starts, it runs completely out of hand and blows up the entire star. Contrary to normal hydrogen-rich stars like our sun, white dwarfs lack a built-in "safety valve" that prevents nuclear fusion to run out of control. Nuclear fusion in a white dwarf will, however, not be ignited unless the mass of the white dwarf reaches the *Chandraskekhar limit* of about 1.4 solar masses. Since white dwarfs are born with masses below this value, an increase of mass to this limit can occur only if matter is dumped onto the white dwarf. When a white dwarf is single, this is very unlikely to happen. Even if moves through a dense interstellar cloud the amount of mass it can capture is so tiny that it is highly unlikely to reach the limiting mass. An entirely different situation arises if the white dwarf has a companion star in a binary system (see Fig. 13.5b). In this case it may be able to grow in mass by capturing matter from its companion star. It is for this reason that it is nowadays generally thought that Type Ia supernovae occur only in binary systems. Observations show that these explosions do not always reach the same intrinsic brightness: it appears that in the explosions not always the same amount of nuclear fuel (carbon and oxygen) is consumed. In some Type Ia supernovae only 0.2 solar masses of fuel is burned[1], while in others cases as much as 1.4 solar masses is burned. The first-mentioned ones are therefore less bright than the second ones, but the observations show something very important: there is a clear correlation between the real peak brightness that the supernova reaches, and the rate at which *after* the peak the brightness of the supernova decreases: the brighter the peak, the slower the decrease in brightness following the peak brightness. The nice thing is that the rate of decrease of the brightness after the peak can be accurately measured for every supernova, and this rate then tells us immediately how bright the supernova really (intrinsically) was at its peak. This means that the Type Ia supernovae are excellent *standard candles*, of which we know the real brightness. One can therefore—since they are extremely luminous—use them to measure the curvature of the universe, and see whether we are in a closed, open or flat universe.

Comparing the *true brightness* of the supernova at its peak with the *observed peak brightness* gives us the "light-distance" of the supernova: the distance obtained by assuming that the observed brightness decreases inversely proportional to the square of the distance, as explained in Fig. 3.10. A measure of this light-distance is the so-called *distance modulus* (m−M) in Fig. 13.7. One can also take the spectrum of the supernova, which gives us its redshift. For nearby objects—where the universe is still flat—the distance modulus (m−M) is, according to the Hubble law proportional to the redshift (Fig. 6.13). However, for large distances this proportionality no longer holds, and the precise relation between (m−M) and the redshift depends on the type of universe in which we live: open, closed, or flat or

[1] Astronomers often use the word "burn" for nuclear fusion which, of course, is not correct, but this has become a habit.

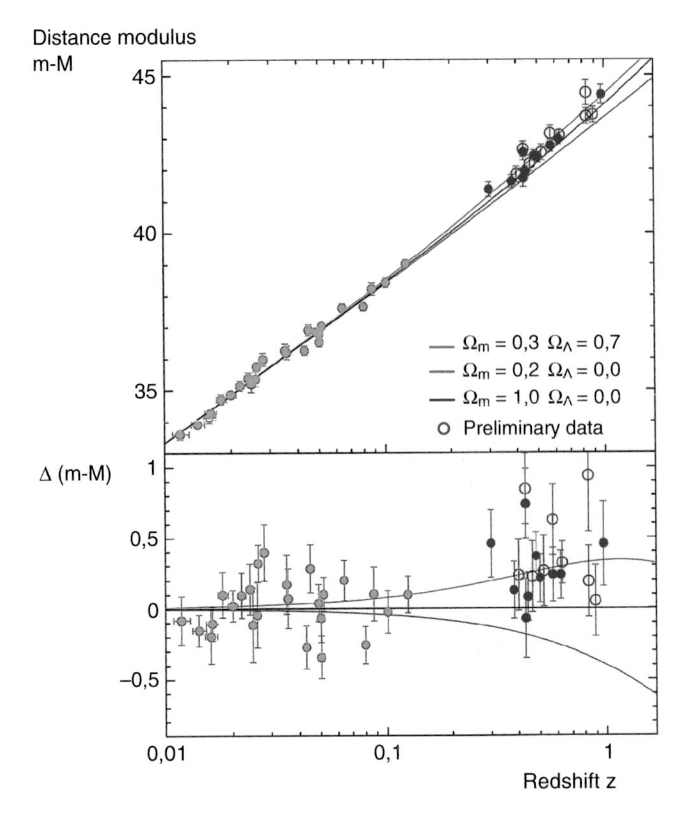

Fig. 13.7 Light distances of a number of supernovae of type Ia, plotted against their redshifts, and compared with different models of the universe. The observations are by the team of Saul Perlemutter of Berkeley, California. The light distance is the distance one obtains if one assumes that the observed brightness of an object decreases inversely proportional to the square of the distance. The 'distance modulus' (m–M) is a measure for this light distance. The quantity Δ(m–M) in the lower figure is the difference of the observed (m–M) relative to the relation expected for an open universe with $\Omega = 0.2$ and no cosmological constant (the horizontal line in the lower figure). The *blue* curve in the lower picture is for a flat model universe with 30 % matter and 70 % "Dark Energy": $\Omega_m = 0.3$ and $\Omega_\Lambda = 0.7$, and the lower curve is for a flat universe with only matter and no cosmological constant: $\Omega_m = 1.0$. One notices that only the blue curve fits the observations well

whether or not there is a cosmological constant Λ. This real form of the universe becomes detectable only for redshifts larger than 0.5. One therefore needs very distant, and thus: faint, supernovae to discover the true nature and shape of the universe. During the 1990s two teams of scientists started to apply this method. One of these was led by Saul Perlemutter of the Lawrence Berkeley Laboratory, who was the first to suggest that Type Ia could be used in a systematic way to measure the state of the universe. The other team was led by Australian-born Brian Schmidt (nowadays at the Australian National University in Canberra) under the guidance of Robert Kirshner at Harvard University. Both teams discovered that the very distant supernovae, with redshifts larger than 0.5, are systematically fainter than expected for a universe that would only contain matter with gravity. In such a universe—no matter whether it is open, closed or flat—the speed of expansion decreases gradually with time, and one does not expect very distant supernovae to be fainter than corresponding to their redshifts. This is shown in Fig. 13.7 for two cases of a matter-dominated universe: and open one in which the Ω-value produced solely by matter is $\Omega_m = 0.2$ and a flat one in which it is $\Omega_m = 1.0$. One sees in the figure that the (m−M) values of the observed supernovae at redshifts >0.5 are higher than the curves for these universes that are solely dominated by matter. Here the difference $\Delta(m-M) = 0.5$ means that the distant supernovae are about 1.6 times fainter than expected and $\Delta(m-M) = 1.0$ means that they are about 2.5 times fainter than expected. The young member Adam Riess of the team of Brian Schmidt was the first who in 1998 realized that the only model of the universe that fits these supernova observations is one that contains a cosmological constant Λ, which represents a *repulsive gravity*. And this repulsive gravity accelerates the expansion of the universe, just like we saw in the late-time expansion in the Lemaître model of Fig. 8.7. This was a very strange discovery, since a positive Λ value implies that empty space (vacuum) contains energy. The mass equivalent of this energy contributes to the value of Ω, and the Ω-value that corresponds to the Λ that fits best with the observations is about $\Omega_\Lambda = 0.7$. Riess found that the best fit to the observations is obtained for the combination $\Omega_m = 0.3$ and $\Omega_\Lambda = 0.7$ such that the total value of Ω becomes unity: $\Omega = 1.0$. This means: a flat universe, just as Dicke already has suspected (see Chap. 12).

We thus see that the Λ-term which Einstein had called his "greatest blunder" made a surprising come-back in the form of a mysterious energy present everywhere in the universe—also in your living room—which is completely independent of matter. It is this energy in empty space (vacuum) which produces the accelerated expansion of the universe, causing distant supernovae to look fainter than expected. By lack of a better name this energy is now called *Dark Energy*. It has the strange property to be an amount of energy per unit volume of empty space: the more space there is the more dark energy. The presence of this dark energy has become noticeable only during the past 5 billion years. Before that time it, of course, also existed, but since there was less space, there was less dark energy, and its effects on the rate of expansion of the universe was so small that it could not be noticed. In these days the gravity of the matter still fully dominated, and slowed down the expansion of the universe. However, the more the universe expanded, the larger the amount of dark energy became and about 4–5 billion years ago the point was reached from where on this energy began to accelerate the expansion of the universe, as depicted in Fig. 13.8b. A remarkable coincidence is that at the moment at which the effects of dark energy became noticeable in the universe is the time at which the solar system and Earth were born. Had our solar system been born 5 billion years earlier, we would not have been able to detect the presence of dark energy in the universe. It thus seems as though Earth was born just at the right time to allow after 4 billon years of evolution of life, to produce beings that could detect the presence of dark energy in the universe. This is one of many "coincidences" concerning the

Fig. 13.8 (**a**) The three discoverers of Dark Energy who shared the 2011 physics Nobel Prize. From *left* to *right*: Saul Perlemutter (born 1959), Brian Schmidt (born 1967) and Adam Riess (born 1969). (**b**) Shift, since the Big Bang, of the balance between the initial deceleration of the cosmic expansion due to the gravitational attraction of the matter, and the acceleration due to the action of the dark energy. In the first 8 billion years the gravity of the matter dominated, but since about 5 billion years the dark energy is beginning to manifest itself and causes the expansion of the universe to accelerate

nature of our universe, that appear to suggest that the universe was just made to produce "intelligent" observers (humans). We will return to this subject in Chap. 16.

It thus appears now that the universe consists for only about 4 % of "normal" matter consisting of atoms and molecules. In addition there is according to the best recent measurements some 26 % of dark matter and some 70 % dark energy. The fact that the matter which makes up our bodies, Earth, the sun, stars and galaxies, represents only 4 % of the total mass-energy of the universe, is one of the most extreme illustrations of the Copernican Principle, which states that we do not occupy a special (central) position in the universe. Even the matter of which we consist is not the most important ingredient of the universe! As an English colleague once remarked: "we now even have to give up matter-chauvinism".

In 2007, Saul Perlemutter and Brian Schmidt were awarded the Gruber Cosmology Prize for the discovery of Dark Energy, and in 2011 they, together with Adam Riess were awarded the physics Nobel Prize for this discovery (see Fig. 13.8a).

Vacuum Energy

The nature and origin of the dark energy is still a complete mystery. In Chap. 10 we already mentioned that the vacuum (space in which no particles or radiations are present) is not as empty as one would expect: it consists of a sea of short-lived particles and their anti-particles which are continuously created and almost immediately again annihilate one another. According to quantum mechanics the energy of the vacuum fields that create these pairs of "virtual" particles is not constant but fluctuates with all kinds of frequencies. These fluctuations are so small and so fast that they escape detection. However, Dutch theoretical physicist Hendrik Casimir (1909–2000) in 1948 conceived an experiment aimed at detecting their existence. This experiment is based on what nowadays is called the *Casimir effect*. Casimir proposed placing two perfectly flat parallel metal plates in vacuum, at a small distance from each other (Fig. 13.9). If these plates are not electrically charged they will, according to classical physics, not exert any force on each other (their mutual gravitational attraction is negligible). However, the small distance between the plates poses restrictions to the *wavelengths* of the particles produced by the quantum fluctuations of the vacuum between the plates (according to quantum mechanics, particles also behave as *waves* and have their own wavelength, the DeBroglie-wavelength, called after French physicist Louis DeBroglie). Between the plates, only waves are allowed for which the distance between the plates is an entire number of wavelengths, whereas outside the plates all wavelengths are allowed (see Fig. 13.9). For this reason there are fewer virtual particles in the vacuum between the plates than outside the plates. As all particles collide with the plates and in this way exert a pressure on the plates, the pressure on the outside of the plates will be larger than the pressure on the inside (the side facing the space between the plates). Already in 1949 Casimir's collaborator Dirk Polder in the Philips Research Laboratories in

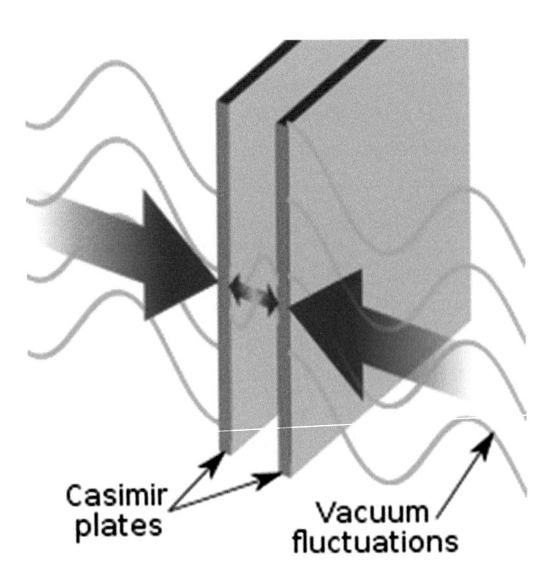

Fig. 13.9 The *Casimir-effect*. Due to the restriction of the wavelengths of quantum fluctuations between the two plates, fewer vacuum fluctuations can occur in the space between the plates than on the outside of the plates. As a result, the two plates, placed in vacuum, will be gently pushed towards each other. Careful measurements have shown that this 'vacuum energy'-effect really exists

Eindhoven (Netherlands), of which Casimir at the time was director, succeeded in measuring this small force with 15 % accuracy. With modern techniques the Casimir effect has been measured with much greater precision by among others: Marcus Sparnaay in 1957 and more recently, Steve Lamoreaux in 1997. The measured force has exactly the value predicted by Casimir. This proved beyond doubt the reality of the existence in the vacuum of the particle pairs that are continuously created by quantum vacuum fluctuations. From this existence one can, however, not find out whether or not the total energy of the vacuum is zero (or more precisely: if the fluctuations take place around a mean energy level equal to zero) or whether it has a small positive value. If it is not zero, the amount of energy will depend on the size of the volume that one considers: the more volume, the more energy. This has the strange consequence that when the space expands, the amount of vacuum energy in the universe increases. During the expansion of the universe the total amount of vacuum energy then becomes larger and larger. In this respect vacuum energy behaves exactly as the dark energy (Fig. 13.10).

From the standpoint of quantum mechanics, vacuum energy must be produced by a quantum field, and one would expect that this field in some way is related to gravity, since according to Einstein's General Theory of Relativity, it is gravity that produces space. One therefore would need a quantum theory of gravity to understand vacuum energy. However, when theorists try to construct such a theory, for example by means of *string theory* mentioned in Chap. 10, the calculations produce a value for the cosmological constant Λ that characterizes vacuum energy, which is a factor

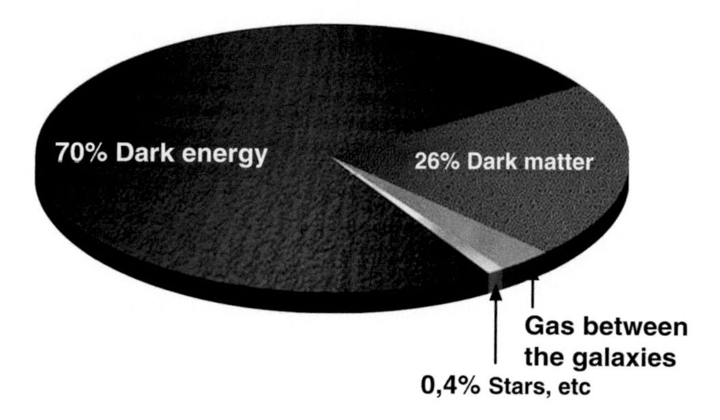

Fig. 13.10 The different ingredients of the universe. Normal matter, consisting of atoms (stars, gas and dust), is only about 4 % of all matter and energy. The remaining 96 % consists of dark matter and dark energy

10^{120} times larger than the observed value of Λ. With such a cosmological constant the universe would expand so extremely fast that never stars or galaxies could have formed. Nobody understands the reason for this colossal difference between the theoretical prediction and observations. It is clear from this that, if anything, string theory in its present form is not (yet) the correct theory for quantum gravity. As we will describe in Chap. 16, some string theorists try to explain their embarrassingly wrong prediction of the cosmological constant by invoking the *anthropic principle*. But as we will explain there, it is to be feared that this is a poor man's solution.

The Perfect Free Lunch

We have seen that the value of Ω, the mean density of the universe divided by the critical density, is very close to $\Omega = 1.00$. The present best observations yield a value between 0.99 and 1.01. As we saw in Chap. 12, this means that shortly after the Big Bang the value of Ω must have been extremely close to exactly unity. It is therefore highly likely that also today Ω is exactly equal to unity, which would mean that the total energy of the universe is exactly equal to zero.

As mentioned by Alan Guth in his book *The Inflationary Universe* (pp. 12–15), in the late 1960s American physicist Edward P. Tryon, then at Columbia University in New York, put forward the idea that the universe could be the result of a (large) quantum fluctuation. As we saw in Chap. 10, Heisenberg's uncertainty principle implies that the vacuum can spontaneously create particles. We cannot exactly predict what particle will emerge, although the probability that an elementary particle is created is, of course, many times greater than that spontaneously a Volkswagen emerges from empty space. But the probability of this last-mentioned event is not exactly zero. Tryon published his idea in 1973 in the British scientific journal Nature with as title "Is the universe a vacuum fluctuation?" His idea is that the universe can have originated from a vacuum fluctuation, that is: out of nothing—because the large positive energy of all the masses in the universe (including their energy of motion and the cosmological constant, although Tryon did not know about the latter) could be exactly compensated by the negative energy of all the gravitational attractions in the universe. According to Tryon the fact that the universe would have been created out of nothing is not in disagreement with the laws of physics. Tryon does not explain why a vacuum fluctuation with the enormous size of the universe would arise. In fact, the vacuum fluctuations that we know always take place only on the level of elementary particles, not of macroscopic objects. The only argument that Tryon put forward against this objection is that the laws of nature do not set a limit on the size that vacuum fluctuations may have. He did not attempt to calculate the probability of the occurrence of a universe-sized vacuum fluctuation, but just wrote: "In answer to the question to why it happened, I offer the modest proposal that the Universe is simply one of those things that happen from time to time". Since according to this theory the formation of the universe did not require any energy, and the total energy of the universe is still equal to zero, Tryon described our universe as "*the perfect free lunch*".

The same idea of a universe created out of a vacuum fluctuation was further elaborated in the early 1980s by others, such as Alex Vilenkin of Tufts University, Steven Hawking of Cambridge University and Jim Hartle of the University of California in Santa Barbara.

Indeed it is a nice and elegant idea that all that we see around us—Earth, sun, Milky Way, galaxies, plants and animals, the entire universe—in fact can have been created out of nothing, and that this Creation did not require any energy. The observations of the last decades that show that the universe is perfectly flat ($\Omega = 1$), appear to fully support this idea.

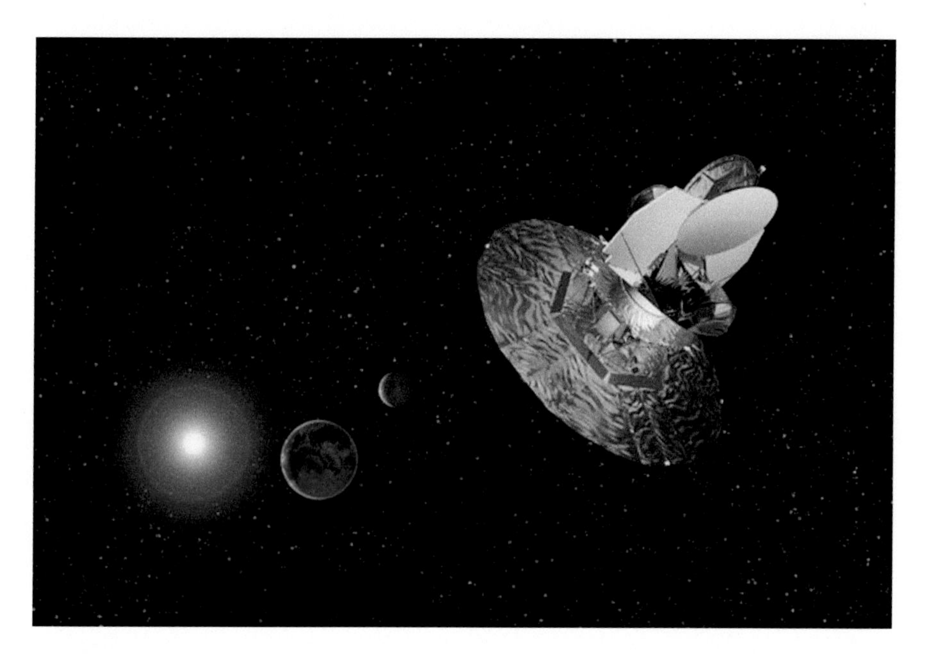

The WMAP satellite of NASA

Chapter 14
Ripples in the Cosmic Microwave Background Radiation

If you are religious, it is like seeing the face of God

George Smoot, American astrophysicist,
2006 Physics Nobel laureate

The Cosmic Microwave Background Explorer satellite (COBE) of NASA, launched in 1989, measured with unprecedented precision the intensity of the cosmic microwave background radiation (CMB) at many wavelengths and in all directions over the sky. The spatial resolution (sharpness) of the microwave optics in the satellite was 7°—14 times the diameter of the full moon. This means that the temperature of the background radiation was measured in separate areas of the sky with this size. The result is a sky map on which one can see that (small) temperature differences between such areas on the sky exist. These are depicted in Fig. 14.1. Before this map could be obtained, the observations first had to be corrected for the Doppler effect produced by the motion of the sun with respect to the background radiation, with a speed of about 390 km/s (see Fig. 9.14). This Doppler effect causes the

E. van den Heuvel, *The Amazing Unity of the Universe*,
Astronomers' Universe, DOI 10.1007/978-3-319-23543-1_14

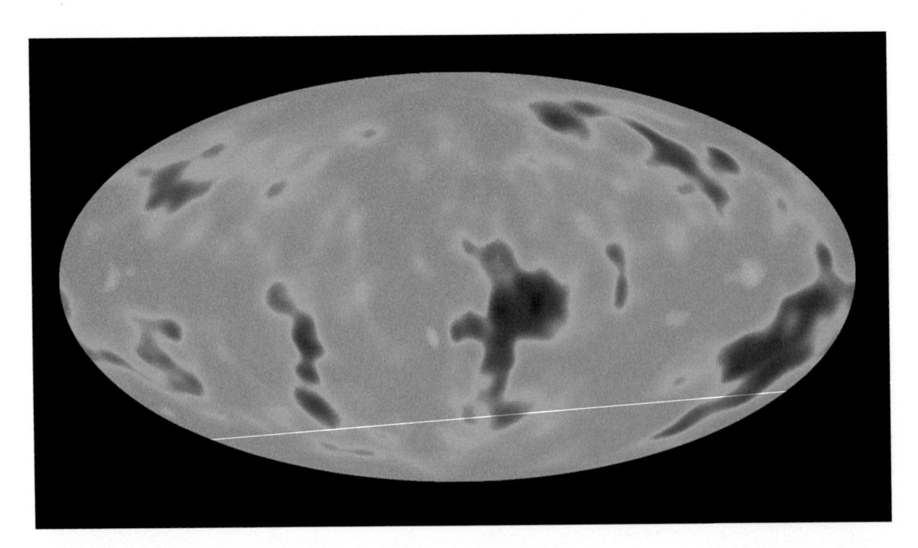

Fig. 14.1 Sky map of the deviations from the mean temperature of the cosmic microwave background radiation as measured in 1992 by NASA's COBE satellite, with an angular resolution of 7°. The deviations are smaller than one ten-thousandth of a Kelvin. The map is in galactic coordinates: the sky is pictured on a plane, with the centre of the Milky Way system as centre, and the horizontal axis along the Milky Way. One sees here the deviations that remain after the Doppler effect of the sun's motion—depicted in Fig. 9.14—has been removed and the radiation of the dust in the Milky Way has been subtracted. In the *blue* areas the temperature is slightly lower than average, in the *green* and *yellow* regions slightly higher than average

temperature on one half of the sky to be slightly higher than average and on the other half to be slightly lower than average. The motion of the sun relative to the background radiation is a combination of the motion in its orbit around the Galactic centre and the motion of our Galaxy relative to the average background of distant galaxies. The last-mentioned motion, with a velocity of a few hundred km/s, is probably caused by the attraction of the other galaxies in the Local Group and of the Virgo Cluster (see Chap. 6), or of the local super cluster of which the Local Group and the Virgo Cluster are members.

After subtraction of this Doppler effect the average Planck curve of the CMB could be determined. As we saw in Fig. 9.6, this results in a curve that deviates less than 0.03 % from a perfect Planck curve for a temperature of 2.725 K. Looking at the sky distribution of the remaining temperature differences of this radiation between different regions on the sky (Fig. 14.1) one observes that the temperature differences between 7°-size areas on the sky never exceed a few hundred-thousandth of a Kelvin.

These temperature differences are due to tiny local enhancements and reductions of the matter density in the universe at the time when it became transparent, 380,000 years after the Big Bang. At places of higher density we look less deep into the universe than elsewhere and the redshift of the radiation will therefore be slightly

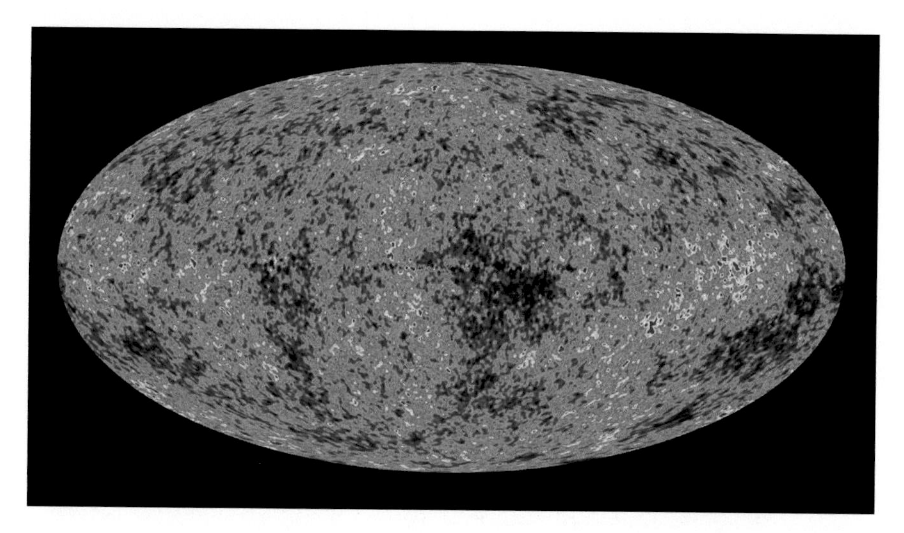

Fig. 14.2 Sky map of the temperature deviations of the cosmic microwave background radiation, as measured between 2003 and 2005 by NASA's WMAP satellite, with an angular resolution of 0.2°. Also here the Doppler effect of the sun's motion and the radiation of the dust in the Milky way have been removed. The over-all areas with higher and lower temperatures are the same as those in Fig. 14.1, but show here much more detail. Again the maximum deviations are of order one ten-thousandth of a Kelvin. Figures 14.1 and 14.2 show that at the moment when the cosmic background radiation was emitted, 380,000 years after the Big Bang, there were already small density enhancements in the gas that filled the universe. These were the 'seeds' that later grew into the super-clusters and clusters of galaxies

lower than the average value of about 1100, resulting in a slightly higher temperature. Similarly, in regions where the density was lower than average, we will see a slightly lower temperature. These tiny density enhancements, which already formed before the universe became transparent, were the "seeds" that later on gave rise to the formation of large structures, such as the clusters and super-clusters of galaxies that we nowadays observe in the universe.

These tiny temperature differences in the cosmic background radiation were at the limits of what the COBE instruments could detect, but they were beautifully confirmed by the measurements with NASA's next and more accurate satellite for measuring the cosmic background radiation, the Wilkinson Microwave Anisotropy Probe launched in 2003, abbreviated as WMAP (it was called after David Wilkinson, mentioned in Chap. 9, who died before its first results became available). The instruments of WMAP had an angular resolution on the sky of 0.2°, 35 times better than that of COBE, and could measure the radiation temperatures with an accuracy of a few hundred-thousandths of a Kelvin. Comparison of the sky distributions of the temperature measured by COBE and WMAP (Figs. 14.1 and 14.2) shows that WMAP found the same large-scale (global) temperature deviations that COBE had discovered, which beautifully confirmed that the COBE team was the first one ever to discover the existence of temperature differences in the distribution of the cosmic microwave background radiation over the sky. This, together with the beautiful measurement of the Planck curve of the microwave background radiation (Fig. 9.13) earned COBE team leader John Mather and George Smoot, who had built the instru-

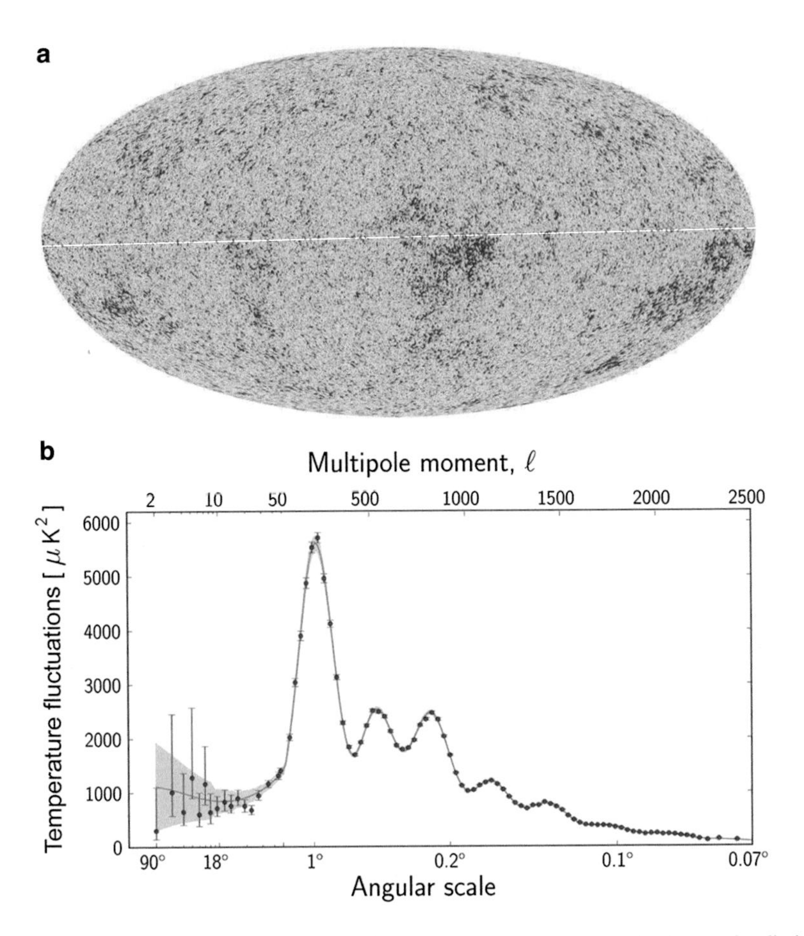

Fig. 14.3 (**a**) Map of the temperature deviations of the cosmic microwave background radiation, as measured between 2009 and 2013 by ESA's Planck satellite, with a highest angular resolution of 0.08° (2.5 times higher than that of WMAP). This is the most detailed snapshot of the infant universe ever made. On the basis of these measurements the age of the universe has now been established as 13.8 billion years. (**b**) The distribution of the dependence of the temperature deviations of the cosmic background radiation on the angular sizes of the deviating regions in the map of (**a**), as measured with the Planck satellite. The drawn curve is the theoretical prediction of this distribution for a universe with 26.0 (±0.5) percent matter 4.8 (±0.5) percent of ordinary matter, and 69.2 (±1.2) percent dark energy. This is called the ΛCDM model, where Λ is the cosmological constant, and CDM stands for 'Cold Dark Matter'. The dots are the observations with their error bars. The *bluish* hue indicates the uncertainty of the place of the curve that best fits the observations. One sees that only for very large angular scales the curve is still somewhat uncertain

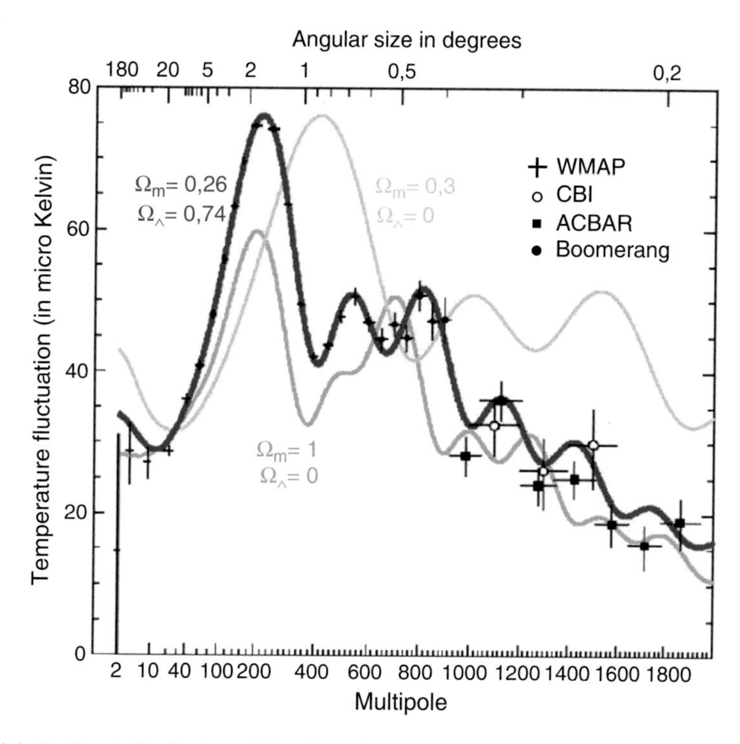

Fig. 14.4 Predicted distribution of the dependence of the temperature deviations of the background radiation on the angular sizes, for three different models of the universe. The *red* curve is the observed distribution for the WMAP results: 26 % matter and 74 % dark energy ($\Omega_m = 0.26$; $\Omega_\Lambda = 0.74$), which was later slightly revised by Planck (see Fig. 14.3b), the *yellow* one is the prediction for an open universe with only matter ($\Omega_m = 0.3$) and the *green* one for a flat universe with only matter ($\Omega_m = 1.0$). One observes that the *green* and *yellow* curves do not at all fit the observations, but the one for $\Omega_m + \Omega_\Lambda = 1.00$ does fit the WMAP observations very well; the same is true for the Planck observations

ment that discovered the temperature differences (the "ripples") over the sky, the 2006 physics Nobel Prize[1].

After Penzias and Wilson in 1978 this was the second Nobel Prize awarded for cosmological research, followed in 2011 by the third Nobel prize, for the discovery of Dark Energy (see Chap. 13).

A still further improvement on the WMAP results are those of the Planck satellite of the European Space Agency ESA, which was launched in May 2009 and functioned in space until October 2013. Like WMAP, Planck made an all-sky survey of the Cosmic Microwave Background radiation with still further enhanced spatial resolution, as can be seen in Fig. 14.3a.

[1] A nice overview of these discoveries is given in "The Very First Light" by John Mather, Basic Books, New York 2008.

Theories for the formation of the temperature differences in the cosmic microwave background radiation predict that the value of the temperature difference varies with the size of the area in which this temperature difference is found. This is depicted by the drawn curve in Fig. 14.3b, where Planck's observed distribution of the temperature differences as a function of area size is shown. In fact, the theoretically predicted curve for this distribution varies in shape for different combinations of the amounts of dark matter and dark energy in the universe, and also for other models of the universe, as shown in Fig. 14.4. The curve in Fig. 14.3b is the one for a flat universe with 26.0 (±0.5) per cent of cold dark matter and 69.2 (±1.2) per cent of dark energy, plus 4.8 (±0.5) per cent of ordinary matter. Thanks to Planck these are now the most accurate measurements of these quantities (March 2015). These percentages fit well with the percentages of these ingredients derived from gravitational lensing in galaxy clusters and from the type Ia supernovae discussed in the last chapter. Planck's data fit best with an age of the universe of 13.8 billion years.

Structure Formation in a Static Universe

As we noticed above the tiny density differences that manifest themselves as temperature differences in the cosmic microwave background radiation must have arisen already before the universe became transparent when the universe had an age of 380,000 years. What could have been the origin of these density perturbations? In the late seventeenth century Isaac Newton already realized that if the universe started out with a perfectly smooth distribution of the matter, with everywhere exactly the same matter density, the effects of random motions of particles will by pure chance cause at some places the density to become slightly larger than average and at other places slightly lower than average. The places of enhanced density will then exert a higher gravitational pull than average on the matter surrounding them, causing matter in their neighbourhood to start moving towards the density enhanced region, thus further increasing the already enhanced density in this region. Newton suggested that this *snowball effect* is the cause why the matter of the universe, which he had assumed to originally have been smoothly distributed throughout space, had become clumped into stars and planets. Since gravity is the cause of this fragmentation of an initially smooth matter distribution, one calls the reason for this fragmentation the *gravitational instability* of the smooth distribution. Newton in his time was not able to give a good mathematical solution for gravitational instability, as the precise physical laws for a gas as a function of temperature and density were not yet known. Early in the twentieth century British astronomer James Jeans (1877–1946) was the first to discover an exact criterion for the occurrence of gravitational instability. He discovered that if somewhere in a homogeneous gas a density enhancement occurs, this enhancement can only start growing if the *size* of the enhanced region (assumed to be spherical in shape) exceeds a critical value, the so-called *Jeans length*. The corresponding mass of a sphere of the gas with a diameter equal to the Jeans length is called the *Jeans mass*. It turns out that for an ideal gas (without photons) the Jeans mass depends only on the density and temperature of the gas: it is proportional to the temperature to the power 3/2 divided by the square root of the density (see Appendix E). The Jeans mass is the smallest-mass condensation that can form due to gravity, in a gas of a given density and temperature. This means that according to the classical Jeans theory, for a static universe of very low density, the Jeans mass is very large, such that only very large condensations can form: *proto-galaxies*. If such a condensation begins to contract under its own gravity, its density and temperature will increase (when a gas is compressed, its temperature rises). If the contracting gas sphere remains transparent, it can radiate away its excess heat and cool itself such that it can maintain a low temperature. Due to its increase in density, the value of the Jeans mass inside this contracting sphere will then become smaller, such that inside this contracting sphere, condensations of smaller mass can form: *proto star clusters*. When these contract and are able to stay transparent and radiate away their excess heat and cool themselves, then inside these condensations again the Jeans mass will decrease, such that again smaller condensations can form: the *proto stars*. If now the density of these smallest condensations has become so large that they no longer can stay transparent and cool, further fragmentation cannot occur. This then terminates the process of fragmentation of the original smooth low-density gas. It was originally thought that galaxies and globular cluster and stars could have formed through this *hierarchical fragmentation process*, on the basis of Jeans' theory for a static universe. However, in an expanding universe, in which further also radiation plays a role, the situation becomes very different. This does not mean that Jeans' theory is not valuable. It still remains very valuable for understanding the formation of stars inside the clouds of gas and dust in present-day galaxies like our own. Inside the cloud only condensations can form that are larger than a Jeans mass, and these are the *proto star clusters*, inside which then, in the hierarchical way described above, the *proto stars* form. So, inside galaxies, Jeans' theory remains valid. But for the formation of galaxy clusters and galaxies in an expanding universe, it can no longer apply and has to be modified.

Structure Formation in an Expanding Universe

Before the universe became transparent 380,000 years ago, it was very hot. The very intense heat radiation present in the early universe prevented the growth of density enhancements by gravitational instability. The reason for this is that in this opaque universe the photons of the radiation were completely coupled to the matter: already after travelling a very short distance, photons were absorbed again by the matter, so the matter dragged the photons along with it, and as a result, regions of enhanced matter density also contained more photons. However, photons always stream from regions of high photon density to regions of lower photon density—in regions of higher photon density the radiation pressure is higher than in regions of lower photon density, so that these regions want to expand. The photons therefore have the tendency to stream away from the regions of higher matter density, and they will in their turn drag the matter along. The presence of the intense radiation in the early universe will therefore cause density enhancements in the early universe to be washed out again, and prevents gravitational instability to develop. On the other hand, if there occurs somewhere a density enhancement by chance, this enhancement will cause a pressure enhancement, and we know that pressure enhancements in a gas propagate through the gas with the speed of sound, in the form of sound waves. These sound waves are what we observe as the density ripples in the cosmic microwave background radiation dating from 380,000 years after the beginning of the universe. British cosmologist Joe Silk was in 1967 the first to calculate the propagation of sound waves in a radiation-dominated early universe. In these days the importance of dark matter had not yet been realized, and so the presence of this matter was not included in his calculations. The same was true in the work of Russian astrophysicists Rashid Sunyaev and Yakov Zeldovitch who, in 1969, building forth on Silk's work were the first to make an attempt to calculate the spectrum of the sound waves of the early universe. They realized that, since no signal can travel faster than light, the size of a density enhancement that may arise at a time t after the beginning of the Big Bang cannot be larger than the size $c.t$ of the horizon at time t. So, sound waves that arise at time t cannot have a wavelength longer than $c.t$, waves that arise at $2t$ cannot have wavelengths larger than $2c.t$, etc. Thus, the earliest arising sound waves have the shortest wavelengths, the later ones have longer wavelengths. The fact that a sound wave (pressure enhancement) arises at a certain time, implies that at that time the pressure had its largest value, so the *phase* of the wave at that time is known. This implies that automatically one can calculate the phase also for any later time, so also for the moment at which the universe became transparent. For some sound waves at that time the phase will have been such that it reached its highest pressure (and density) at that moment, for others that it reached its lowest density when the universe became transparent. This reasoning allowed Sunyaev and Zeldovitch to calculate the spectrum of the density fluctuations at the time the universe became transparent, 380,000 years after the beginnings. This spectrum is the distribution of the sizes of the density enhancements one expects to observe today in the distribution of the cosmic microwave background radiation over the sky. These density enhancements (and reductions) translate into temperature reductions (and enhancements), as we saw earlier in this chapter. The discovery of these fluctuations by the satellites COBE, WMAP and Planck (Figs. 14.1 and 14.2 and 14.3a) qualitatively confirmed the predictions of Sunyaev and Zeldovitch, although quantitatively the observed distribution of the sizes of the fluctuations (Fig. 14.3b) differs considerably from their original prediction. This is due to a number of factors, the most important one being that in the original computations the presence of dark matter was not yet included. The presence of dark matter considerably changes the shape of the spectrum, depending on the precise ratio of the amounts of ordinary matter and dark matter. Also, in the late 1960s nobody was aware of the importance of the cosmological constant Λ (although Zeldovitch in these day already thought that it might play a role in the real universe). In these days the universe was assumed to be *open*, because not enough matter was known to close it or make it flat. In an open universe the rays of the cosmic microwave background radiation are deflected less by the gravity of the matter than in a closed or flat universe. The sizes of the fluctuations will therefore in an *open* universe *nowadays* look *larger* than they actually were. In a *closed* universe they will look *smaller*. Only in a flat universe their dimensions remain unchanged. Figure 14.4 depicts the predicted spectrum of the presently observed sizes of the temperature fluctuations for an open, flat and closed universe. Amazingly, the spectrum observed by the WMAP and Planck satellites agrees perfectly with what one expects for a *flat* universe ($\Omega = 1$). This is possible only if, apart from normal matter and dark matter, there also is a large amount of dark energy in the universe. While WMAP gave a best fit for the proportions of dark matter and dark energy of 23 % and 73 %, respectively, and an age of the universe of 13.7 billion years, Planck gives 26 (±0.5) per cent dark matter, 69.2 (±1.2) per cent dark energy and an age of the universe of 13.8 billion years. The latter values are now the best determined ones.

Figures 14.3b and 14.4 show that this so-called ΛCDM-model (the model that implies a large contribution to the present universe of dark energy (symbolized by Λ) plus slow-moving (Cold) Dark Matter particles) perfectly fits the observations. It should be noticed here that 380,000 years after the beginning, when the universe became transparent, dark energy did not yet play any part in the production of the spectrum of the density fluctuations. At that time, the dark matter represented 63 % of the energy of the universe, ordinary matter 12 %, radiation energy 15 %, and neutrinos 10 %—neutrinos were produced by the Big Bang in numbers comparable to the number of photons. Since that time the relative energy fractions of these constituents have changed enormously: nowadays neutrinos and photons (radiation) play a negligible role in so far as their energy is concerned—although their numbers are huge (of order one billion per proton or neutron). Dark energy has now become the dominant constituent with dark matter as second.

The Cosmic Web: Further Confirmations of the Existence of Dark Matter and Energy

Dark matter played a crucial part in the formation of galaxies and galaxy clusters. After the universe became transparent, the photons no longer were coupled to the matter. From that moment on the already present density enhancements in the universe started to contract due to their own gravity. The increase in density of these collapsing regions was partly counteracted by the expansion of the universe. Computations show that if the gravity of the contracting regions would solely have been due to the ordinary matter of the universe, consisting of atoms and molecules, the density in these regions would have grown so slowly that the expansion of the universe would finally have "won" from the tendency of these denser regions to contract, and the density enhancements would have been erased. In that case no galaxies and galaxy clusters could have formed. Only if the existence of dark matter—some 5–10 times more than normal matter—is included in these computations, the contraction of the regions of enhanced density will win from the expansion of

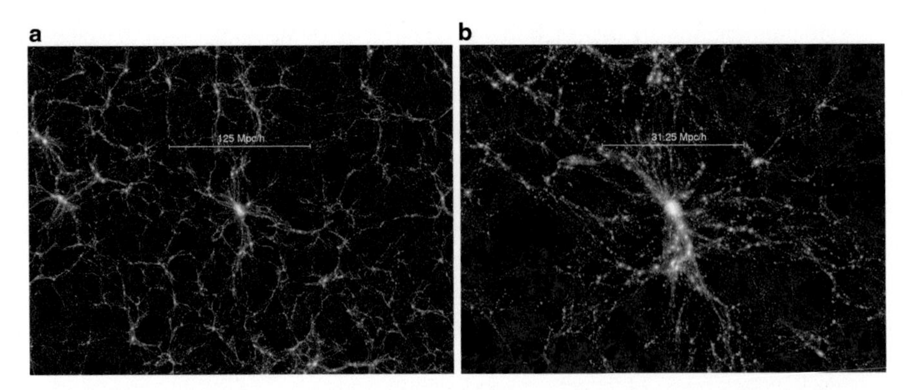

Fig. 14.5 (**a**) Structure formation in a universe with 25 % dark matter, 4 % normal matter and 71 % dark energy, as calculated in the 'Millennium simulation' carried out by German researcher Volker Springel of the Max Planck Institute für Astrophysik in Garching. The scale indicates a region of 125 Megaparsec (397.5 million light years). One observes here the clustering of dark matter as predicted for the present age of the universe. In the *yellow* regions the density is 1000 times higher than average, in the *black* 'empty' regions it is 10 times lower than average. It is expected that the 'normal matter' will concentrate in the regions where the density of dark matter is highest. One thus expects that the *yellow* regions indicate the places of clusters and super-clusters of galaxies. The predicted 'foam-like' structure of the distribution of galaxies corresponds well with the observations, as presented in Figs. 14.6 and 14.7. (**b**) Close-up of a small part of (**a**), where a very large cluster of galaxies formed. The scale here is 31.25 megaparsec (100 million light years)

the universe, and galaxies and galaxy clusters will form. Dark matter is therefore an indispensable ingredient of the universe, without which galaxies and clusters could not have formed.

Structure formation in the expanding universe can nowadays be simulated with help of powerful computers (see Fig. 14.5). Starting from the tiny density enhancements observed in the cosmic microwave background radiation, in these computations one follows the growth of the density enhancements. One includes in such computations different amounts of "hot" and "cold" dark matter and compares the obtained results with the observed distribution in space of millions of galaxies. Such distributions have been mapped in the past decades (e.g. see Figs. 14.6 and 14.7), and from this comparison one finds which combinations of assumed inputs for the computations does best fit the observations. In these computations also the effects of dark energy can be included, as this energy has become important during the past 5 billion years. Dark energy slows down the contraction of clusters and super-clusters of galaxies in comparison with a universe without dark energy.

Figure 14.5 shows the results of such a computer simulation for $\Omega = 1$, a flat universe, which apart from 4 % ordinary matter contains 25 % cold dark matter and 71 % dark energy. The result is a foam-like structure in which the dark matter has gathered on the surfaces of imaginary soap bubbles that enclose large empty spaces, so-called "voids". The dominant gravity of the dark matter has dragged the ordinary

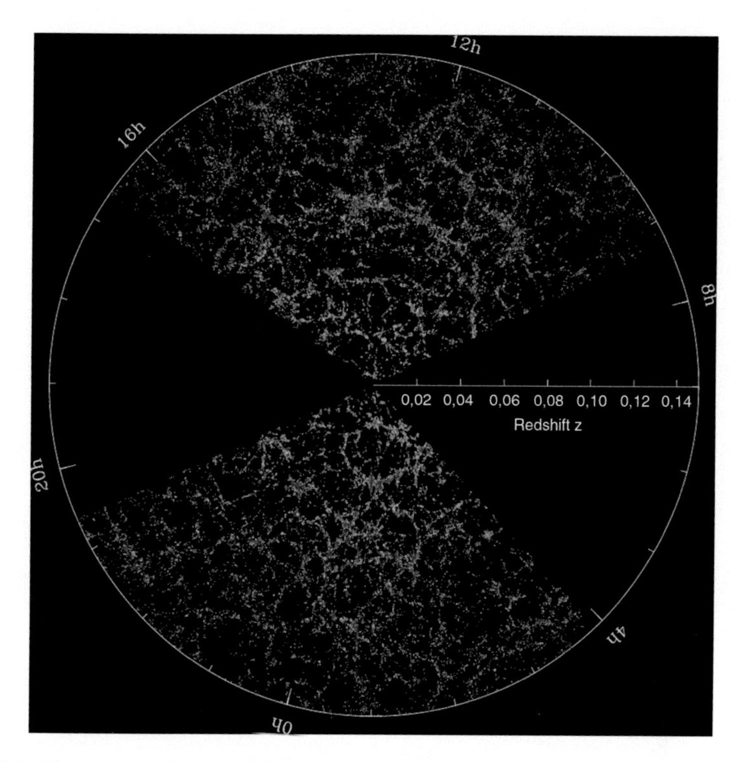

Fig. 14.6 Observed distribution of the positions in space of 930,000 galaxies, for a slice of the universe measured in the Sloan Digital Sky Survey. The redshifts of all these galaxies were measured, such that their distances are known. One clearly observes that the galaxies are mostly found along the walls of colossal 'soap bubbles', the interiors of which are practically empty

matter along, such that the ordinary matter is also concentrated on the soap-bubble-like surfaces surrounding the voids. This therefore is the structure that one expects for the distribution in space of galaxies. That the real galaxy distribution in space has indeed this foam-like structure, had already been known prior to these computer simulations. This was thanks to the pioneering work of Princeton astronomer Jim Peebles—whom we already met in Chap. 9—in the 1970s. This work was followed in the past decades by huge galaxy surveys using large telescopes. Figure 14.6 shows, for example, the space distribution of 930,000 galaxies measured with the Sloan Digital Sky Survey (SDSS). This survey was made with a telescope that was especially built for this purpose, located in New Mexico. It measured the redshift (which gives the distance) and the brightness of every galaxy in the survey. One clearly observes the foam-like structure of the distribution, which is called the *cosmic web*. Figure 14.7 depicts the result of an earlier survey of 211,414 galaxies in the southern sky made with the Anglo-Australian Telescope, and again shows the foam-like structure. If one carries out the structure-formation computations with different assumed amounts of cold dark matter, or with a value of Ω different from

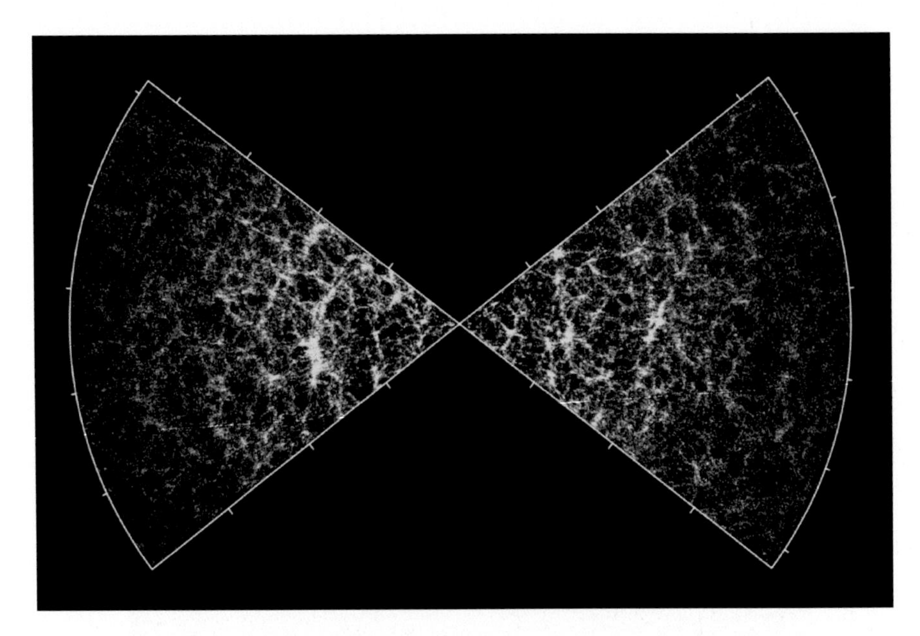

Fig. 14.7 Distribution in space of 221,414 galaxies in the southern sky, as measured in the 2dF-survey with the Anglo-Australian Telescope, by a team of Australian and British astronomers. Like in Fig. 14.6, the structure of the cosmic web is clearly visible

unity, the resulting structure of the cosmic web no longer fits the observed structure. Therefore, also these computer simulations tell us that we live in a flat universe that has some 4 % of ordinary matter, about 6 times more dark matter and over 70 % of dark energy.

As we saw in Figs. 14.3b and 14.4, completely independently, the distribution of the temperature fluctuations of the cosmic microwave background radiation also yield these percentages of ordinary matter, dark matter and dark energy. These different types of observations therefore provide completely independent confirmations of the findings about dark matter and dark energy that originally been derived from motions and gravitational lensing in galaxy clusters, and from distant Type Ia supernovae, as described in the foregoing chapter. It is thanks to these two completely independent confirmations, that the Nobel prize committee in 2011 decided to award the prize for the discovery of dark energy.

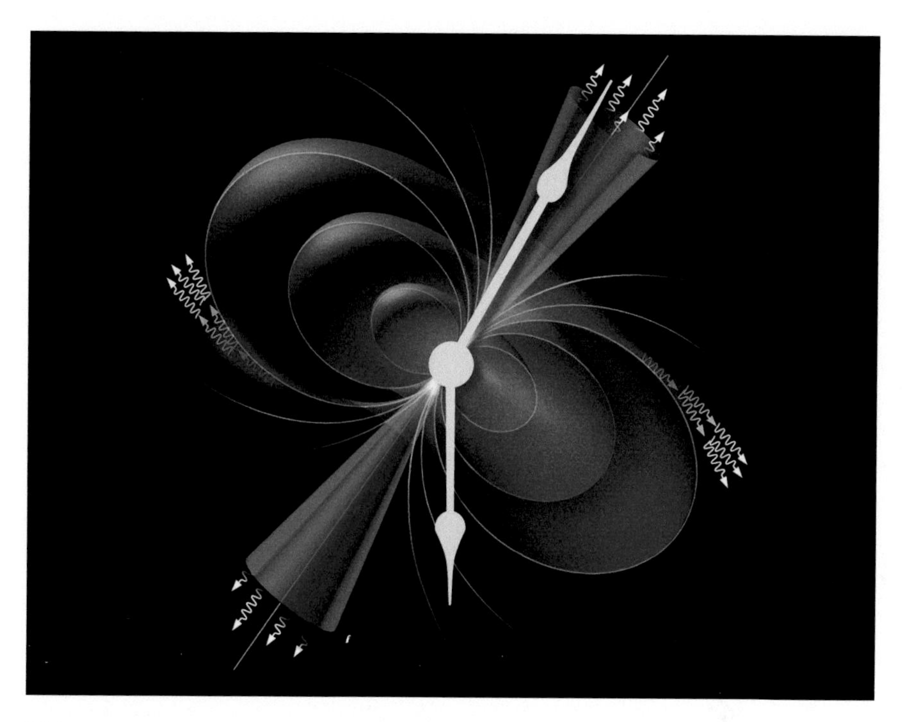

Which clock indicates the time for the entire universe?

Chapter 15
Time in the Universe

Time is what prevents everything from happening at once

John Archibald Wheeler (1911–2008)—American physicist

In the past chapters we have, as though it was completely evident, spoken about *time* and *development of the universe in the course of time*. However, since Einstein' relativity theory, the idea of one time for the entire universe is not at all obvious. We must therefore look in more detail at the basic concepts and assumptions underlying models of the universe.

Scientists and philosophers have for many centuries been thinking about the meaning of time. Time keeps ticking away, from the past to the future. This makes time different from space. While in three-dimensional space one can go forward and backward, and up or down, or to the left or the right, in time one can only go forward. While in space I can say: "I go backward", in time I cannot say: "I go back to yesterday". Time goes only in one direction. We will return later to this problem of

E. van den Heuvel, *The Amazing Unity of the Universe*,
Astronomers' Universe, DOI 10.1007/978-3-319-23543-1_15

the arrow of time. Here we will first concentrate on whether one can define one time for the entire universe.

Since antiquity, the passing of time is measured by observing phenomena that recur with great regularity, such as the daily motion of the sun and moon along the sky. When the sun has completed one full revolution along the sky (from reaching its highest point on the sky, in the South, till its next passage through the South), exactly one solar day has passed. When the moon has returned to the same phase—for example: the full moon—1 month has passed. When the sun, in its yearly course along the constellations of the zodiac, has returned to the same point in the same constellation (e.g. in Aries) exactly 1 year has passed, etc. Already the Babylonians and the Ancient Greeks had realized that the path of the sun along the sky is not fully regular: in January it moves faster along the stars than in July, and this fact also makes that the length of the solar days is not exactly the same throughout the year: in January it is slightly longer than in July. Therefore, sun and moon are not completely regular clocks.

Looking for a better clock, Galilei used the swinging of a pendulum, but was not yet able to turn this into a real clock. This was done by Dutch physicist Christiaan Huygens (1629–1695), who was the first to construct an accurate pendulum clock. Nowadays, the very rapid oscillations of the electromagnetic radiation produced by electronic transitions between the two hyperfine ground states of caesium-133 atoms are used to control the output frequency of a so-called *atomic clock*. These clocks can keep time accurately within one ten-millionth of a second over the course of a year, and are now used as the standards of time. Thanks to these clocks we now know that neither the daily rotation of the Earth, which determines the length of the stellar day (which is about 4 min shorter than the average solar day), nor the orbital motion of the moon, which in our calendars, determines the length of the month, are really accurate clocks. Due to the friction of the oceanic tides against the coasts and shallow ocean floors, Earth's rotation gradually slows down, leading to an increase of the length of the (stellar) day by over one-thousandth of a second per century.[1] At the same time the moon, which is the main cause of the ocean tides, gradually spirals away from the Earth in a wider and slower orbit. It moves away nowadays from Earth by about one inch per year.[2]

Newton thought that time is universal and that for everyone in the universe time advances at the same pace. At the opening of the *Principia* he wrote: "Absolute, true and mathematical time, of itself, and from its own nature, flows equably without relation to anything external." This means that his idea was that there is an *absolute time*. This immediately implies that there is absolute causality: if one knows for one point in time the positions of all objects in the universe, one will, with the laws of Newtonian mechanics, be able to exactly calculate the further development of the

---

[1] This increase looks almost negligible, but it can be easily measured, because this is an *acceleration*. After 1 year, the length of the year has already increased by 0.6 s, and a century lasts 62 s longer than an earlier century.

[2] This has been measured very accurately by reflection of laser signals from Earth by reflectors which the Apollo astronauts have placed on the moon.

positions and velocities of all objects, and thus of the entire universe, for the entire future. It was Newton's idea that God had set the universe—all of creation—in motion as a clockwork, and that from the moment of creation onwards, the entire future development of the universe was already determined. After the creation of the universe, God could just rest, and did not have anything further to do.

Einstein's discoveries of relativity of time destroyed this simple and beautiful picture. He showed that the pace at which time proceeds depends on the speed of motion (special relativity) and on the strength of the gravitational field that one experiences (general relativity). As a result, time was degraded from an absolute and universal concept to something relative and local. Thus the ticking of the clock depends on how one moves and on local gravity. Then the question arises: how can one define *one time* for the entire universe? And how can one speak of the development of the universe *in the course of time*? We follow here now the reasoning of J. Narlikar (in *The Structure of the Universe*, Oxford Univ. Press 1977).

Every galaxy might have its own time, and the universe might therefore be very complicated, with different galaxies following space-time tracks that cross each other and form a tangled bunch, as depicted in Fig. 15.1a. But one could also imagine a simpler situation, as depicted in Fig. 15.1b, where galaxies flow along regular tracks that do not intersect, except in the point X where all space-time tracks come together. In Fig. 15.1a time does not have a universal physical mean-

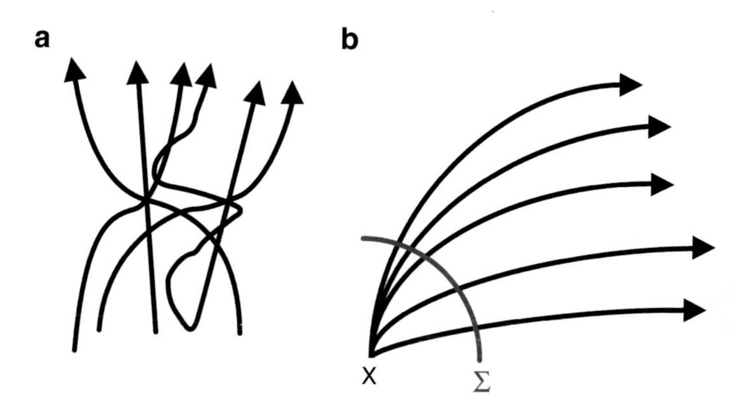

Fig. 15.1 Examples of: (**a**) disordered space-time trajectories for different parts of the universe, which may intersect and cross one another, and (**b**) motions of different parts of the universe according to *Weyl's postulate*. In case (**a**) it is impossible to define one unique time for the entire universe. On the other hand, in case (**b**) such a unique time, on which everyone in the universe can agree, can be defined. In case (**b**) the galaxies stream along ordered space-time trajectories that do not intersect, except at the starting point: the Big Bang. At a time t the galaxies are located on a space-time surface Σ. At any time such a surface can be defined in four-dimensional space-time, and the surface then corresponds to one unique well-determined time, valid for the entire universe

ing: each galaxy has its own time (clock) and it is not possible to synchronize the clocks of different galaxies. On the other hand, in the situation of Fig. 15.1b such a synchronization can be achieved, thanks to the regularity of the motions of the galaxies in space-time. Here one can draw a surface Σ that perpendicularly cuts the space-time tracks and then make sure that on this surface the clocks of all the galaxies are synchronized, that is: show the same time. In this way, one obtains one universal cosmic time, about which all galaxies agree, and which can be used as a reference time for the entire universe. When the galaxies move through space-time, the surface Σ moves along and all clocks advance by the same amount, thus keeping their synchronization. Mathematician Herman Weyl has derived the precise conditions for how regular the different space-time curves of the galaxies must be in order to define such a universal cosmic time. This is the so-called *Weyl postulate*. With this postulate as a boundary condition the study of the large-scale evolution of the universe is considerably simplified. This postulate allows one to speak of one time coordinate relative to which every change in the universe can be indicated. The expanding universe with a Big Bang nicely fulfils the Weyl postulate, since here the space-time motions of the galaxies are completely regular: all galaxies move away from each other, and all space-time trajectories intersect each other in the Big Bang. As a matter of fact, Friedman and Lemaître had—without realizing this—already intuitively included the Weyl postulate in their dynamical models of the universe. Weyl gave their work a more solid mathematical basis.

As mentioned before, Einstein as well as Friedman and Lemaître, had assumed that on every surface Σ in space-time, as depicted in Fig. 15.1b, the universe is homogeneous and isotropic. We cite here Narlikar (in the reference given above): "In physical terms this can be interpreted as follows. Suppose we fix the value of the cosmic time t and look at the universe from any of the galaxies that are moving according to Weyl's postulate. Then the universe will look the same, no matter which galaxy we choose to observe it from. Moreover, it will present the same view in all directions from any galaxy. To put it in another way: if you were blindfolded and left in any part of the universe, upon opening your eyes you could not tell where you are or in which direction you are looking."

Thanks to this cosmological principle, of homogeneity and isotropy, in combination with Weyl's postulate, the cosmologist is able to study the evolution of the universe. All models of the universe are based on these two, very logical and simple, assumptions. Without them, it would not be possible to construct models of the evolution of the universe. The extreme smoothness of the cosmic microwave background radiation beautifully proves that the universe fulfils the conditions of homogeneity and isotropy, and the many lines of evidence that indicate that the universe originated from the Big Bang prove that it fulfils the Weyl postulate. For these reasons we are able to define a *cosmic time* and one universal *cosmic timescale*. We will later see how this timescale can be measured.

What Happened Before the Big Bang?

When an astronomer gives a public lecture about the history of the universe, there almost always is a person in the audience that asks this question. The answer is: everything in the universe, including time itself, originated in the Big Bang, so one cannot speak about "*before the Big Bang*", because time did not exist. Counting backwards from the present time, one can always calculate backwards to earlier times: one can always find an earlier time, but in doing so, one never reaches a real beginning. The laws of physics show that before the Planck time of 10^{-43} s everything, including time, becomes undetermined, such that no longer one can speak about time.

Stephen Hawking has compared the question "what was there before the Big Bang?" with the question "what places lie North of the North Pole?" Both questions are equally senseless. On Earth, as long as your Northern geographical latitude is smaller than 90°, there always is a point Northward of you, and you can always answer the question "what lies North of you?". However, once you are standing on the North Pole, this question has become senseless.

This reasoning strongly reminds of that of Saint Augustine (354–430) who deeply thought about the question "what was God doing before he created the World?" He concluded that this is an incorrect question, since God exists outside of time. It is remarkable how closely this reasoning resembles our present views on the question about "what happened before the Big Bang?"

As an aside, according to George Gamow, when someone asked Saint Augustine "What was God doing before he created the World?" he would have replied: "He was creating Hell for persons who ask such questions." Gamow, who was a great practical joker and gave many popular lectures on the universe, always referred to this alleged answer of Saint Augustine, when he was asked what happened before the Big Bang.

About the Direction of Time

According to the theory of relativity we live in a 4-dimensional space-time, which has three spatial dimensions (two horizontal ones, indicated as x and y, and one vertical, indicated as z) plus time, t.

As mentioned already at the beginning of this chapter, what makes time so totally different from the three spatial dimensions is, that while one can move in all directions (backwards and forwards) in the three spatial dimensions that surround us, in time one can only go forwards. But one cannot say: "I go now forward to next year". One does go forward to the future, but one cannot influence the speed with which this happens. One just has to wait until a future moment arrives. It is as though we are carried forward in a train whose speed we cannot influence. This sole forward direction of time is in physics often called *the arrow of time*, and why such an arrow exists is one of the great puzzles of physics. The reason why this is so puzzling is that all physical equations for the motions of objects and of electromagnetic radiation are *symmetrical* with respect to time. For example, in the equations of motion of the planets around the sun, one can reverse the sign of the time (give it a negative value) without any problem. The planet will then orbit the sun in the reverse direction with respect to its present direction of motion, and with such an orbit is nothing wrong: it is fully allowed by physics (e.g. we can without any problem send a spacecraft in an orbit which moves in a 180° opposite direction of the directions in which the planets orbit the sun). The only reason why the planets describe their orbits in their present direction around the sun is that they obtained their velocities during the formation of the solar system, when they condensed out of a disk of gas and dust that was orbiting around the sun in this direction. Had the disk orbited in the opposite direction, all planets would now be moving in that opposite direction. Their present orbital directions are the result of the arbitrary formation conditions of the solar system.

The strange thing therefore is that, while the laws of physics are perfectly symmetric with respect to time, just as they are symmetric with respect to space, still we experience time as a phenomenon that has only one direction.

The reason why time has a direction has been ascribed to the phenomena of *probability* and *entropy*. This is called the *thermodynamic arrow of time*. Entropy is a thermodynamic quantity that was introduced by nineteenth century German physicist Rudolf Clausius, in connection with steam engines. Entropy is a measure of the disorder of a physical system. When a system is highly ordered, the entropy is low, if it is very disordered, the entropy is high. With "ordered" we mean here that the different components of the system have been arranged in a neatly ordered state with the same types of objects grouped together, while in a disordered state they are fully randomly mixed through each other. We may illustrate this by the following two examples. A teacup is a nice and orderly object. But if we drop it on the floor it becomes a random collection of pieces of broken china. The cup then goes from low entropy to high entropy. A somewhat more complex example is that of a rectangular glass container that has a left- and right-hand compartment separated by a removable wall. One fills the left compartment with blue ink and the right one with clear water. This is a nicely ordered state, with all the ink molecules at left and the water molecules at right. If one removes the wall that separated the two compartments, the ink and the water will mix, and after some time the entire container is filled with a light-blue liquid. The ink molecules are now completely mixed with the water molecules. The system has gone from a completely ordered state to a totally disordered state: from low entropy to high entropy.

The ink molecules are now moving randomly criss-cross among the water molecules, and the probability that, due to these random motions of billions of molecules one will ever reach a situation in which all the ink molecules are in the left-hand half of the container and all the water molecules in the right-hand half, is negligibly small: it will basically never happen again. We thus see that systems always have a tendency to go from low entropy to high entropy, and the way back is basically impossible.

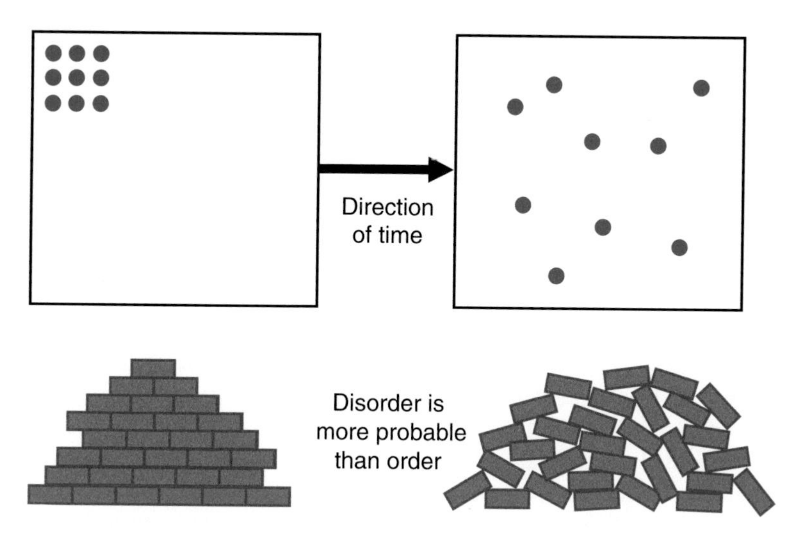

Fig. 15.2 Examples of order and disorder. In the *upper-left* picture the nine air molecules are grouped nicely in an ordered pattern in the *upper-left* corner of the container. When released, they will, as a result of their thermal velocities, start roaming through the container in all directions, leading to the much more disordered situation pictured *upper-right*. It is extremely unlikely that as a result of their random motions they will ever return to the ordered *upper-left* situation. Similarly, it is extremely unlikely that if one randomly throws a number of bricks together, a nicely ordered pile will result, as depicted at *lower-left*; practically always a disordered pile will result, as depicted at *lower right*. A system that is left to its own, will in the course of time always evolve from ordered to less ordered. One therefore can define the direction of time as the direction of increasing disorder

The completely mixed situation of ink and water is statistically seen the most probable outcome of the system, as a result of the random motions of the molecules, and the completely separated situation has an extremely low probability to arise by pure chance. We thus see that the entropy is basically a measure of the probability (likelihood) that a certain situation of the system is produced by pure chance processes. Leaving the liquid to itself, it will always strive to the most likely, completely mixed, configuration. One can also illustrate this with the two somewhat simpler systems depicted in Fig. 15.2. At room temperature, molecules of a gas move around with velocities of order a kilometre per second. It is highly unlikely that these random motions will result in all the nine molecules in Fig. 15.2 to be gathered in the upper left corner of the container, as depicted in the upper left picture. It is much more likely that they get evenly distributed throughout the container as depicted in the upper right-hand picture. The same holds when one throws bricks on a pile. The likelihood that a nicely ordered wall results, as depicted in the lower-left picture of Fig. 15.2, is very much smaller than the likelihood that a messy pile results, as depicted in the lower-right picture.

Austrian physicist Ludwig Boltzmann, whom we already encountered earlier, was able to mathematically demonstrate that the thermodynamic quantity entropy of a system in a certain state is exactly related to the probability (likelihood) that this state of the system will occur by random processes. The ordered situations depicted on the left-hand side of Fig. 15.2 are much less likely to occur by random processes than the situations depicted on the right-hand side. The situations on the left have therefore much lower entropy than the systems depicted on the right. If one leaves a system to itself, it will, due to random processes, always evolve to its most likely state: it will evolve from low entropy to high entropy. This fact is expressed by the *second law of thermodynamics*: "the entropy of a closed system (that is: a system that is left alone, and does not exchange energy with its surroundings) can in the course of time only increase." [this means: it can never decrease]. The evolution of a system towards its most likely state takes place simultaneously with the advancement of time, and therefore the advancement of time might be seen as a result of the evolution of a system towards its most likely state. As a system can never evolve back to a less likely state (that is: to lower entropy), the arrow of time seems to be connected with the fact the evolution of systems towards higher entropy is irreversible. The universe being a closed system then forces the arrow of time to run only into one direction. This is called the *thermodynamic arrow of time*. This complicated term can be nicely illustrated by making movies of irreversible processes. For example: a movie of dropping a china teacup on the ground, such that it fragments in random pieces. If one would run the movie backwards, one sees the pieces reassemble and form again a complete tea cup. Everybody immediately sees that this film has been run backwards and shows an impossible situation, that never happens in real life. Similarly, if one films the mixing of ink from the left container half with water from the right container half, one observes that the result is a container filled with a smooth light-blue liquid. Turning the movie backwards, one sees the ink and the water separate with the ink on the left and the clear water on the right. One sees immediately that this is a film run backwards, as this will never happen in nature. The arrow of time therefore seems to be due to the irreversibility of the striving of real physical systems towards disorder.

Universal Clocks of the Universe

Already some 80 years ago British cosmologist Edward Arthur Milne proposed that the cause of the arrow of time is the expansion of the universe. The universe started from a highly ordered state (the Big Bang) and evolves towards a more and more disordered state. Milne pointed out that while the Newtonian universe has a clock that keeps ticking forever, the expanding universe itself is a clock. The passing of time can be read on this clock by looking at the increase of the distances between the galaxies: the larger their distance the later it is on this cosmological clock. Nowadays this explanation for the arrow of time is called the *cosmological arrow of time*, and is seen by many physicists as a very likely explanation for the arrow of time. It should be noticed here that since the universe is moving towards a more disordered state, this arrow of time probably identical with the thermodynamic arrow of time.

The expansion of the universe started at time zero: the Big Bang. For every later time t one can measure the distance r between two galaxies, and also the velocity v at which they are moving away from each other. One then finds the age t of the universe from $t = r/v =$ the Hubble time. (This is the simplest approximation of the time t. In reality the calculation of the value of t is a bit more complex and depends on the model of the universe that one has chosen, see Appendix C.) Everywhere in the universe one measures with this method at age t of the universe on average the same value of t, by choosing a pair of galaxies, measuring the distance between them and the velocity at which the move away from each other (we say "on average" because the velocity difference between the galaxies also contains a slight local random component, due to small local deviations from the average expansion of the universe. This is seen, for example, from the fact that the Andromeda Nebula is approaching our Galaxy, instead of moving away from us, due to a local velocity deviation of a few hundred km/s). We thus see that there is indeed one universal age of the universe on which all observers throughout the universe can agree.

Even simpler and still more straightforward this universal age of the universe can be measured by measuring the temperature of the Cosmic microwave background radiation (CMBR). As we saw in the last chapter, after subtraction of the Doppler effect due to the local velocity deviation of our Galaxy, the distribution of the CMBR over the sky is completely smooth (apart from the small ripples of order a few hundred-thousandths of a Kelvin produced in the epoch of inflation), and beautifully fits a Planck curve of temperature 2.725 K (Fig. 9.13). This temperature is the one for the present age of the universe of 13.8 billion years, and everywhere in the universe observers will today measure this same temperature and therefore conclude that the universe has an age of 13.8 billion years. But in the past this temperature was higher and in the future it will be lower, everywhere in the universe. One can therefore use the temperature of the CMBR to measure the age of the universe, and everywhere in the universe observers will measure the same temperature at the same time, and therefore will agree with each other on the age of the universe. As the expansion proceeds in one direction, and the temperature drops in the course of time, also the time advances in one direction. This, of course, holds only for a flat or an open universe. In a closed universe the expansion will at a certain moment halt and then reverse to contraction. This would create a problem with the arrow of time, but we already know that this is unlikely to happen as all available observational evidence indicates that we live in a flat universe, as described in the last chapter.[3]

Directly related with the arrow of time is the concept of *causality*, which is one of the obvious fundamental principles of physics. Causality means that certain processes are the cause of other processes that occur at a later time, such that there is a logical chain of cause and effect. This automatically implies that time has a direction.

---

[3] The fact that the universe is not static but expanding and furthermore is open or flat, is also very important for the emission of light by the stars. If the universe had been static, the stars would not have been able to radiate simply according to the laws of electrodynamics, as we now observe that they do. For an explanation of this fundamental problem, see J. Narlikar: *The Structure of the Universe*, Oxford University Press 1977, p. 186ff.

But physics does not explain *why* time has a direction, and why not all physical processes could take place at once, at the same time. This strange problem about the nature of time was the inspiration for John Archibald Wheeler's aphorism at the beginning of this chapter.

In summary, we can state that the arrow of time concerns a fundamental question closely related to all processes in nature. And that the solution of this question can probably be found in the fact that we live in an expanding universe that started only a finite time ago, and that the thermodynamic arrow of time and the concept of causality in fact result from the cosmological direction of time, that is: from the expansion of the universe. This then would imply that the time which we read from our watches is directly related to the Big Bang and the evolution of the universe.

Multiverse

Chapter 16
From Universe to Multiverse

God not only plays dice. He also sometimes throws them where they cannot be seen

Stephen W. Hawking, British mathematical physicist

In the 1970s British mathematician and astrophysicist Brendon Carter noticed a number of surprising coincidences in the laws of physics and in the properties of our universe, that seem to suggest that our universe was created precisely in such a way that it could produce life and "intelligent" observers, *intelligent* meaning: similar to us, human beings. In the first place there is the fact that the age of the universe, 13.8 billion years, does not differ much from the age of the solar system and Earth of 4.65 billion years, and from the time of about 4 billion years that life on Earth needed to produce observers that can think about the laws of nature and about the evolution of the universe. There is only about a factor of three differences between

E. van den Heuvel, *The Amazing Unity of the Universe*,
Astronomers' Universe, DOI 10.1007/978-3-319-23543-1_16

the latter timescale and the age of the universe. Why is this difference not a factor of a hundred or a million? Actually, we—intelligent observers—appeared already in a very early stage of the evolution of the universe. Of course, we could appear only after sufficient amounts of elements heavier than helium had been produced, in order to enable planets like Earth to form. This required that a sufficient number of earlier generations of massive stars had passed, which produced these elements and terminated their short lives with supernova explosions which distributed these elements throughout the interstellar hydrogen clouds in our Galaxy. Only when a sufficient amount of these elements was present in these clouds, could a solar system with planets like Earth form, and could the development of life begin. It thus seems as though, almost immediately after the universe was born, a development had been set in motion aimed at producing intelligent beings.

The coincidences go even a lot further. All life on Earth is based on the element carbon (^{12}C). The unique property of this element is to form large and flexible chains, together with atoms of other elements, such as hydrogen, nitrogen, oxygen. No other element can do this. This property is essential for the building of cells of living organisms, and thus for life. Life therefore is possible only thanks to the existence of carbon, and if somewhere else in the universe life has developed, it will undoubtedly also depend on carbon. (There has been speculation in the past about life that could be based on silicon. However, chains of compounds of silicon miss the enormous flexibility and the innumerable possibilities offered by carbon for the formation of long chains of compounds that can move in all possible directions.) After hydrogen (H) and helium (He), carbon (C), nitrogen (N) and Oxygen (O) are the most abundant elements in nature, and these three elements can have been produced only in stellar interiors and not in the Big Bang, as explained in Chaps. 9 and 11.

In 1953 British astronomer Fred Hoyle, whom we already met in Chaps. 9 and 11, discovered that there is something special with the properties of the nucleus of the carbon atom, and that if the properties of this nucleus had been slightly different carbon could never have been produced in stars and life could not have originated. To appreciate the importance of Hoyle's discovery we have to look into the nature of the nuclear fusion reactions with which the stars generate their energy. Our sun, and some 90 % of all stars produce their energy (light) by the fusion of nuclei of hydrogen to nuclei of helium, as explained in Fig. 16.1. The core of the star that is hot enough for this process to take place, at temperatures of between 10 and 25 million Kelvin, occupies between 10 and 30 % of the mass of the star, depending on the stellar mass. In stars like our sun it is about 14 %, in stars of 20–30 solar masses it is about 30 %. When all hydrogen in this core is consumed, the core, which now consists of helium, contracts and gets hotter due to the release of gravitational energy by this contraction. When its temperature has risen to about 100 million Kelvin, the fusion of helium nuclei into carbon begins. This is a two-step process. In the first step two nuclei of helium fuse into a nucleus of the element beryllium (^{8}Be), and in the second step the beryllium nucleus captures a helium

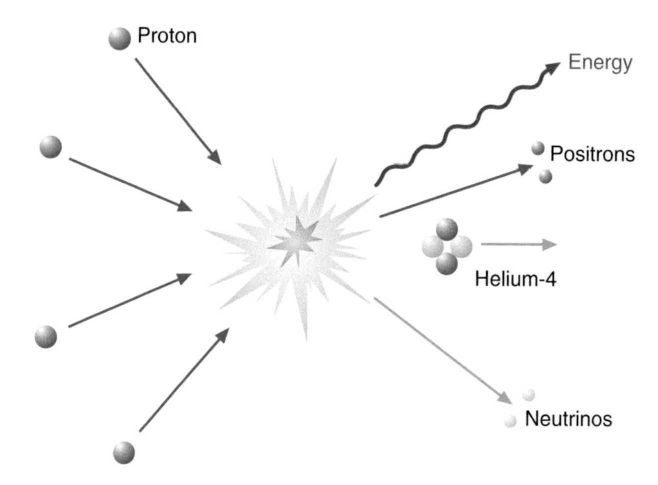

Fig. 16.1 In the central parts of a star the temperature is sufficiently high for nuclear fusion of four hydrogen nuclei into a helium nucleus to take place, as schematically depicted here. This reaction proceeds in a number of steps: first two hydrogen nuclei fuse into a deuterium nucleus, which then again captures a proton to produce a nucleus of helium-3. Two helium-3 nuclei fuse into a helium-4 nucleus, releasing two protons. In order for these fusion reactions to take place, the collisions between the nuclei must be sufficiently strong, such that the electric repulsion of the positively charged nuclei is overcome and the attractive strong nuclear force, which acts at very short distances, causes the nuclei to merge (see also Fig. 16.2). To achieve this in the stellar interior, a temperature above 10 million K is required. As the helium-4 nucleus is 0.7 % less massive than the four protons together, in this reaction this mass difference is transformed into energy (mostly electromagnetic radiation) which slowly propagates through the star, and is emitted as light from the stellar surface. In this way the sun in its interior transforms every second 600 million tons of hydrogen into helium to maintain its energy output of 4×10^{26} W

nucleus and forms a carbon nucleus. These two nuclear fusion reactions are written as follows:

$$^{4}\mathrm{He} + {}^{4}\mathrm{He} \rightarrow {}^{8}\mathrm{Be} - \mathrm{Energy}$$

$$^{8}\mathrm{Be} + {}^{4}\mathrm{He} \rightarrow {}^{12}\mathrm{C} + \mathrm{Energy}$$

The first of these reactions, which in the early 1950s was discovered by American astrophysicist Ed. E. Salpeter (1924–2008), requires extra energy (95 keV) and the produced beryllium nucleus is very unstable, and easily splits back into two helium nuclei. As a result at any time only a very small amount of beryllium is present in the stellar interior. These few nuclei then can capture another helium nucleus and produce a carbon nucleus, the second step in the formation of carbon by helium fusion.

In this phase of helium fusion, the star is a red giant: due to the large temperature difference between its centre and its outer layers, these outer layers have expanded to a giant size. In order to produce enough energy to maintain the output of light of this giant, the star must in its central parts fuse large amounts of helium nuclei into carbon nuclei. In the early 1950s, Fred Hoyle discovered that this is possible only if the carbon nucleus has a very special property, which at the time was not known, but for which he could make a prediction. This prediction later turned out to be completely true, with was a great triumph for Hoyle's work, as I will describe in the next section.

In order to understand what this is all about, I have to tell a little more about atomic nuclei. It turns out that, just as the electrons that orbit around an atomic nucleus are allowed to have only a distinct number of quantified energy levels (see Chap. 5), also the atomic nucleus can only be in a limited number of quantized energy states. The lowest energy state of the nucleus is called the ground state, and above this level are a number of quantized energy states. When the nucleus is in one of these higher energy levels, we say that is in an *excited state*. One can picture the atomic nucleus as an energy *mountain* within its centre a deep *pit*, as depicted in Fig. 16.2. The mountain is due to the electric repulsion of positively charged nuclei. When two positively charged nuclei approach each other, the energy needed to get them closer together increases steeply with decreasing distance, causing the mountain to become very step. Only when nuclei get very close together, they start to attract each other by the strong nuclear force. Since attraction is a negative force, this produces the deep pit in the centre of the mountain. The energy levels are located in the pit. When a nucleus is in an energy level in this pit that is above zero (see Fig. 16.2) the nucleus can easily fragment back into smaller nuclei.

Red Giants, Carbon Nuclei and Fred Hoyle

Nuclear fusion reactions have some strange properties. Fusion reactions in which the new nucleus is formed in an excited energy state have a much higher probability for forming this nucleus than reactions in which this nucleus is formed in its ground energy state. For example, in the second reaction described above, in which a carbon nucleus forms by fusion of a beryllium nucleus and a helium nucleus, the probability for forming the carbon nucleus is much higher if this nucleus is able to be formed in an excited state, than when it is formed in the ground energy state. However, to be able to form this nucleus in an excited state, the energies that play a role in this reaction should be such that their combination matches the energy of one of the energy levels above the ground level of the carbon nucleus, as we will describe below. In 1953 the energy levels in the carbon nucleus were not yet known: they have to be determined by laboratory experiments that had not yet been done at that time. The formation of the nucleus in an excited level is possible only if the two nuclei that are fused can provide an amount of energy that is equal to one of the energy levels in the carbon nucleus. The amount of energy that the two nuclei can provide is the sum of their energies of motion—which are determined by the local temperature—plus the energy released by the fusion—by converting into energy the

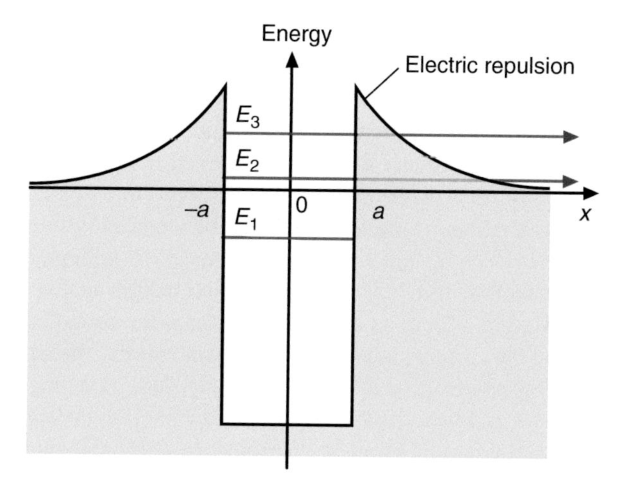

Fig. 16.2 The atomic nucleus can be pictured as an 'energy pit'. At large distances the *repulsive* electric force (represented here as positive energy) can be pictured as an energy mountain around the nucleus, as shown here. The mountain rises when a positively charged particle approaches the nucleus. However, very close to the nucleus, the *attractive* strong nuclear force begins to act, which can be represented as negative energy: in the central part of the positive energy mountain there is a deep negative energy pit (called a 'potential pit'). In this energy pit there are a ground level plus a number of quantized energy levels above this ground level. Particles in the nucleus (helium nuclei, protons, etc.) can occupy energy levels above the ground level. In that case the nucleus is said to be in an 'excited state'. If this energy level is above zero (the *black* horizontal x-line) the nucleus can disintegrate, for example, by emitting a helium nucleus. The helium nucleus then as it were 'tunnels' through the 'potential mountain', as indicated by the red arrows. In 1953 Fred Hoyle concluded from the existence of red giant stars that in the carbon nucleus there must be an energy level at about 7.60 MeV above the ground level. This was later confirmed by nuclear experiments of William Fowler (Fig. 16.3)

mass difference between the carbon nucleus and the sum of the masses of the helium and the beryllium nuclei. If the two fusing nuclei are able to provide the energy needed to form the carbon nucleus in an excited state, then the reaction has a high probability to take place (much higher than for forming the carbon in the ground state) and one calls this reaction then a *resonant nuclear reaction*. In 1953 Hoyle calculated that if the carbon nucleus would emerge in it ground state from the fusion between nuclei of beryllium and helium, far less carbon would have been produced in nature than we nowadays observe. And also that in stars in the stage of helium fusion far too little energy would be produced to power the light of red giant stars. It was at that time already known that red giants are in a stage beyond hydrogen fusion in the stellar core, and that the only stage in which they could still survive long enough to be able to be observable for more than 5 % of the duration of the hydrogen fusion stage, is the phase of helium fusion. Carbon fusion and still later fusion stages produce so little energy that the lifetime of stars in these stages is extremely short, such that the probability of observing a star in such stages is extremely small. Therefore the sole possible explanation for the fact that we see so many red giants in the sky (over 5 % of the stars are red sub-giants or giants) is that they represent a relatively long-lived phase in the evolution of a star, and the only phase for which this is possible is helium fusion. Since this would not be possible if in helium fusion the carbon nuclei would have formed in the ground state, Hoyle came to the conclusion that in the carbon nucleus there must be an energy level precisely such that a *resonant nuclear reaction* between beryllium and helium nuclei can occur. Hoyle knew that masses of the nuclei of helium, beryllium and carbon, and also knew that the temperature at which helium fusion takes place is about 100 million Kelvin. The mass difference between the sum of the helium and beryllium nuclei and the carbon nucleus, converted into energy through the formula $E=mc^2$ is 7.20 MeV (one MeV is a mega-electronvolt). Due to the high temperature the nuclei of beryllium and helium have high random velocities giving them average kinetic energies of a few tenths of an MeV. Hoyle therefore concluded that in the carbon nucleus there must be an energy level at about 7.60 MeV above the ground level.

At that time no one had measured the energy levels in the carbon nucleus, and to measure such levels, one has to bombard carbon nuclei in a particle accelerator with other particles, such as protons or helium nuclei. The bombarded nucleus is then kicked into an excited state, and when falling back to its ground state, it emits the energy difference between the excited state and the ground state in the form of a gamma-ray photon. So, by measuring the energy of the emitted gamma rays, one finds the energy of the excited energy level. Hoyle knew that nuclear physicist William Fowler (1911–1995, Fig. 16.3) at the California Institute of Technology

Fig. 16.3 Willy Fowler (1911–1995), who discovered the energy level in the carbon nucleus predicted by Fred Hoyle

("*Caltech*") was able to carry out such measurements. So, when in 1953 he visited Fowler at Caltech, he told him that the existence of red giant stars had convinced him that in the carbon nucleus there much be an energy level located at an energy of about 7.60 MeV. Fowler first thought that he was dealing with a lunatic: how could the existence of red giant stars tell where the energy levels in the carbon nucleus are located? Nevertheless Hoyle was able to convince him to do the required experiments, and to Fowler's great surprise he discovered an energy level at 7.67 MeV in the carbon nucleus. From this moment on he had no more doubts about Hoyle's abilities and this was the beginning of a life-long friendship and collaboration in which Fowler did the experiments and Hoyle, together with the British-born couple Margaret and Geoffrey Burbidge, developed the theory for the formation of elements in stellar interiors, described in Chap. 11. As we saw in that chapter, Fowler was extremely upset when in 1983 the Nobel Committee awarded him the physics prize for the formation of elements in stars and did not also include Fred Hoyle in this prize. It is unclear what the motives of the Nobel Committee for this decision were. Possibly, the cause was that during his entire life, Hoyle kept opposing the idea that the universe originated in the Big Bang, while since the discovery of the cosmic microwave background radiation 1964/1965—for which in 1978 the Nobel Prize had been awarded—the evidence for the Big Bang had become overwhelming, as described in Chap. 9. Although Hoyle missed the Nobel prize, he was awarded other important distinctions for his work: a knighthood, bestowed by the British Queen, and in 1997 the Crafoord prize of the Royal Swedish Academy of Sciences, which he shared with Ed Salpeter, the other scientist that made a key contribution to our understanding of helium fusion in stars. This prize is awarded in a 3-year cycle for astronomy, mathematics and geo sciences—disciplines for which there is no Nobel Prize—and in terms of prestige and money is comparable to the Nobel prize. I felt honoured that the Swedish Academy invited me and my wife to attend the award ceremony by the Swedish king, of this prize to Hoyle and Salpeter, and to take part in the accompanying dinner presided by the Swedish king and queen.

Another Coincidence: The Excited Energy Levels in the Oxygen Nucleus

The temperature in the stellar core during helium fusion is sufficiently high to, in principle, allow carbon to immediately fuse with helium to oxygen, through the reaction $^{12}C + ^{4}He \rightarrow ^{16}O + \text{Energy}$

In this reaction the energy liberated by the mass difference between the carbon plus helium nuclei on the one hand and the oxygen nucleus on the other, is 7.17 MeV. The danger of this reaction is that in red giants all carbon would be immediately converted into oxygen, such that no carbon would be left—and life would not be possible in the universe. However, nature has provided a second coincidence, to just prevent this. This coincidence is that in the oxygen nucleus there is an energy level *just below* 7.17 MeV, but *no level sufficiently close above it* to allow a resonance reaction. Because of the combined energies of motion of the helium and carbon nuclei, a resonance reaction would have required an energy level at about 7.17 MeV +0.5 MeV = 7.67 MeV to enable a resonant reaction. But no such level is present in the oxygen nucleus. For these reasons, the fusion reaction of carbon and helium into oxygen *is not resonant*, and therefore has a very low probability to occur. As a result, during helium fusion in stars, only a tiny fraction of the carbon is fused into oxygen, and much carbon is produced in the universe.

How Finely Tuned Are These Properties of the Carbon and Oxygen Nuclei?

The locations of the energy levels in atomic nuclei are determined by a number of constants of nature such as: the mass of the electron m_e, Planck's constant h, the velocity of light c, the electric charge of the electron e, etc. A dimensionless[1]) combination of such constants is the so-called *fine structure constant α:*

$$\alpha = e^2 / c = 1/137 \tag{16.1}$$

where $= h/2\pi$.

If the value of α had differed only a few per cent from 1/137, the energy levels in the carbon and oxygen nuclei would have shifted such that no carbon could have been produced in nature. Nobody knows why these constants of nature, including α, have the values that they have. The suspicion is that they have been awarded their values in a very early phase of the Big Bang, before or during inflation, and that during these times, due to random fluctuations, they could just as well have been assigned different values. If that had happened, no carbon could have been formed and we would not have existed.

The Anthropic Principle

The above considerations suggest that during the Big Bang the fundamental constants were assigned just the right values to enable the existence of life in the universe. Astrophysicist Brendon Carter, whom we mentioned in the beginning of this chapter, was the first to realize that, from the fact that we exist, one can in fact derive that the fine-structure constant α must have a

(continued)

[1] *Dimensionless* means that this constant cannot be expressed in units, such as metres, kilograms or seconds. Due to the division that produces α, all such units have dropped out, and α is *dimensionless*.

(continued)
value close to 1/137, and that all of the above-mentioned constants of nature must have values close to the ones they have in our universe. From the fact that man (anthropos) exists, one can therefore derive fundamental information about the values of the constants in the laws of physics! Another predictable parameter is another dimensionless quantity called the *gravitational fine-structure constant* α_G defined as:

$$\alpha_G = Gm_p^2/(\hbar.c) = 10^{-38} \quad (16.2)$$

where G is Newton's gravitational constant and m_p is the mass of the proton.

In the 1930s, Indian astrophysicist Subrahmanyan Chandrasekhar and British theoretical physicist Paul Dirac (both were Nobel prize laureates later in life) already realized that the number of protons of a typical star (like our sun) is $(\alpha_G)^{-3/2} = 10^{57}$.

The value of the gravitational fine-structure constant therefore determines the mass of a typical star, and that, in turn, determines how long a star like the sun will live: about 10 billion years. We know that Earth has an age of 4.65 billion years and that life required about 4 billion years to develop observers like us. We can now reverse the reasoning, as follows: From the fact that life needed about 4 billion years to produce us (anthropos), we see that the sun must be a star that lives for at least 4 billion years, and thus that its mass cannot differ much from the mass of our sun (since the more massive a star is, the shorter it lives: a star of twice the mass of the sun lives only 1 billion years, too short to produce us). This mass tells us the value of $(\alpha_G)^{-3/2}$, and since this parameter is composed of G, m_p, $\hbar$ and c, we can, if m_p, $\hbar$ and c are known, calculate the value of G. If G had been twice as large, gravity would have compressed the gas of the sun twice as strong, and all nuclear fusion reactions would have proceeded much faster, the sun would have lived much shorter and there would not have been sufficient time for life to develop into producing observers. Similarly, if one would have kept G, $\hbar$ and c fixed, we would have found that m_p cannot have differed much from its present value, etc. We thus see that from the time required for the development of intelligent observers (humans) we can derive the values of the constants of nature!

Anthropic Considerations

All of the above seems to suggest that the universe was created with just the right values of the constants of nature to enable the development of us humans as intelligent observers. There are two ways in which one can view this situation—these are known as the *strong* and the *weak* anthropic principle. The base underlying these two viewpoints is that it is completely unknown why the constants of nature have the values which they are observed to have in our universe. We have no physical theory that can predict these values. For example, if we define the mass of the electron to be equal to 1 (one), the mass of the up-quark (see Appendix B) is 8, the down-quark 16, the strange quark 293, the muon (heavy electron) 207, the tau lepton (another kind of heavy electron) 3447, the charm quark 2900, the bottom quark 9200, the W-boson 157,000, the Z-boson 178,000 and the top quark 344,000. These masses are very important as they determine, for example, the masses of protons and neutrons that make up all atomic nuclei, and are the basic constituents of the matter from which stars, galaxies, planets and living beings are made. The standard model of elementary particles—briefly summarized in Appendix B—is an extremely successful theory that in an elegant way explains the existence of the different kinds of elementary particles. But it does not predict the masses of these particles—given above—and also not the numerical values of the coupling constants that regulate the strengths of the various types of interactions between these particles. These masses and coupling constants must be determined by laboratory experiments and then are put into the standard model "by hand". The point of view of the *strong anthropic principle* is that there must exist an underlying theory, which we still do not know, that is able to unequivocally predict the values of the constants of nature. According to this principle, the laws of nature are such that the constants of nature can only have precisely the values that they have in our universe, as these are due to an overarching universal physical theory, which we do not yet know. The fact that we exist is, according to the strong anthropic principle, a direct consequence of the laws of nature that from the moment of the creation of the universe allowed only one possible development of the evolution of the universe. If other universes would exist—which is not necessary in the view of the strong anthropic principle—these will evolve in exactly the same way as our universe, because they have the same laws of nature with all the same fundamental constants of nature.

On the other hand, according to the weak anthropic principle, there are large numbers of universes with the same basic laws of nature but with different values of the fundamental constants, which were assigned to them by random processes very early in the Big Bangs that created these universes.[2]

All available evidence indicates that in a very early phase of the universe, within the first 10^{-40} s after the formation of the universe, all forces of nature—gravitation, strong and weak nuclear force and electromagnetism—were equally strong and unified into one single universal force, through a theory that includes a quantum theory of gravity (see Chap. 10). We do not have such a theory yet but we do know that subsequently, when the universe expanded and cooled, the first force to separate from the others was gravity, which from here on went its own way. After this, at the end of the epoch of inflation, at about 10^{-33} s, the strong nuclear separated from the

[2] There are also models that adopt one huge expanding universe that is divided into a large number of independent sub-regions that forever remain outside each other's horizons, each having randomly assigned values of the fundamental constants of nature.

other forces, and the baryons were created. Finally, at a relatively late time, 10^{-10} s, the weak force separated from the electromagnetic force. The separation of these last-mentioned two interactions, occurred at sufficiently low energies to be studied with the largest particle accelerators on Earth. From these studies we know that this separation was the result of *symmetry breaking*, a process in which chance plays a important role. Symmetry breaking is a process that occurs in a so-called phase transition. Examples are the freezing of water or the crystallisation of salt out of a saturated salt solution. In the liquid state, matter is completely symmetric, homogeneous and isotropic (the same in all directions), but when a liquid solidifies (freezes) the result can be a crystal that has a clear preferred main axis, in one direction in space, and therefore is no longer isotropic. The original symmetry, which it had in the liquid state, is now *broken*. The direction that the main axis of the crystal adopted when the liquid froze, is determined by chance processes at the moment of the freezing. This can be the presence of a dust particle, or a slight draft of air passing over the liquid, and the axis can freeze, in principle, in any arbitrary direction. The present view in theoretical physics is that during the separation of the different forces in the early phases of the Big Bang, such symmetry breaking took place. The precise forms that the strong and weak nuclear forces and the electromagnetic force nowadays have in our universe, would then have been determined by symmetry breaking, and the "*directions*" in which their "*axes*" of symmetry arbitrarily froze during this symmetry breaking, take here the form of the precise strengths that these forces adopted at the moment of symmetry breaking, the precise strengths being characterized by the values of the fundamental constants of these interactions. These were, in this view, therefore determined by random effects at the moment of symmetry breaking, and could have taken any arbitrary value. Theories developed by theoretical physicists that implement this idea of symmetry breaking are the various forms of so-called Grand Unifying Theories that attempt to unify the different fundamental physical forces. Examples are various versions of so-called supersymmetric theories, as well as superstring theory (mostly just called string theory)—of which the supersymmetric theories are a low-energy approximation. As mentioned in Chap. 10, string theory is thought by many theoretical physicists to be a very promising theory for unifying all fundamental physical forces, and for obtaining a quantum theory of gravity. In string theory elementary particles, such as quarks or electrons, are thought to be represented by vibrations of tiny strings, in a space of higher dimensions than our four-dimensional space-time. Such vibrating strings in higher dimensions would in our three-dimensional space be observed as elementary particles. According to string theory quantum processes could continuously lead to the creation of new universes out of vacuum in higher dimensions space. These universes would then, like our universe, be four-dimensional space-time structures. In this higher dimensional space these universes are distinct from each other and will not be able to observe one another. Each universe develops out of its own Big Bang, created by a vacuum quantum fluctuation. Such a collection of universes that continuously keep forming in a higher dimensional space is called a *multiverse*.

As the values of the fundamental constants of nature in the different universes were determined by the chance process of symmetry breaking, by far the largest number of these universes will have values of the fundamental constants that will not allow life to develop. For example, universes with a much larger value of the gravitational constant G than in our universe will—even if all other constants of nature are the same as in our universe—soon collapse after their emergence, and also stars, if they would be able to form, would burn their nuclear fuel so fast that there is insufficient time for the development of intelligent life. Universes that are born with a cosmological constant Λ that is larger than in our universe, will expand so fast that galaxies and stars have no time to form. And universes with different values of the quark masses may be able to solely produce hydrogen and no other elements, such that also stars and life cannot emerge. According to the newest versions of string theory as many as 10^{300} or even 10^{500} universes should form in order to produce—by chance—just one that obtained the right values of the fundamental constants of physics to allow the formation of stable long-lived stars, carbon and *intelligent life* with observers like us. Thus: in order to produce one universe like ours, nature should—according to the weak anthropic principle of grand unified theories like string theory—have produced more than 10^{300} sterile universes in which no life could develop.[3]

The great difference with the strong anthropic principle is that according to the strong anthropic principle the values of the fundamental constants of nature are due not to chance, but to an underlying set of fundamental laws of nature that have not yet been discovered, and that already from the start of the universe had in it the potential for the development of intelligent life. If there would be more than one universe, for which according to the strong anthropic principle there is no need, then all these universes would have exactly the same fundamental constants as our universe, and in all these universes intelligent life would be able to arise. The strong anthropic principle therefore suggests that the universe *was made with a purpose*, namely: to develop intelligent life. This strongly suggests "the hand of God" in the creation of our universe. For this reason, the strong version of the anthropic principle is popular among religiously inclined persons.

This holds particularly for the pseudo-scientific fundamentalist religious movement of *intelligent design,* which attempts to provide an alternative for Darwin's theory of biological evolution.[4]

On the other hand, the weak anthropic principle has the disadvantage that it does, in fact, not explain anything, and leaves everything to chance processes. This implies that we no longer have to search for deeper underlying laws of nature. This actually would mean the end of science. This is one of the reasons why quite a number of prominent physicists are sceptical about anthropic "explanations". Examples are Nobel laureates David Gross of UC Santa Barbara, Gerard 't Hooft of Utrecht

[3] See for example the book "*The Cosmic Landscape, String Theory and the Illusion of Intelligent Design*" by Leonard Susskind, Little Brown and Company, New York, 2006.

[4] Contrary to Darwin's theory, *intelligent design* is not a scientific theory, as there is no way for it to be tested by experiments or observations. For a critical review of this "theory" see the book "*Intelligent Thought*" (editor John Brockman), Random House, New York, 2006, which also gives a good overview of the anthropic principle.

University and Steven Weinberg of the University of Texas at Austin.[5] According to David Gross the use of the anthropic principle to explain our laws of nature would be a sign of defeat. In fact, string theorists that use this principle as an explanation have given up the search for deeper underlying causes for the structure of our world. According to Gross, physicists should adopt the motto of Winston Churchill: "Never, never, never give up".

In Gross's opinion, at this moment string theory indeed provides the most promising way to achieve further understanding of unification of the forces of physics: "it is the only game in town", but it requires still much further development and refinement.

Also among cosmologists, which often are supporters of the weak anthropic principle, there are sceptics about the "explanation" that string theorists propose for the origins of the properties of our universe. Prominent cosmologist Paul Steinhardt of Princeton University remarked that the anthropic principle makes a number of assumptions about the existence of a vast number of universes… "Why to assume an infinite number of universes, all with different properties, just to explain the existence of *our one universe*?"[6]

Weinberg gives an interesting fictive example of how one might be able to predict the values of the constants of nature, if one would know the more fundamental underlying laws of nature that determine the values of these constants. In this example, he pictures a civilisation that developed on a planet called Earth2, that is an almost exact copy of Earth, the only difference being that on Earth2 the sky is always completely clouded. The physicists on Earth2 have discovered the same laws of physics that we know and just like us they have concluded that all constants of nature have been extremely fine-tuned to allow the development of intelligent life. They have, however, one more fundamental constant of nature than we, which they called K2. This constant is the input of energy into their atmosphere from outside which, for to them unknown reasons, nature provides, and is absolutely indispensable for life to exist on Earth2. They have measured the value of K2 and found it to be 1400 J per square meter per second. They came to the conclusion that if K2 had been only slightly larger or smaller than 1400 J per square meter per second, life on Earth2 would have been impossible. Then, one day the cloud cover on Earth2 breaks open and the inhabitants see that K2 is not a fundamental constant of nature, but has an underlying cause which until then was not known to them: the existence of Sun2, at a distance of one astronomical unit, which provides the atmosphere of Earth2 with an energy input of 1400 J per square meter per second. According to the strong anthropic principle, every constant of nature could have such an underlying cause, not yet known to us. Interestingly, with a small adaptation, Weinberg's exam-

[5] However, Weinberg, in a 1987 publication and in his book "*Dreams of a Final Theory*" notices that he can think of no other explanation for the small value of the cosmological constant Λ of our universe (Physical Review Letters 59, 1987).

[6] See, for example, the critical book about string theory by mathematician Peter Woit: "*Not even Wrong*", Basic Books, New York, 2006, and here particularly the chapter "*The Landscape of String Theory.*"

ple can also be used to support the weak anthropic principle. This is because the distance of a planet from the sun in our solar system is the result of chance processes during the formation of this system. Life on a planet can develop only if the planet is located at the right distance from the sun, such that there is liquid water, and the temperature is not too hot or too cold, the planet has an atmosphere and its gravity is not too strong, etc. It is therefore not a coincidence that we live on Earth, and not on Venus or Mars or Mercury or Jupiter. Mercury and Venus are too hot, Mars is too cold and the gas giants like Jupiter or Saturn have no solid surface, and extremely stormy atmospheres consisting of methane, ammonia and hydrogen and are a very unpleasant environment for the development of life. Earth is the only planet in the solar system with just the right conditions for the development of life. Someone who, following eighteenth century clergyman William S. Paley (1743–1805)[7] says: "How incredibly beautiful and precise the Good Lord has tuned the temperature, gravity and other circumstances on Earth to allow the development of humans", uses the wrong arguments. If the temperature, gravity, atmosphere, etc. had not been suitable for the development of life, we would not have existed. The properties of Earth were determined by chance processes that occurred during the formation of the solar system, and by good luck happened to be just right for enabling the development of intelligent life. Without any doubt there are billions of planets in our Milky Way system that are just as badly placed, and therefore sterile, as the other seven planets of our solar system. On such planets no intelligent life will be found. But near some stars there will be planets that, just like Earth, have the right size and atmosphere and oceans, as well as the right distance to their stars, and a star that is sufficiently long lived to allow intelligent life to develop. The string theorists extend this chance process for selecting habitable planets even further, by postulating that there is a similar self-selection process among a large number of universes: only in universes that by chance obtained in their Big Bangs the right values of the fundamental constants of physics, life and intelligence will have been able to develop.

It is, of course, at present impossible to say which of these two variants of the anthropic principle is the correct one. The problem with both variants is that they are not real scientific theories. A real scientific theory should, according to the criterions put forward by philosopher Karl Popper, be falsifiable: if the theory is wrong, there should be experiments or observations possible that enable to show that it is wrong. If a theory does not allow such experiments or observations to possibly falsify it, it is, according to Popper's criterion, not a real scientific theory. If all predictions of the theory are later confirmed, as was the case so far with the Standard Model of particle physics, then this gives confidence that this theory provides a good framework for understanding nature. However, string theory already

[7]William S. Paley, in his 1802 book "Natural Theology or Evidences of the Existence and Attributes of the Deity", uses the watchmaker analogy to argue that God has created all living beings (if you find somewhere in the grass a watch, you will immediately conclude that this watch has been *created* by a watchmaker, and did not originate by some random process). He had taken this watchmaker analogy from Dutch philosopher Bernard Nieuwentyt who published it in 1750.

fails with the first testable prediction: the value of the cosmological constant Λ, for which string theory predicts a value 10^{120} times larger than the value observed in the universe (see Chap. 13). Unfortunately, we have only one observable universe and so far no ways have been found to get a glimpse of the presence of other universes. We therefore perhaps may never know whether the weak anthropic principle is right, as many string theorists believe. The same is also true for the strong anthropic principle, as we do not have a theory that predicts the values of the fundamental constants, although according to this principle such a theory should exist, but still has to be discovered. Therefore, concerning the strong anthropic principle, there is still hope that in the (perhaps distant) future, if we are smart enough, the situation may change and the postulated underlying deeper layer of fundamental physical laws will be found.

The Multiverse Theory: A New Form of the Continuous Creation Theory of Hoyle, Bondi and Gold?

As mentioned in Chap. 11, Hoyle, Bondi and Gold in the 1950s developed their theory of continuous creation as an alternative to the Big Bang theory of Lemaître and of Gamow and his students Alpher and Herman. Hoyle kept fighting the Big Bang theory until his death. The reason why Hoyle could not accept it was philosophical: for Hoyle, who as a humanist did not believe in a God that created the universe, the Big Bang model smelled too much like a creative *act of God*. If the universe was created a finite time ago, and had not existed infinitely long, as in *the continuous creation theory*, one would have to accept that there had been a clear *act of creation* which, he feared, implied the existence of a God. The fact that the Big Bang idea had been put forward by a Jesuit priest, Abbé Lemaître, made Hoyle even more sceptical of this idea.

It is interesting to notice that with the emergence of multiverse ideas—fitting with the weak anthropic principle—the situation has now become quite analogous again to that of Hoyle's continuous creation theory. In Hoyle's theory the continuous creation (of hydrogen atoms in the space between the galaxies in the expanding universe) took place *inside our universe*. In the new multiverse theories, the continuous creation has been moved up one step: here entire universes are continuously being created in the *cosmic landscape* of string theory, as described in Leonard Susskind's book "*The Cosmic Landscape, String Theory and the Illusion of Intelligent Design*", mentioned above. The title of this book already shows that the motivation of Susskind—and presumably of all supporters of the multiverse idea—is to get rid of an *act of creation*, that smells like the need for a God. In these multiverse models, the higher dimensional Cosmic Landscape is continuously producing new universes due to quantum fluctuations, a process that goes on continuously and therefore has no need for an act of creation. The multiverse idea—exactly like Hoyle's continuous creation model—fits with the humanistic philosophical ideas that there is no need for the concept of a God. One may, however, wonder whether this is not a prejudice just like the belief in a God.

In any case, we do live in a universe with values of the fundamental constants that have been accurately tuned to allow the emergence of intelligent life, once there is a suitable planet, as our own Earth shows. This knowledge is sufficient to continue with Chap. 17, which discusses the likelihood that somewhere else in our universe intelligent civilisations may have developed.

Complex molecules from space raining down on Earth

Chapter 17
Intelligent Life Elsewhere in the Universe

Where is everybody?

Enrico Fermi (1901–1954), Italian-American physicist

One of the most fundamental questions of all science is whether elsewhere in the universe *intelligent* civilisations have developed, *intelligent* being defined as having consciousness and a scientific/technological level at least comparable to that of present humanity. A positive as well as a negative answer to this question will have far-reaching consequences, for philosophy as well as religion. If we are the only ones in the universe, then the universe has apparently been created especially to allow us to come into being, and then we are extremely special: in a way "at the centre of the universe". On the other hand, if intelligent life is found elsewhere, we are much less special. The latter case will fit well with the Copernican principle, which states that Earth, Sun and Milky Way do not occupy special privileged positions in the universe: Earth is just one of the eight planets orbiting the Sun, the Sun

E. van den Heuvel, *The Amazing Unity of the Universe*,
Astronomers' Universe, DOI 10.1007/978-3-319-23543-1_17

is just one of some 200 billion stars orbiting around the centre of the Milky Way, and our Milky Way galaxy is just one among hundreds of billions of galaxies in the observable part of the expanding universe. We do, however, not know if the Copernican principle also holds with respect to intelligent life. In this chapter we discuss the probability that elsewhere intelligent life could exist, and possible ways to get into contact with such life.

The Origin of Life

Since the discovery in 1995 of the first planet orbiting another star, by Swiss astronomers Michel Mayor and Didier Cheloz, thousands of planets have been discovered outside our solar system. We now know that two out of every ten sun-like stars have a planetary system. This means that there are of order 40 billion planetary systems in our Milky Way galaxy. We also saw in Chaps. 2 and 11 that almost as soon as the large bombardment of the inner planets by asteroids had terminated, primitive life appeared on Earth (Fig. 17.1). Apparently, when the chemical ingredients required for life are present, such as water, carbon, nitrogen, phosphor, iron, etc., and also circumstances such as temperature, presence of an atmosphere, lightning, etc. are in the right range, primitive life seems to arise spontaneously on a relatively short timescale of a few hundred million years. There is no way to find out how this precisely did happen on the primitive Earth. It is known, however, that meteorites may contain dozens of different amino acids, which are the building blocks of proteins. It is possible, therefore that in the early solar system fairly complex organic molecules were created already in the solar nebula from which the meteorites condensed, and that these organic molecules were deposited on Earth by meteorites and comets. And that this contributed to life getting started on Earth. But it is also possible that the origin of life was a purely terrestrial affair in which the complex organic molecules were produced by the actions of lightning and a variety of chemical reactions. Already in the 1950s, American chemists Stanley Miller and Harold Urey experimented with a "primeval soup" consisting of a mixture of the above-mentioned elements which they exposed to electrical discharges ("lightning"). To their surprise they found that in this way all kinds of amino acids were produced. In addition, on Earth spontaneously complex organic molecules can be produced by catalytic chemical reactions on the surfaces of a variety of substances, such as clay in water. This suggests that oceans, lakes and rivers, in combination with the atmosphere, provided a suitable environment for the creation of complex organic molecules which, through certain forms of self-organization succeeded, after a few hundred million years, to produce structures that were able to maintain themselves, and to make copies of themselves by feeding on their surrounding "soup" of complex organic molecules. This is, in fact, what defines life: the ability to feed on one's environment, in order to maintain oneself, and to produce offspring, that is: to make copies of oneself. The discovery of large tube-worms and many other strange forms of life surrounding the hot volcanic vents at mid-ocean ridges at a depth of several miles below the ocean surface, shows us that in complete darkness, at temperatures exceeding the boiling

Fig. 17.1 Life on Earth started between 3.7 and 4.0 billion years ago. The solar system and Earth formed some 600 million years earlier from an interstellar cloud, and the large bombardment of the planets with asteroids—part of their formation process—had terminated about 4 billion years ago

point of water (100 °C), life can abundantly flourish. This life, that does not depend on sunlight or oxygen may give us a glimpse of what the first organisms on Earth may have looked like, when the atmosphere not yet contained oxygen and when our entire planet was surrounded by a thick cover of clouds. At the bottom of the food chain of this extreme life near mid-ocean ridges are the Archaea, primitive bacteria that as energy source use hydrogen-sulfide expelled by deep-ocean volcanism.[1]

[1] There is a large amount of literature about how life might have originated in these circumstances. See, for example, Lynn Margulis and Dorion Sagan: "*What is life?*", University of California Press, Berkeley, 1995, and Simon Conway Morris: "*Life's Solutions*", Cambridge University Press 2003.

The fact that life was there already so soon, within a few hundred million years after Earth had become more or less "quiet", appears to tell us that the probability is not at all small that life will arise under the conditions that reigned on the primitive Earth. It therefore seems very likely that if somewhere else in our universe similar conditions are present for several hundred million years, also life will arise. We therefore expect that on millions of other planets in our Milky Way galaxy at least simple lifeforms, resembling bacteria, will be present.

Where Is Everybody?

A big question is: how often will such primitive lifeforms have evolved into complex or even *intelligent* organisms? Italian-American physics Nobel laureate Enrico Fermi (1901–1954, Fig. 17.2) was among the first ones to ask this question. He already knew that there are of order 100 billion stars in our Galaxy. Applying the Copernican principle that our solar system is not special, he reasoned that most stars will have a planetary system. Even if one assumes that only one per cent of these planetary systems have a planet with about the same conditions as Earth, one still has of order a billion Earth-like planets in the Milky Way on which life could have developed. And, since our solar system and Earth appeared more than 8 billion years after our Galaxy was born, among these planets there must be many where life has already developed further than on Earth. One would expect then that on many

Fig. 17.2 Enrico Fermi (1901–1954), born and educated in Italy and later working in the US, was one of the greatest physicists of the twentieth century

Fig. 17.3 Frank Drake (born 1930) was the first to make an estimate of the number of possible civilizations in our Galaxy with whom we might be able to communicate. He also started the first program, using radio telescopes, to search for possible signals from extra-terrestrial civilizations. He was founder of the SETI Institute in Mountain View, California

planets very highly developed civilisations exist, that could, for example, have invented techniques for interstellar space travel, and therefore could have visited us. Nevertheless, despite many exciting stories by people that claim to have observed flying saucers and other UFOs, so far serious researchers never have found a shred of evidence for visitors from outer space. These considerations inspired Enrico Fermi in 1950 to pose the question: "Where is everybody?"

The first one who seriously tried to find an answer to Fermi's question was radio astronomer Frank Drake (born 1930, Fig. 17.3), who in the 1950s worked at Cornell University. In 1958/1959 Drake started a program that used radio telescopes for listening to possible radio signals from extraterrestial civilisations. Drake reasoned that radio waves provide by far the cheapest and energetically most efficient way for communicating over very large distances. To illustrate that radio communication requires hardly any energy, one may consider how much energy all radio telescopes in the world together have received from all radio sources in the universe that have been studied since the beginning of radio astronomy more than 80 years ago. These millions of sources include: tens of thousands of galaxies, hundreds of supernova remnants, gaseous nebulae, thousands of radio pulsars (spinning neutron stars), etc. The total combined radio energy received from all these sources is similar to the energy released when a few-inch size shred of paper is dropped on the ground from a height of 6 ft. This miniscule amount of energy has allowed to accurately map the hydrogen clouds in our entire Milky Way, measure the spins of thousands of neutron stars, the radio emission of tens of thousands of quasars and other radio galaxies, etc. Drake reasoned that if there are high civilisations elsewhere, they will undoubtedly have discovered that radio waves provide by far the most efficient way of communicating with other civilisations.

In 1961 Drake published his famous *equation* (one better should call this a *formula*, instead of equation), which enables to estimate how many higher civilisations one may expect to exist in our Milky Way Galaxy, with which one would, in principle, be able to communicate. We give here this formula in a form that is somewhat simplified, but accurately follows Drake's original reasoning when he put it forward.[2]

In fact his formula deals with a simple probability computation: the probability F that another sun-like star will have a planet on which a higher, communicative civilisation has come into being, and presently exists, such that we can nowadays communicate with it. This F simply is the product of a number of factors and probabilities, as follows:

$$F = f_p.n_e.f_L.f_i.f_c.t \qquad (17.1)$$

Here f_p is the fraction of solar-like stars that have planets, n_e is the number of earth-like planets in a planetary system that are suitable for the development of life, f_L is the probability that indeed life will develop on such a suitable planet, f_i the probability that this life reaches the level of intelligence at least comparable to that of humans, and f_c the probability that these intelligent civilisations will be interested in communicating with other civilisations. Finally, t is the lifetime of an intelligent civilisation, expressed as a fraction of the lifetime of our Galaxy during which civilisations can have developed, which we take to be 10 billion years. Here we still have to define what we call intelligent. For this we take simply that the extraterrestrial civilisation has at least reached the same level as humans in the fields of mathematics, natural sciences and technology. An implicit assumption in setting up Drake's equation is, as Barrow and Tipler have noticed,[3] that on every habitable planet it will take life, like on Earth, of order 4 billion years to develop to the stage of "intelligence".

Optimistic and Pessimistic Estimates

In estimating the various factors in the right-hand member of Eq. (17.1) one can be optimistic and pessimistic. Some optimistic early scientists originally estimated the values to be quite high. Because of the Copernican principle, they assumed, like Fermi, f_p to be equal to 1. For n_e they adopted the value 2, since in our solar system both Earth and Mars are basically in the habitable zone. Since life emerged on Earth almost immediately after the asteroid bombardment had terminated, they estimated the probability for life to emerge to be very high $(f_L = 1)$, just like the probability that in the course of time life will become intelligent (f_i=0.5). Since some civilisations might be introverted and not eager to communicate, the factor f_c was put equal to 0.1. Based on these estimates, the probability that at some time an intelligent and communicative civilisation will develop will be 0.1 (10 %). However, in order to know what fraction of such civilisations are still present such that we can

[2] For an excellent overview of Drake's work and his search for extraterrestrial civilisations, see Frank Drake and Dava Sobel: *"Is Anyone Out There?"*, Delacorte Press, New York, 1992.

[3] J. D. Barrow and F. J. Tipler: *"The Anthropic Cosmological Principle"*, Clarendon Press, Oxford, 1986, Chap. 9.

communicate with them, we still have to know the average lifetime t of such a civilization. If we put it as 10,000 years then, expressed in the lifetime of the Galaxy of 10^{10} years, $t = 10^{-6}$. This then leads to a value of $F = 10^{-7}$. Assuming 100 billion solar-like stars in the Galaxy, this value of F would then lead that at present there are 10^4 such civilisations with which we would be able to communicate. (The underlying assumption in these calculations is that since 10 billion years ago, stars where such civilisations can develop have been born continuously, such that such civilisations continuously have arisen and then lived for 10^4 years; the ones that have arisen in the last 10^4 years are then still present today.)

If these 10,000 present intelligent and communicative civilisations are spread evenly throughout the Galactic disk, which has a diameter of 100,000 light years, the nearest such civilisation will be at a distance of about 900 light years. This distance can be easily bridged with radio communication. Drake himself assumed for f_p, n_e and f_L the same optimistic values given above, but assumed $f_i = 0.01$ and $f_c = 0.01$. The value of F then comes out 5000 times smaller than given above. If that is true there will at present only be two intelligent communicative civilisations in our Galaxy, one of them being us. The nearest other civilisation with which we would be able to communicate is then some 50,000 light years away.

The Search for Extra-Terrestrial Intelligence

Assuming the real situation to be somewhere between the optimistic estimate and Drake's estimate, Drake and his colleagues in the early 1960s initiated a systematic "listening program" in which nowadays many radio telescopes in the world participate.

One may wonder in what language an extraterrestrial civilisation would be able to communicate. Because of the universal validity of mathematics, one would expect that any advanced civilization will have discovered the same mathematical theorems and logical structures that mathematicians here on Earth have discovered. For example, a theorem like that of Pythagoras is valid everywhere in the universe. It has been found that on the basis of the universally valid theorems of mathematics one can develop a language that any other advanced civilization will be able to decipher and understand. In the 1950s, Dutch mathematician Hans Freudenthal (1905–1990) of Utrecht University developed precisely such a language, which he named LINCOS (Lingua Cosmica). The listening programs developed by Drake and his collaborators are aimed at finding signals expressed in such a universal language. Since 1984 the SETI program (SETI = Search for Extra Terrestrial Intelligence) has its own institute, the SETI-Institute in Mountain View, California, of which Drake was one of the founders. The budget of the institute is provided by scientific institutions, companies and rich sponsors. Drake was its first director. Apart from its scientific research, the institute is also an important centre for scientific outreach to the general public. Thanks to a large gift from Paul Allen, co-founder with Bill Gates of Microsoft, the SETI Institute has its own large radio telescope, the Allen Telescope Array (ATA), consisting of 42 dishes of 20 ft diameter

Fig. 17.4 The Allen Telescope Array (ATA) in California was built for the SETI institute to search for extra-terrestrial civilizations. The telescope was funded by a donation of Paul Allen who, together with Bill Gates, founded Microsoft. The telescope is also available part of the time for research by astronomers from the University of California, Berkeley

spread over a circular area with a diameter of 1000 ft (Fig. 17.4). The telescope is managed by the Astronomy Department of the University of California in Berkeley, which also is granted a part of the observing time for its own research projects.

Why Were We Never Visited?

Even if we adopt the conservative estimate by Drake that at present there are only two intelligent civilizations present in the Milky Way, there must, in the past billion years already have been many such civilizations. Adopting the above-assumed typical lifetime of 10,000 years of such a civilization, there will, in a billion years have been 200,000 such civilizations, and in the past 5 billion years: one million. It is then very surprising that—in so far as we know—our Earth has never been visited by representatives of such a civilization. American mathematician Frank Tipler (born 1947) concludes from this that that even Drake's pessimistic estimate of the presence of two such civilizations in the Milky Way at any time is far too optimistic. In his view the emergence of an intelligent, communicative civilization on a habitable planet is far less likely than most people had thought before. According to Tipler, Earth and humanity are unique in our Milky Way system, and nowhere else in this system a lifeform of comparable intelligence has arisen.

Fig. 17.5 John von Neumann (1903–1957), born and educated in Hungary and later working in the US, was one of the greatest mathematicians of the twentieth century. He developed one of the first digital computers and was the first one to realize that computers should be programmed by means of an *operating system*

The argument on which Tipler bases his belief is that a technically highly advanced civilization will at some time always start to develop so-called *von Neumann machines* (also called *von Neumann probes*), and that if in the past somewhere in the Milky Way an advanced civilization had arisen, thc von Neumann machines constructed by this civilization would already long ago have reached us and have colonized Earth. Since this has not happened, Tipler concludes that such civilizations have never existed.[4]

Von Neumann machines were conceived by the great Hungarian-American mathematician John von Neumann (1903–1957; Fig. 17.5). He put forward the idea of these robot-like machines the 1950s. It is a machine that is able, by using materials it takes up from its environment, to make exact copies of itself. In fact, every living organism is a von Neumann machine. Tipler has in mind a real machine: a kind of space probe, which when it arrives on another planet, is able to make two exact copies of itself, which it then launches into space to reach two other planets, where they land and each of them makes two copies of itself, which are launched again to reach four other planets, make copies, that are sent to eight other planets, etc. If one does this one can calculate that within a relatively short time, say a few hundred million years, all planetary systems in our Milky Way will have been colonized by von Neumann machines. It is easy to see why. In order to launch them into space, a machine must have a speed of at least a few tens of km/s. Let us simply assume a speed of 100 km/s, which is not unrealistically high, even with present space technology. With this speed the distance of 4.2 light years to the nearest star is covered in about 13,000 years. The trip to the next two stars will again take about

[4] See, for example, F. J. Tipler, "*Extraterrestrial intelligent beings do not exist*" in "*Extraterrestrials*" (ed. E. Regis Jr.) , Cambridge University Press, 1987, pp. 133–150.

13,000 years. In this way the von Neumann machines will have reached the opposite edge of the Milky way in about 300 million years, and have doubled their numbers 20,000 times. Even if the majority of the von Neumann machines had been lost by accidents and malfunctions on the way to the opposite side of the Milky Way, or during their landing on a planet, still all habitable planets in the Milky Way will have been visited. A highly developed civilization that has arisen longer than 300 million years ago would therefore already have colonized all planets in the Milky Way with von Neumann machines. The great advantage of sending von Neumann machines instead of living beings (humans) is that such machines much more easily will be able to survive the enormously long travel to other stars. Apart from the psychological, sociological and physical problems involved in a space trip of 13,000 years—longer than the entire human history since the invention of agriculture—there is, for example, the great danger of cosmic rays in interstellar space: charged atomic nuclei and electrons with relativistic speeds. On Earth our atmosphere and magnetic field shield us from the bulk of these dangerous particles, but already on a trip to Mars an astronaut will receive 5–10 % of the maximum dose of this radiation allowed for humans. A 10 year space trip is therefore already close to lethal. Von Neumann machines are much better astronauts than living beings, and one expects that also other higher civilizations in our Milky Way will have come to this conclusion. From the fact that on Earth no trace of a von Neumann machine has been found, Tipler concludes that no colonization of the Milky Way system has taken place and that therefore nowhere else in the Milky Way a civilization has arisen that has reached a similar or higher level of intelligence than ours.

American astronomer and novelist Carl Sagan (1934–1996, Fig. 17.6) was not convinced by Tipler's argument. Sagan said: "The absence of evidence is not the evidence of absence." Together with Russian astrophysicist Josif Shklovskii (1916–1985) Sagan wrote in the 1960s the first book about possible intelligent life elsewhere in the universe ("*Intelligent Life in the Universe*", Delta Books, 1966).

It is very well conceivable that the only von Neumann machines that can be realized in practice are simple living organisms such as bacteria. These indeed are able to use the raw materials available on a planet to feed on and make copies of themselves and thus multiply. If later on an impact by an asteroid or comet occurs, rocks containing these organisms can be launched into space, and after a long travel, fall on another planet, causing life to spread to this other planet. Evidence that this is possible comes from the discovery on Earth of meteorites that originated from Mars and from the moon. These have been launched long ago from Mars and the moon due to the impact of an asteroid or comet. In this way life may, once it has arisen on one planet, gradually be spread by *random walk* to other planets, and finally spread throughout the entire Milky Way. This spread would then take place purely randomly: this form of *sowing life* on other worlds one might call *random panspermia*. The idea of panspermia was already proposed early in the twentieth century by Swedish Nobel laureate Svante Arrhenius, who believed that the spores of bacteria can survive interstellar travel and therefore could be spread throughout the Milky Way. It has since been found that unprotected spores of bacteria cannot survive the intense bombardment by cosmic rays in interstellar space. They can survive when they are deeply buried inside a piece of protecting rock. Nobel laureate Sir Francis

Fig. 17.6 Carl Sagan (1934–1996) and Frank Drake at Cornell University in 1994; Sagan in the 1960s wrote the first book, together with Russian astrophysicist Josif Shklovskii, about *Intelligent Life in the Universe*. Apart from being a great researcher, he also was an outstanding populariser of science

Crick (1916–2004), who in the 1950s together with Jim Watson discovered the double helix structure of DNA, has proposed a slightly different form for the spread of life throughout the Milky Way and for the origin of life on Earth. His theory is called *directed panspermia*. (Francis Crick: "*Life Itself: Its Origin and Nature*", Simon and Schuster, 1981.) Crick's point of view is that the intelligent inhabitants of a planet will always finally discover that the spores of bacteria are the only life-form that can survive long-duration space travel, however, only if they have been stored in capsules that protect them against cosmic rays. These intelligent extraterrestrials may then decide to direct spaceships with these capsules to planetary systems which these extraterrestrials suspect to harbour habitable planets. When such a capsule lands on a planet, it opens and sows the spores in this new environment. The spores awaken and become bacteria, which multiply and mutate and so the evolution of life gets started on this planet. Finally, in this way, life would have been spread to all habitable planets in the Milky Way. Also life on Earth might, according to Crick's idea, be due to the initiative of an intelligent civilisation elsewhere. The consequence of this idea, and also of the above-mentioned random panspermia (if that is at all possible), would be that all life in the Milky Way is based on the same DNA that we find here on Earth. This makes it possible to, in principle, test this theory. If one discovers that life on another planet is based on exactly the same DNA as ours, this would confirm panspermia. If, on the other hand, life elsewhere is based on another type of complex molecules, we will know that this life is not due to panspermia, but most probably originated independently on that planet itself, without exchange with the life on other planets.

It should be noted here that panspermia is in fact the spread of von Neumann machines, and thus Tipler's argument that we have not been visited by von Neumann machines, is not 100 % sure. If life arrived here by a directed panspermia, the von Neumann machines are already in our midst.

Possible Stumbling Blocks for the Existence of Intelligent Extraterrestrials

It might well be that the 10,000 year lifetime of an advanced intelligent civilization assumed by Drake is much too optimistic. It is very well possible that soon after the invention of advanced technology and science, particularly nuclear science, the civilization destroys itself and never reaches the point of developing von Neumann machines of the kind Tipler envisaged. Possibly, the inborn aggression that was required to produce a hunter-type being like us, will cause advanced civilizations to keep fighting wars between different parts of the planet, such that within less than a 1000 years after the emergence of an advanced civilization, this civilization destroys itself in a planet-wide nuclear war.[5] Also, short-term and religious thinking may lead a civilization, shortly after inventing medical science, to explosive population growth and overpopulation, exhausting the planet's natural resources and environment, destroying all other larger species on the planet—as is already happening on Earth at accelerating speed—and leading to the collapse of the civilization within a few centuries. Man appears at this moment to be precisely following the last-mentioned path, and it is quite possible that our civilization has reached its final century, as British astrophysicist Sir Martin Rees has argued.[6]

It is therefore very well possible that in the past already many intelligent civilizations have arisen. In our Milky Way system, that have destroyed themselves before reaching a high enough level of technology required for producing von Neumann machines or to have started directed panspermia. If that is true, it becomes extra important that we on Earth do not make the same mistakes.

Earth Could Be Quite Unique

In recent years, more and more the awareness has grown that intelligent life may be muchless common than Drake and Sagan in their optimism thought in the 1960s. For this reason it is important to once more look carefully at the possible values of the different factors in Drake's Equation (17.1). We know nowadays that some 20 % of sun-like stars have planets, so $f_p = 0.2$. Also it has become clear that many

[5] See M. H. Hart and B. Zuckermann (editors): "*Extraterrestrials, where are they?*", Pergamon Press 1982.

[6] Martin Rees: "Our Final Century", Heinemann, London, 2003.

planetary systems have one or more Jupiter-like planets. Computer simulations[7] show that systems with such large planets have a much lower probability to produces planets like Earth. Such simulations suggest that n_e is probably less than 0.1, which means: 20 times lower than assumed by Drake and others before. The value of f_L, the probability that simple life forms arise on a planet, remains about 1. But the probability f_i that simple life forms develop into much more complex ones and finally become *intelligent*, might very well be much smaller than thought previously. The reasons for this come both from astronomical considerations that had, so far, not been included, as well as from the field of evolutionary biology. We now consider these two groups of reasons in more detail.

Refined Astronomical Considerations

Earth has, as described in Chap. 2, an abnormally large moon. This fact may have deeply influenced the evolution of life on Earth. This big moon is crucial for maintaining the stability of the angle between the rotation axis of Earth and its orbital plane. Computations of the various motions of Earth produced by the gravitational attractions of Sun, moon and all other planets of the solar system show that if Earth would not have such a large moon the angle between Earth's rotation axis and its orbital plane would have wildly and erratically varied in the course of time, just like the rotation axis of Mars.

In that case, Earth would not have had stable climate zones, which would particularly have hampered the stable and undisturbed evolution of the higher animals, like our apelike ancestors. As explained in Chap. 2, the formation of our moon was an event of pure chance: the collision of Earth with a Mars-sized planet very early in the history of the solar system. The probability that an Earth-like planet in another solar system also underwent such a chance collision with another small planet is probably very small. From the fact that Earth is the only one with such a large moon among the four rocky planets of the solar system, this probability is certainly smaller than 0.25, but more likely it is of order one per cent or less. On the other hand, while stability is over-all good for the evolution of life, a too stable situation may lead to stagnation, since in such a stable situation the organisms are no longer stimulated to adapt to new circumstances. This was, for example, the case in the epoch of the dinosaurs, which lasted from 245 million years ago to 65 million years ago. During these 180 million years, the mammals, which were already around 250 million years ago, were prevented from evolving into larger-sized species. Only small mammals, which could easily hide, were able to survive. Every time a somewhat larger mammal appeared it was eaten by fiercely carnivorous dinosaurs. Only after an impact of an asteroid that killed-off the dinosaurs 65 million years ago, could the small mammals evolve into a large variety of species of all kinds and sizes, that filled the ecological niches vacated by the dinosaurs. This resulted in countless numbers of new species, including the primates to which humans belong. American evolutionary

[7] E. W. Thommes, S. Matsumura and F. A. Rasio, *Science* 321, 814–817, 2008.

biologist Stephen J. Gould (1941–2002) put forward his theory of *punctuated equilibrium*, which implies that, in order for evolution to advance, from time to tie a catastrophe should happen that disturbs the state of equilibrium into which life had settled for a long period of time. This disturbance, in which many existing species are wiped out, allows the remaining species to evolve further and thus gets evolution going again. This resembles the principles applied by present-day management theorists: they recommend to, after a number of years, completely reorganize the working structure of a company, such that new and stimulating forms of collaboration between workers arise, allowing the company to advance and flourish. In order to achieve this for the evolution of life on a planet there should from time to time, say every 100 million years, be a gigantic catastrophe, such as the impact of an asteroid, that wipes out a large number of species and kick-starts the evolution again. In order for an asteroid impact, like the one that wiped out the dinosaurs, 65 million years ago, to happen, the planetary system should have an asteroid belt, like the one in our solar system. We have no idea what the probability is for a planetary system to have such a belt. It might be quite low, perhaps 10 %. If we include all the above reduced probabilities into the Drake equation, the probability for life to evolve to the stage of intelligence may become much lower than thought before. The multiplication of the above estimated probabilities reduces f_i by a factor 1000. Combining this with the reduction of f_p to 0.2 and of n_e to 0.1, the value of F becomes 100,000 (one hundred thousand) times smaller than originally estimated by Drake. According to Drake's estimate two million intelligent and communicable civilizations formed during the history of the Milky Way, each of which lasting for 10,000 years. Reducing this by a factor 100,000, there remain only 20 civilizations that formed during the entire history of the Milky Way system. In fact when only 20 civilizations arise in 10 billion years, one will have that, on average, only one such civilization will arise every 500 million years, and then lives for 10,000 years, a negligible amount of time compared with the time when the next civilization will arise, and also compared to the times that there is no such civilization around. In this case we will at this moment certainly be the only civilization present in the Milky Way.

Considerations from Evolutionary Biology

The probability for the development of intelligent higher forms of life might even be smaller than estimated in the last section, if factors from evolutionary biology are taken into account. On Earth eukaryotic cells arose some two and a half billion years after the origin of the first primitive prokaryotic lifeforms. Eukaryotes did form by symbiosis of different bacteria, as mentioned in Chap. 11. Without the emergence of eukaryotic cells, more complex organisms could never have developed. Mark Ridley suggests in his book "Mendel's Dream" (published in America with title "The Cooperative Gene")[8] that this transition from prokaryotes (bacteria) to

[8] See for example, Anton Pannekoek: "*The Origin of Man*", Wereldbibliotheek, Amsterdam 1957 (in Dutch language).

eukaryotes was a most unlikely (symbiotic) step in the evolution of life. On Earth this happened 2.5 billion years after the origin of prokaryotic bacteria. Perhaps this was extremely fast, and does this occur on other habitable planets on average only after 10 billion years, or after 100 billion years. If that would be the case, the probability that somewhere else in our Milky Way higher lifeforms developed will be minimal, and Earth would be really unique.

Another question is whether biological evolution, also if eukaryotes have arisen, inevitably leads to a form of intelligence that produces a scientific and technological culture like that of humans. While many astronomers, like Carl Sagan and Frank Drake are (or were) relatively optimistic about this, many evolutionary biologists have serious doubts, for a number of reasons. In rest the brains of humans use twenty per cent of the total energy consumption of our bodies. This is an absurdly large fraction compared to any other animal. Normally, this would be a huge disadvantage in the struggle for life in nature. Another disadvantage of these large brains is that childbirth proceeds much more difficult for humans than in other mammals. The enormous information-processing brains of Homo Sapiens are the result of a unique evolution, driven by climatological and ecological variations in North-East Africa 2–3 million years ago. The rise there of wide, largely treeless grasslands induced our ancestors to start walking upright, freeing their hands for making tools, particularly for hunting. The invention and later refinement of tool making and the accompanying evolution of the fine muscles in our hands, stimulated the evolution of the corresponding parts of our brains. This simultaneous coordinated evolution of our refined tool-making hands and our brains resulted in our large brains and our hands that are able nowadays to perform the most refined tasks (see Footnote 8).

As a chance by-product of these large brains, language emerged at some stage in the evolution, as a unique means of communication. The development of language no doubt originated from the need for efficient collaboration and coordination during hunting parties and other activities, and it highly enhanced the success of humans as a species. Some 13,000 years ago, the invention of agriculture and the domestication of animals were other huge steps which allowed humans to adopt a sedentary lifestyle, in villages and cities. In city states the need for administration led to the development of script, which stimulated further cultural development and the rise of science and technology.

Thanks to the written recording of the discoveries and inventions of earlier generations, knowledge could be passed on to the following generations. These did not have to re-invent important achievements but could build forth on the knowledge of their ancestors. This—together with the concept of learning about these past achievements in schools—has enormously accelerated the development of civilization and of science and technology, an acceleration that still continues at an ever increasing pace, thanks nowadays to computers and new means of communication.

However, there is no indication in nature that the development that we humans experienced in terms of intelligence was unavoidable. Evolutionary biologists such as Stephen J. Gould have pointed out that evolution has no goal and does not strive for ever higher levels. High-school biology books often appear to suggest this by drawing evolutionary trees with humans at the top. This is, however, just a human interpretation. Darwinian evolution has no pre-determined goal, apart from the survival of the species. Those species survive that are best adapted to their environments. The simple prokaryotes that already existed 3 billion years ago, are still around and still are excellently able to survive, just like many other primitive organisms that developed later, such as cockroaches and numerous kinds of insects. Dolphins, which are highly intelligent and have large brains, already reached their present stage tens of millions of years ago. They never evolved further because in their environment there was no need for further development. Also elephants obtained their large brains and high intelligence long ago, and did not develop these facilities further, because in their environment they could excellently survive with their achieved level of abilities. After all, large brains are not a requirement for survival. The dinosaurs survived for 180 million years despite having only small brains. Evolutionary biologists George Simpson and Ernst Mayr have noticed[9] that the evolutionary path that led to the large brains of humans was determined by a number of highly unlikely circumstances that are highly improbable to ever occur again. This was not a set of evolutionary steps that inevitably was leading to a higher level, but a collection of random events that just as well could have led to a totally different outcome. For this reason most of my evolutionary biologist colleagues consider it extremely unlikely that on other planets, where life originated, a similar series of highly unlikely developments would occur leading to large brains, consciousness and intelligence. In their view, the factor f_i in Drake's equation, which represents the probability that life will become intelligent, is extremely small, and the number of intelligent civilizations that have occurred in the history of the Milky Way system is even much smaller than the 20 that we estimated in the foregoing pages from the refined astronomical considerations.

Our Rare Earth

From combining the thoughts from evolutionary biology with the above mentioned refined astronomical considerations, it appears likely that Earth is the only planet in the Milky Way system where ever intelligent life has arisen. This is absolutely not in disagreement with the Copernican principle, because there are hundreds of billions of galaxies in the universe, and it is well conceivable that on average in each galaxy (or in every ten or hundred galaxies) just one planet occurs on which at some point in time an intelligent civilization arises, with a level at least comparable to that on Earth. This would mean that the nearest planet with the potential to develop an intelligent civilization may be in the Andromeda Nebula. If in each galaxy only one intelligent civilization arises that survives for, say, 10,000 years (without spreading over the entire galaxy) the probability of finding a galaxy that

[9] See the references in Chaps. 3 and 9 of the book "*The Anthropic Cosmological Principle*", by Barrow and Tipler, and E. Mayr in "*Extraterrestrials*", ed. E. Regis Jr., Cambridge University Press 1987, pp 23–30.

Fig. 17.7 Our rare Earth. As argued in this chapter, it is very well possible that Earth is the only planet in the Milky Way on which ever an 'intelligent' civilization has developed. This makes our planet unique and invaluable

has an intelligent civilization at the same time as us, is only one in a million. Among the over 100 billion galaxies in the observable universe there will then, at this moment be some hundred thousand intelligent civilizations. These are, however, so far away that it seems impossible to ever communicate with them. An enormous obstacle to the communication will be that it will take a radio signal tens of millions of years to reach them. And by the time the answer arrives here, our civilization will have gone long ago, unless much faster ways of communication would be invented, such as the science-fiction-like use of wormholes. The fact that Earth with its intelligent life is most probably unique and rare in our Milky Way galaxy[10] urges us to make all possible efforts to conserve our beautiful planet and all its wonderful forms of life.

[10] See the book "Rare Earth" by P. D. Ward and D. Brownlee, Copernicus Books, New York, 2000. A number of the arguments given above for the uniqueness of the development of intelligent life on Earth were borrowed from this book.

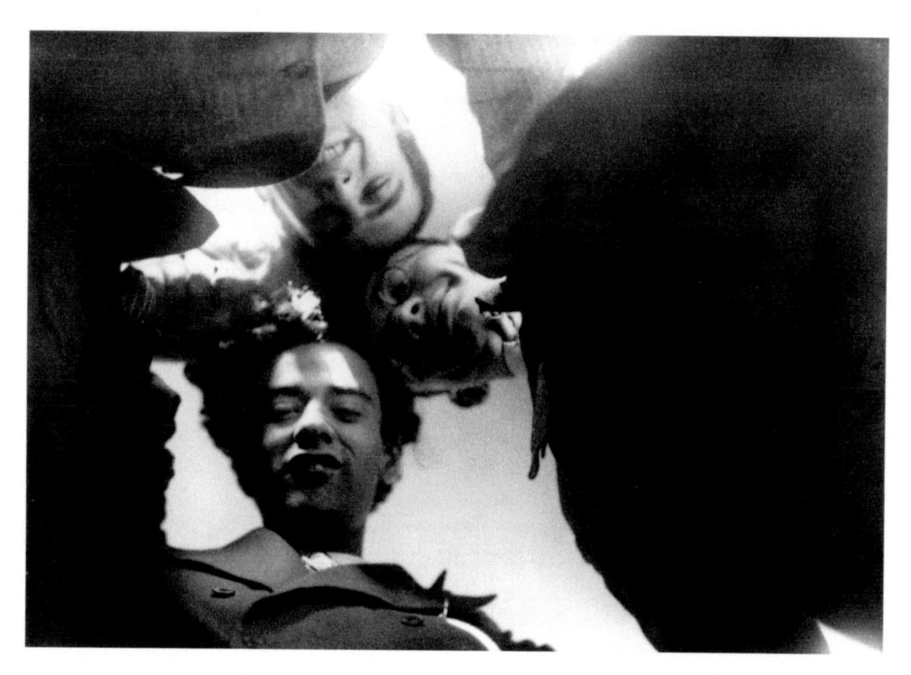

Edward Teller (N), George Gamow (E), Lev Landau (S) and Hendrik Casimir (W), around 1930 in Copenhagen

Chapter 18
Epilogue

> *One thing that I have learned in a long life—that all our science, measured against reality, is primitive and childlike—and yet it is the most precious thing we have.*
>
> **Albert Einstein (1879–1955), German-Swiss-American physicist**

When we look back to the important discoveries that have shaped our image of the universe and its evolution, we cannot escape from wondering why and how certain developments took place, and how it could have happened that researchers sometimes were not aware of earlier developments in their field, even when these were abundantly available in the literature. We saw this, for example, with Friedmann and Lemaître and with the prediction and discovery of the cosmic microwave background radiation (CMBR). Particularly this last example is an interesting case history of how discoveries in science may take place. A very important question is:

E. van den Heuvel, *The Amazing Unity of the Universe*,
Astronomers' Universe, DOI 10.1007/978-3-319-23543-1_18

how is it possible that Dicke and Peebles were not aware—as they claimed—of the predictions of 1949 and later years of this radiation by Gamow's collaborators Alpher and Herman?

In the 1950s Margaret and Geoffrrey Burbidge, together with Fred Hoyle and Willy Fowler had very successfully explained how the elements heavier than helium were formed inside stars. Because of this, the work of Gamow and his collaborators, that originally had been aimed at explaining the formation of these elements in the Big Bang, had around 1960 been largely forgotten. The dominance of the *steady state cosmology* of British astrophysicists Hoyle, Bondi and Gold was so great at the time that, when in 1958 the famous Solvay Conference of Physics in Brussels was devoted to the 'Structure and Evolution of the Universe', the leader of this conference, British Nobel laureate Bragg, decided to *not invite* Gamow, even though he had been recommended by Wolfgang Pauli, another Nobel laureate, who was known to be hyper-critical of almost everybody, but had a very great respect for Gamow.

As often happens in science, a new generation of researchers appears and seems to know nothing, and discovers the same things that others have discovered 10 or 20 years earlier.

This was the case with Jim Peebles and Robert Dicke in Princeton, who around 1964 started to make computations about the hot Big Bang universe. Just like Gamow and his collaborators 15 years earlier, they reached the conclusion that the heat radiation from the time when the universe became transparent, about 400,000 years after the beginning, should still be present in the universe in the form of microwave radiation. When shortly later this radiation was discovered, they claimed to never have heard about the earlier work of Gamow and his collaborators Alpher and Herman. Although this is not impossible, it still is quite strange and for me almost unbelievable. This is because in the 1950s Gamow, in his popular books and in an article in Scientific American, read by hundreds of thousands of people, had described the high temperature of the Big Bang and the expected present temperature of the radiation it produced. Gamow's popular books were translated and sold worldwide in large numbers. In the 1950s, when I was in high school in the Netherlands, my friends and I read these books in Dutch translation. In his 1952 book *The Creation of the Universe* (reprinted as a pocket in 1957) Gamow describes in the text as well as in the appendix the evolution of the temperature of the universe with time. As a 15 year old boy I already had read this. In this book Gamow wrote that he believed in a cyclical universe, which after reaching its greatest size collapses in what he called the *Big Squeeze*, in which the temperature and density rise to gigantic values. After this the universe would in his view explode again in a hot Big Bang. Amazingly, Dicke had envisaged exactly the same cyclical model to explain why the universe started from a hot Big Bang. I must say that I find it very hard to believe that neither Dicke nor Peebles had ever seen or read Gamow's popular books, which were sold in such large quantities worldwide, nor his popular articles in journals such as Scientific American. Also Gamow himself did not believe this, as one can see from his 1965 letter depicted in Fig. 18.1, in which he drew the attention of the discoverers of the Cosmic Microwave Background Radiation to the

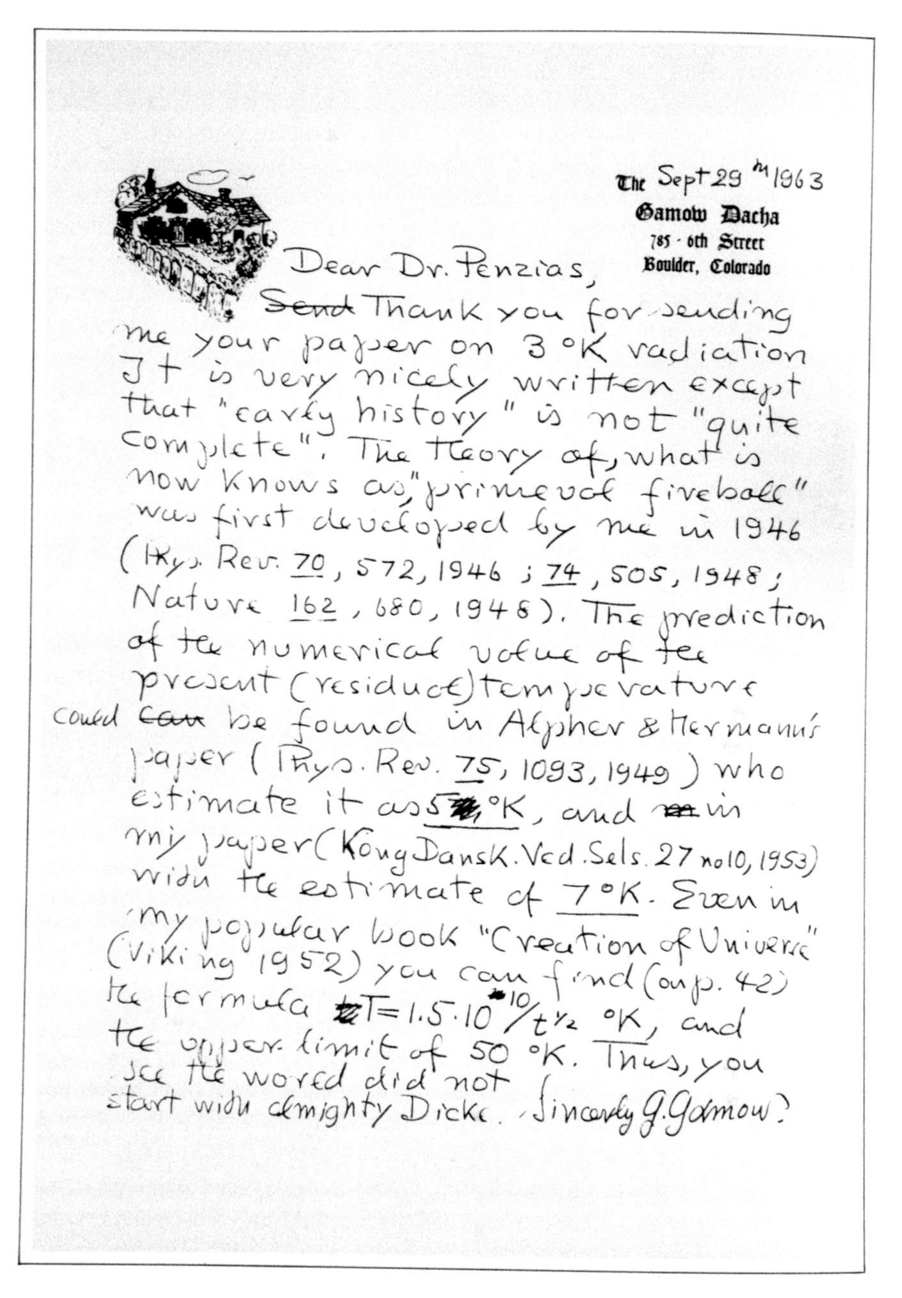

The Gamow Dacha
785 · 6th Street
Boulder, Colorado

Sept 29th 1963

Dear Dr. Penzias,

~~Send~~ Thank you for sending me your paper on 3 °K radiation. It is very nicely written except that "early history" is not "quite complete". The theory of, what is now known as "primeval fireball" was first developed by me in 1946 (Phys. Rev. 70, 572, 1946; 74, 505, 1948; Nature 162, 680, 1948). The prediction of the numerical value of the present (residual) temperature could ~~can~~ be found in Alpher & Herman's paper (Phys. Rev. 75, 1093, 1949) who estimate it as 5 °K, and in my paper (Kong. Dansk. Ved. Sels. 27 no 10, 1953) with the estimate of 7 °K. Even in my popular book "Creation of Universe" (Viking 1952) you can find (on p. 42) the formula $T = 1.5 \cdot 10^{10}/t^{1/2}$ °K, and the upper limit of 50 °K. Thus, you see the world did not start with almighty Dicke.

Sincerely G. Gamow

Fig. 18.1 Letter from George Gamow to Arno Penzias of 29 September 1965, about the 1946/1949 predictions by himself and his students Alpher and Herman, of the microwave background radiation from the Big Bang (By mistake Gamow wrongly dated the year of the letter: 1963)

earlier work of Alpher, Herman and himself, and finishes with stating: 'Thus you see the world did not start with almighty Dicke'.

A well-known psychological phenomenon, that also often occurs outside of science, is that one has read an idea and likes it very much, then forgets about it, until at a later time, when it emerges from memory, one has the feeling that this is one's own original idea, and forgets completely that one has read or heard it before. I cannot keep but thinking that this is what happened to Dicke and Peebles. Peebles has often declared that he and Dicke knew nothing about the earlier work of Alpher, Herman and Gamow. How is this possible if I as a 15 year old kid in another continent did know about it? Did he live in the middle of the wilderness with no contacts with the civilized world? He also says that that from the moment he heard about this work, he always in his publications cited the earlier articles with the predictions of the temperature of the CMBR. But this is not fully correct. In his 1966 article in the Astrophysical Journal Peebles lists the 1953 article of Alpher, Herman and Follin in his reference list, but nowhere mentions it in the text of the article. (This 1966 article was a revised version of his 1965 article that was refused by the Physical Review, of which Alpher and Herman had been the referees, without Peebles knowing about this at the time.)

These facts all strongly suggest that Peebles better liked to forget about the work of his predecessors as it, of course, cast a shadow on his own work. This idea is strengthened by the fact that while Gamow, Alpher and Herman received hardly any recognition for their pioneering work, Peebles made the quite arrogant remark: 'Everything considered, I think that Bob and Ralph have been given the credit they deserve' (see J. Mather, "The Very First Light", 2008, Basic Books, p. 61). This while Peebles himself received many prestigious distinctions such as the Shaw Prize of Hong Kong and the Swedish Crafoord Prize. Only in 2007, a few weeks before he died at age 86, Ralph Alpher was awarded the American National Medal of Science, a high distinction. Better late than never. Bob Herman, who died earlier, and also Gamow, never received a distinction for their work.

Be it how it may, Dicke and Peebles were the ones to realize that this radiation could be measured and they stimulated their colleagues Roll and Wilkinson to build instruments for this. This was an important initiative that clearly went a step further than what Gamow and colleagues had done. It is a pity for them that even before they were able to carry out their measurements, Penzias and Wilson had already discovered the background radiation from the Big Bang, such that they, and not the Princeton group were awarded the 1978 physics Nobel prize. Also Gamow was not included in that prize, as he then had already died. It is a pity that Ralph Alpher, who in his 1948 Ph.D. thesis work discovered that the Big Bang must have been very hot, and therefore rightfully should be called the *father of the hot Big Bang theory*, was not included in this prize. He certainly would have deserved it.

Another interesting subject concerns the formation of the elements heavier that helium. We saw in Chap. 11 that Hoyle, because he did not believe in the Big Bang, between 1946 and 1957 developed the theory for the formation of these heavier elements in the interiors of stars, by nuclear fusion processes and neutron capture. Independently, in 1946 Dutch astronomer Bruno van Albada, who later became my predecessor as a professor of astrophysics at the University of Amsterdam, presented the same theory. We now know that Hoyle and van Albada were right, and that the Big Bang made only hydrogen and helium (plus tiny amounts of a few other light isotopes). However, in his 1952 popular book *The Creation of the Universe*, Gamow mocks these ideas of Hoyle and van Albada. He writes: "What Hoyle and van Albada demand, sounds like the request of an inexperienced housewife, who wanted three electric ovens for cooking a dinner: one for the turkey, one for the potatoes and one for the pie. Such an assumption of heterogeneous cooking conditions, adjusted to give the correct amounts of light, medium-weight and heavy elements, would completely destroy the simple picture of atom-making, by introducing a complicated array of specially designed 'cooking facilities'".

Nevertheless, it has turned out that Gamow here was wrong and van Albada and Hoyle were right. Gamow had assumed that apparently the simplest model—such as one in which all elements were made in the Big Bang—always would be the right one. This is the "Occam's razor" principle, which implies that in general the simplest solution, with the least assumptions, is the correct one. And in science indeed this is very often true. But apparently sometimes the razor cuts deeply and nature is more complicated than had been expected.

Finally, an interesting phenomenon that we encountered in Chap. 8 was Einstein's conviction between 1917 and 1929, that the universe is static and that therefore an extra lambda (Λ) term should be added to his original equations, to keep the universe static. At first, Einstein had even thought that Friedmann's equations were mathematically wrong, but later he realized that this was not the case. And when Lemaître independently had found the same solutions and showed these to him, Einstein mentioned Friedmann's work to him and remarked that he did not believe in these solutions and that "mathematically correct solutions often do not correspond to situations that are realized in the physical world". Finally, Hubble's discovery of the expansion of the universe led to Einstein's confession to Gamow that the introduction of this lambda-term had been his greatest blunder. But now we see that, after all, this lambda was not a blunder, and is back to explain the accelerated expansion of the universe. All of this teaches us how haphazard the path of science is and also: how enormously important the observations are to keep theorists on the right track. On the other hand, observations cannot do without a theoretical framework that allows them to be understood. Hubble's observations appeared to fit the framework of Friedmann's theory and led to Lemaître's theory of the Big Bang. This theory, elaborated by Gamow and collaborators, predicted the Cosmic Microwave Background radiation, and the discovery of this radiation in turn conformed the Big Bang theory. Further work on the Big Bang theory in the late 1960s led to the prediction by Silk, Sunyaev and Zeldovich of little wrinkles in the temperature distribution of the background radiation over the sky—the "seeds" from which later the clusters and super-clusters of galaxies formed. Thirty years later these wrinkles were discovered with the COBE, WMAP and Planck satellites.

We thus see that science proceeds along an alternating path of observation and theory: observations lead to new theoretical insights and the resulting theories lead, in turn, to predictions of new observable phenomena. Observers then go out to search whether these predictions can be confirmed, or should be rejected. The new observations then lead to new or refined theoretical models, which may or may not be confirmed by further observations. It is this continuing spiral of observation and theory that drives the progress of science.

Astronomy is a natural science. This means: we study phenomena that we observe in nature, and try to understand these phenomena. In the end of the day always the observations decide whether or not a theory is correct. It is thanks to the spiral of observations and theories that, step by step, over the past few thousand years, we have gained more and more insight in the structure of the universe and the history of space and time.

Appendix A: Some Data About the Solar System

The Sun, planets, the dwarf planet Pluto and the Moon

Object	Equatorial diameter (km)	Equatorial diameter (Earth = 1)	Mass (Earth = 1)	Mean density (kg/dm^3)	Acceleration of gravity (Earth = 1)	Escape velocity (km/s)
Sun	1 392 530	109	330,000	1.41	28	617
Mercury	4 878	0.382	0.055	5.43	0.38	4.25
Venus	12 104	0.95	0.820	5.24	0.90	10.36
Earth	12 756	1	1	5.52	1	11.18
Mars	6 794	0.53	0.107	3.93	0.38	5.02
Jupiter	142 800	11.2	318	1.32	2.69	59.6
Saturn	120 000	9.5	95.1	0.70	1.19	35.6
Uranus	51 000	4.0	14.5	1.2	0.93	21.1
Neptune	49 500	3.9	17.2	1.76	1.22	24.6
Pluto	2 290	0.18	0.002	2	0.05	1
Moon	3 476	0.27	0.0123	3.34	0.165	2.4

E. van den Heuvel, *The Amazing Unity of the Universe*,
Astronomers' Universe, DOI 10.1007/978-3-319-23543-1

The orbits of the eight planets

Planet	Semi-major axis of orbit (in millions of km)	Semi-major axis of orbit (in astronomical units)	Orbital period	Orbital eccentricity	Angle between orbital plane of planet and Earth
Mercury	57.9	0.387	87.97 days	0.2056	7.0
Venus	108.2	0.723	224.7 days	0.0068	3.39
Earth	149.6	1	365.26 days	0.0167	0
Mars	227.9	1.524	1.881 years	0.0934	1.85
Jupiter	778.3	5.203	11.86 years	0.0485	1.3
Saturn	1427	9.539	29.46 years	0.0556	2.49
Uranus	2870	19.19	84.01 years	0.0472	0.77
Neptune	4497	30.06	164.79 years	0.0086	1.77

Appendix B: The Structure of Atoms and the Standard Model of Elementary Particles and Forces

The Structure of Atoms: Protons, Neutrons and Electrons

Atomic nuclei consist of electrically positively charged protons and uncharged neutrons. These two kinds of particles are about 1840 times heavier than the negatively charged electrons that describe orbits around the nucleus. The values of the electric charge of proton and electron are the same, but have opposite signs. The number of electrons in an atom is the same as the number of protons in the nucleus, such that the atom as a whole is not electrically charged (also called: *electrically neutral*). The *size* of the atom is determined by the size of the orbits of the electrons. In the case of the hydrogen atom, which has a proton as nucleus, and thus has one electron, the smallest allowed electron orbit has a diameter of about 50,000 times the diameter of the proton. The volume of the atom is therefore $50{,}000 \times 50{,}000 \times 50{,}000 = 1.25 \times 10^{14}$ times larger than the volume of the proton. As the volume of the electron is negligible, the atom consists largely of empty space. This holds for all the atoms that make up the matter that surrounds us and of which we consist.

E. van den Heuvel, *The Amazing Unity of the Universe*,
Astronomers' Universe, DOI 10.1007/978-3-319-23543-1

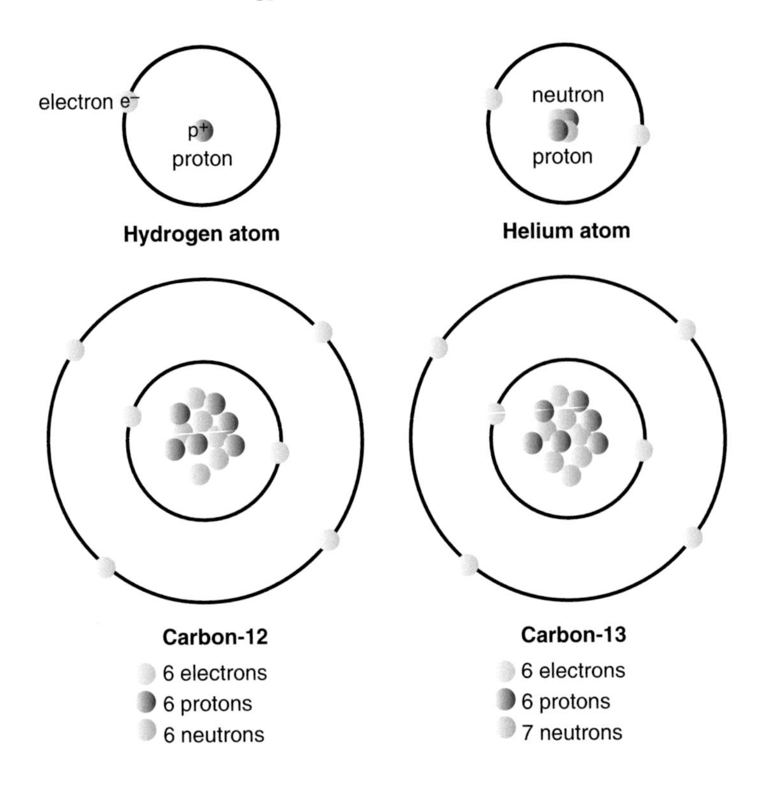

The nucleus of the helium atom consists of two protons and two neutrons, around which two electrons are orbiting. There also exists a stable isotope *helium-3*, with a nucleus that consists of two protons and one neutron, and there consists a stable isotope of hydrogen, called *deuterium*, with a nucleus that consists of one proton and one neutron. The chemical properties of an element are determined by the number of electrons that is orbiting the nucleus. Therefore, deuterium is chemically identical to hydrogen, and helium-3 is chemically identical to helium-4. The different *isotopes* of an element have always the same number of protons in the nucleus, and thus: the same number of electrons around the nucleus. But the nuclei of the isotopes do differ in the number of neutrons. For example, carbon-12 has six protons and six neutrons in its nucleus, but carbon-13 and carbon-14 have seven and eight neutrons in their nuclei, respectively, while their number of protons remains six. Carbon-12 and -13 are stable and are present in nature. Carbon-14 is radioactive and its nucleus decays into a nitrogen-14 nucleus by emitting an electron. The timescale of this decay—the so-called half-life of carbon-14—is 5730 years. Bombardment of nitrogen nuclei in Earth's atmosphere by cosmic rays continuously produces new carbon-14 nuclei in the atmosphere, which are taken up in living organisms like trees, etc. Old wood, for example from Roman or Egyptian times, can therefore be accurately dated by measuring how much carbon-14 is still present in it.

The heaviest nucleus found in nature is that of uranium, which has 92 protons. There are two long-lived isotopes found in nature: uranium-235 with a half-life of 700 million years and uranium-238 with a half-life of 4.5 billion years. They have 143 and 146 neutrons in the nucleus, respectively (see Fig. 2.15 of Chap. 2).

The Quarks and the Standard Model

Experiments with large particle accelerators have shown that the proton and the neutron are not really *elementary* particles, but themselves are composed of other particles with weird properties, the so-called *quarks*, a name invented by American physicist Murray Gell-Mann, who discovered their properties. All evidence suggests that these quarks as well as the electrons are truly *elementary* particles, which means that they belong to the family of the smallest building blocks of matter. Strange enough, the electric charge of the quarks is only a fraction of the charge of an electron: they have either 1/3 or 2/3 of this charge.

Apart from the quarks, which are called *baryons*—which means: heavy particles—the standard model of elementary particles also includes a series of particles called *leptons*, literally meaning "light things". These are the electron and the electron-neutrino, and their anti-particles, the positron and the electron-anti-neutrino, plus two kinds of heavier electrons, called the *muon* and the *tau*-particle, with their neutrinos and their sets of anti-particles. In total, there are therefore 12 different kinds of leptons. Also the number of different kinds of quarks and anti-quarks is 12.

Table B1 The particles and forces of the standard model

	First generation	Second generation	Third generation
Quarks	Up	Charm	Top
	Down	Strange	Bottom
Leptons	Electron	Muon	Tau
	Electron-neutrino	Muon neutrino	Tau neutrino
Force carriers	Weak nuclear force: W^+, W^-, Z°		
	Electromagnetism, photon		
	Strong nuclear force, eight gluons		
Cause of rest mass	Higgs particle		

The weak nuclear force, which binds together a proton and an electron in a neutron, has as its force carrier the bosons W^+, W^- and Z°. In fact this force is a close relative of the electromagnetic force. The carriers of the electromagnetic force are the photons.

The carriers of the strong nuclear force, which binds together the quarks in atomic nuclei, are the *gluons*. The *Higgs particle*, that assigns *rest mass* to certain particles, was discovered at CERN in 2012, for which Scotsman Higgs and Belgian physicist Englert, who predicted this particle in the 1960s, were awarded the Nobel prize. Particles such as electrons, quarks, neutrons, protons and neutrinos, have rest mass. This means that they can be brought completely to rest, and can be weighed. On the other hand, other kinds of particles, such as photons, have no rest mass. They always move with the speed of light. It has been found that, apart from their electric charge, quarks also have another kind of "charge", which is called *colour*. While there are only two kinds of electric charge (+ and −), there are three kinds of colour-charge, which have been given the names R (red), G (green) and B (blue). Every kind of quark, for example the up-quark (U) can occur in every of these three colour varieties, in this case: U(R), U(G) or U(B). Just as electric charges create photons—carriers of the electromagnetic field (and force)—the colour charge creates a *colour field*, the carriers of which are called the *gluons*. Thus the gluons are the carriers of the strong nuclear force, which binds the quarks together in the protons and neutrons. And just like the quantum theory of the electromagnetic interaction is called *quantum-electrodynamics*, the quantum theory of colour interaction is called *quantum-chromodynamics*. It appears that only particles that consist of quarks of three different colours are stable. For example, a proton consists of a red and a green up-quark and a blue down-quark. But also the combination U(B) with U(R) and D(G) is possible, just as U(B) with U(G) and D(R). In fact, the quarks in a proton continuously fluctuate between these three combinations. The up-quark has electric charge 2/3 and the down-quark −1/3, such that the combination of two up-quarks and one down-quark has net electric charge +1. On the other hand, the neutron consists of one up- and two down-quarks, and therefore has electric charge $2/3-2/3=0$. In the same way one can make anti-protons and anti-neutrons from combinations of the corresponding anti-quarks (see the illustration at the right). The work of the physicists Gell-Mann and Nambu shows that all the kinds of particles that showed up in particle accelerator experiments since the 1960s, produced by bombarding atomic nuclei with high-energy particles, are composed of combinations of quarks. Thanks to the theory of quantum-chromodynamics all these particles were found to nicely fit with such quark combinations. Already for over 30 years this standard model of the elementary particles appears to be the last word in our understanding of the building blocks of all matter and forces in our universe. However, to make this model complete, the Higgs particle, predicted in the 1960s by above-mentioned Scotsman Higgs and his Belgian colleagues Braut and Englert, should have to be found, which indeed happened at CERN in 2012. The relative strengths of the above-mentioned natural forces is given in Table B2.

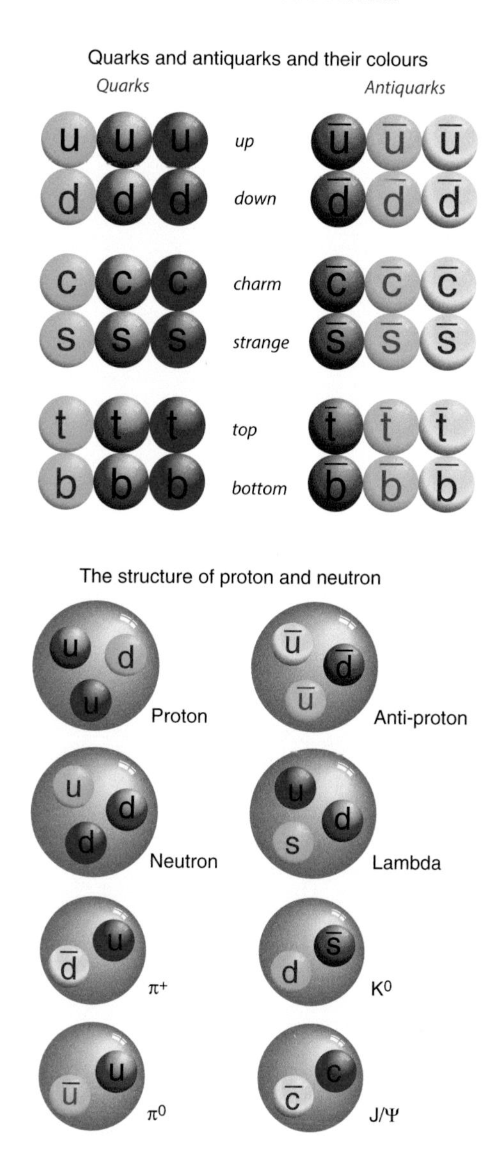

Table B2 Some properties of the fundamental physical forces

Type of force	Strength (relative to the strong nuclear force)	Characteristic distance over which this force acts
Strong nuclear force	1	10^{-15} m
Electromagnetic force	1/137	Infinite
Weak nuclear force	10^{-5}	10^{-17} m
Gravity	8×10^{-39}	Infinite

Appendix C: About the Parameters of the Universe

The Critical Density and the Hubble Constant[1]

After Friedmann's discovery of the different solutions for the evolution of the universe (open, closed and flat) it was found that the same solutions can also be obtained with Newton's law of gravity. In his time, Newton did not know how to treat the dynamics of an infinite homogeneous and isotropic universe. It appears that this is much simpler than most researches might have thought. One can understand this thanks to two theorems that Newton himself had proven in the seventeenth century. To show this, we take a homogeneous and isotropic universe and consider the forces experienced by a galaxy A that is located, as indicated in Fig. C1, at a distance R from our Milky Way Galaxy, that is located at M. In order to calculate the gravitational attraction exerted on galaxy A by our Milky Way Galaxy, and the galaxies in our neighbourhood, we consider the sphere with radius R around our Galaxy. Spherical shells with our Galaxy at its centre, with radius larger than R, do not exert any attraction on galaxy A. This is a theorem proven by Newton, as depicted and described in Fig. C2: an object *inside* a spherical shell of matter experiences no gravitational attraction from this shell: the forces exerted by the different parts of the shell, when all are added together, just cancel. A second theorem that Newton derived, also depicted and explained in Fig. C2, is that an object *outside* a spherical shell of matter experiences a gravitational attraction from this shell that is equal to the force of attraction experienced from a point mass with the mass of this shell, placed at the position of the centre of this shell. This second theorem means that if one wishes to calculate the gravitational attraction exerted on A by all spherical shells with a radius smaller than or equal to R and centred on our Milky Way Galaxy M, one may just replace all these shells by point masses of the same mass, located at M. Together, all spherical shells with radius smaller than or equal to R form a sphere with radius R around M.

[1] See also the mathematical supplement of the book "The First Three Minutes" by Steven Weinberg, Basic Books, New York, 1993.

E. van den Heuvel, *The Amazing Unity of the Universe*,
Astronomers' Universe, DOI 10.1007/978-3-319-23543-1

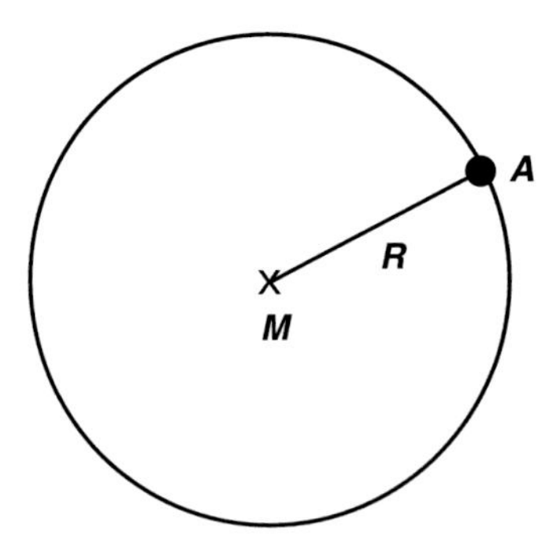

Fig. C1 The gravitational attraction experienced by galaxy A from the direction of galaxy M at distance *R* is determined only by the matter inside the sphere with radius *R* around *M*. This is explained in the text

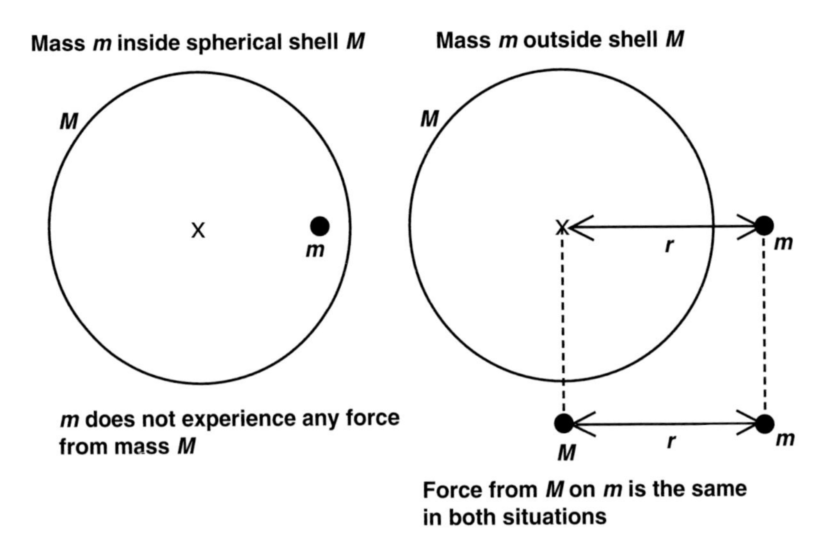

Fig. C2 Around 1680, Newton proved the following two theorems: (1) A spherical shell of matter with mass *M* exerts no gravitational attraction on a point mass *m* *inside* this shell, and (2) A spherical shell of matter *M* exerts on a point mass *m* *outside* the shell a gravitational attraction equal to that of a point mass of size M placed at the position of the centre of the shell

This means that A experiences a gravitational attraction from the direction of M equal to the attraction which this sphere exerts on A.

If the speed with which the galaxy A moves away from M is larger than the escape velocity from the surface of this sphere, the universe will be open. When it is less than this escape velocity it will be closed, and when it is exactly the escape velocity, the universe will be flat. If we imagine the sphere with radius *R* to be Earth and the galaxy to be a stone that is thrown upwards vertically, one will have a situation that is exactly analogous to that of the universe: if the stone has a velocity larger than the escape velocity from Earth, it will never come back and will keep moving away from Earth forever (open universe). If its velocity is less than the escape velocity, it will reach a highest point and then fall back to Earth (closed universe), and when it has exactly the escape velocity, it will keep moving away forever, but will, at infinite distance reach velocity zero with respect to Earth.

The fact that the combined gravitational attraction of the entire part of the universe outside the sphere of radius R around M is exactly equal to zero may seem surprising. It turns out, however, that this is mathematically correct, even for an infinitely large universe. This is called Birkhoff's theorem, after the mathematician who proved it.

We now calculate how the critical density of the universe depends on Hubble's constant, which characterizes the rate of expansion of the universe. One can cast this all in a rather simple mathematical form, as follows. The gravitational potential energy of the galaxy A relative to the mass of the sphere with radius R around M is equal to—GM(sphere). M(A)/R, and the kinetic energy of A relative to M is 0.5M(A).v^2. Here M(sphere) is the mass of the sphere and M(A) is the mass of galaxy A and v is the velocity of A relative to M. The value of v is given by the Hubble relation: $v = \mathrm{H}.R$. The total energy is the sum of the potential and kinetic energy. Using the fact that $M(\mathrm{sphere}) = (4/3)\pi R^3\rho$, where ρ is the mass density [in kg/m^3], one finds that the total energy of galaxy A is:

$$E_{tot} = M(\mathrm{A}).R^2.\left[-(4/3)\pi \mathrm{G}\rho + \tfrac{1}{2}\mathrm{H}^2\right] \quad \text{(C1)}$$

Where $\mathrm{G} = 6.67 \times 10^{-11}$ [Newton.m^2/kg^2] is the gravitational constant in Newton's law.

From this equation it follows that the total energy is zero if the density is equal to the so-called critical density ρ_{crit} given by:

$$\rho_{crit} = 3\mathrm{H}^2/8\pi\mathrm{G} \quad \text{(C2)}$$

For ρ larger than ρ_{crit} the total energy is negative (*closed universe*) and for ρ smaller than ρ_{crit} the total energy is positive (*open universe*). In the latter case one has that even when the radius R has expanded to become infinite—which means: potential energy *zero*—there still is kinetic energy left, so the universe keeps expanding.

Although this result was derived from Newtonian dynamics, it is still valid for a universe with Einstein's General Relativity Theory, provided that the density ρ includes the total energy density divided by c^2. Substituting for Hubble's constant H the presently best estimate of 20 km/s/million lightyears, one finds a critical density of 8×10^{-27} kg/m^3. As one kilogram holds about 6×10^{26} nucleons (protons, neutrons), the critical density corresponds to about 4.8 nucleons per cubic metre.

The Timescale of the Expansion

We now consider how the various characteristic parameters of the universe, such as the density and the Hubble time, change in the course of time t. We again take the galaxy A of mass $M(\mathrm{A})=m$ at a distance $R(t)$ from an arbitrarily chosen galaxy, which we choose as the centre. For the latter we again take the Milky Way M. We saw in the last paragraph that the total energy of A is equal to:

$$E_{tot} = m.R(t)^2\left[\tfrac{1}{2}\mathrm{H}(t)^2 - (4/3)\pi \mathrm{G}\rho(t)\right] \tag{C3}$$

where $\mathrm{H}(t)$ and $\rho(t)$ are the values of the Hubble constant and the density of the universe, at time t. Because of the law of conservation of energy, this total energy of the galaxy A is conserved. Observations have shown that, in fact, the universc is flat (see Chap. 12), which means that this constant is in very good approximation, equal to zero. So, one finds:

$$\mathrm{H}(t) = \left[(8/3)\pi \mathrm{G}\rho(t)\right]^{\frac{1}{2}} \tag{C4}$$

And the characteristic expansion time ("age") of the universe is:

$$t_{\exp} = 1/\mathrm{H}(t) = \left[(8/3)\pi \mathrm{G}\rho(t)\right]^{-\frac{1}{2}} \tag{C5}$$

We see here, that in the early times, when the density was high, the expansion time was very short and Hubble's "constant" $\mathrm{H}(t)$ was very large. Therefore, Hubble's "constant" is not a constant at all! Nowadays, the characteristic expansion time is about 13 billion years. In order to find how $\mathrm{H}(t)$ changes with time, we have to know how the density $\rho(t)$ changed with the characteristic dimension (*scale factor*) $R(t)$ of the universe. When the universe is dominated by matter and not by radiation, as is the case today, the mass inside the expanding sphere stays the same. So, in this case $(4/3)\pi R(t)^3\rho(t)$ is constant, which means that $\rho(t)$ changes proportionally to $1/R(t)^3$. On the other hand, in the very early universe, until about 380,000 years after the beginning, when matter and radiation were still tightly coupled, the energy of the radiation dominated. In these times the density was dominated by the mass-equivalent of the radiation-energy density. This mass equivalent is proportional to the fourth power of the temperature T (Stephan-Boltzmann law, see "Appendix D"). The cosmological redshift changes proportional to the scale factor $R(t)$. The maximum wavelength of the Planck radiation curve is therefore proportional to $R(t)$. Since Wien's law ("Appendix D") states that the maximum wavelength of the Planck curve is inversely proportional to the temperature $T(t)$, one finds that the temperature $T(t)$ is proportional to $1/R(t)$. From this it follows that in the radiation-dominated era the density is proportional to $1/R(t)^4$. In summary:

The density is proportional to $1/R(t)^n$, where n=3 in the matter-dominated era, and n=4 in the radiation-dominated era. For this reason the Hubble constant $H(t)$ is proportional to $(1/R(t))^{n/2}$.

The speed with which an arbitrary galaxy at a distance $R(t)$ moves away from us is $v(t) = H(t).R(t)$, which is proportional to $(R(t))^{(1-n/2)}$.

The velocity $v(t)$ is in fact the change of $R(t)$ with time $= dR(t)/dt$. Using elementary differential calculus we then find, because $v(t)$ is proportional to $(R(t))^{(1-n/2)}$, the following relations between the distances and velocities at times t_1 and t_2 : $t_1 - t_2 = (2/\mathrm{n})\left[R(t_1/v(t_1) - R(t_2)/v(t_2)\right]$, which implies:

$$t_1 - t_2 = (2/\mathrm{n})\left[1/\mathrm{H}(t_1) - 1/\mathrm{H}(t_2)\right] \tag{C6}$$

Since we can express H(t) in terms of the density $\rho(t)$, we see that:

$$t_1 - t_2 = (2/\mathrm{n})\left(3/(8\pi\mathrm{G})^{½}\right).\left[(1/\rho(t_1))^{½} - (1/\rho(t_2))^{½}\right] \tag{C7}$$

We therefore see that the time t_1 that has passed since the beginning $(t_2 = 0)$, when the density $\rho(t_2)$ was practically infinitely large, such that $(1/\rho(t_2)) = 0$, is inversely proportional to the square root of the density $\rho(t_1)$, so one can write that at time t holds:

$$t = (2/\mathrm{n}).(3/(8\pi\mathrm{G}\rho))^{½} \tag{C8}$$

During the radiation era, when n=4, one has $\rho = 1.22 \times 10^{-32} T^4 \left[\mathrm{kg/m^3}\right]$, with T in Kelvins. One then easily calculates that, for example, the time needed for the universe to cool down to 10 million Kelvins is: $t = ½(3/(8\pi\mathrm{G}))^{½} / (1.22 \times 10^{-7})$, which is about 1.9 million seconds=0.06 years.

For the radiation-dominated era, which lasted for about 380,000 years, the time t until the density decreased to $\rho(t)$ is equal to $(1/2)t_{exp}$, and for the present, matter-dominated era, it is equal to $(2/3)t_{exp}$.

As the density is proportional to $R(t)^{-n}$, the scale factor R is, therefore, proportional to $t^{2/n}$. This means, in the radiation-dominated era, $R(t)$ is proportional to $t^{½}$, and in the matter-dominated era it is proportional to $t^{2/3}$.

In Chap. 12 we saw that at any time t after the Big Bang the horizon for everybody in the universe is at a distance of order ct. From outside this horizon we can, at this time, not yet have received information. The above shows that when t decreases, $R(t)$ decreases more slowly than the distance to the horizon. This means that, going to earlier and earlier times, finally every particle in the universe is outside one's horizon. Also the reverse holds: as time increases, one observes that a larger and larger part of the universe appears inside one's horizon.

Redshifts Larger than One

The exact equation for the redshift $\Delta\lambda$ of the rest-wavelength λ, due to the Doppler effect is:

$$z = \Delta\lambda / \lambda = \left(\left(1 + v/c\right) / \left(1 - v/c\right) \right)^{1/2} - 1 \qquad \text{(C9)}$$

Here is v the velocity of the light source and c the velocity of light. For small values of v/c this yields the classical formula for the Doppler effect: $\Delta\lambda / \lambda = v/c$. However, for v approaching the velocity of light, the denominator of the first term on the right-hand side goes to zero, which means that the redshift can become arbitrarily large. For example, for $v/c = 0.9$, one finds $z = 3.36$, which is much larger than 1.

Appendix D: The Radiation Laws of Planck, Wien and Stefan-Boltzmann

In 1900 Max Planck derived the energy distribution of blackbody radiation of temperature T. He discovered that the amount of energy per unit volume in a small wavelength region $d\lambda$ between wavelengths λ and $\lambda + d\lambda$, is given by:

$$u_{\lambda}.d\lambda = \left(8\pi hc / \lambda^{5}\right)\left(e^{hc/\lambda kT} - 1\right).d\lambda \tag{D1}$$

This called a Planck distribution or *Planck's law*. Here k is the Boltzmann constant $= 1.38 \times 10^{23}$ [J/K], c is the velocity of light and h is Planck's constant $= 6.626 \times 10^{-34}$ [J.s].

The Planck distribution reached its maximum at the wavelength λ_{max} given by:

$$\lambda_{\max} = 0.2014052 hc / T \tag{D2}$$

This equation was derived by German physicist Wien and is called *Wien's law*. This law expresses that with increasing temperature, the wavelength where the energy distribution of the radiation of a blackbody reaches its maximum, shifts towards shorter wavelength values.

Finally, if one integrates Planck's formula over all wavelengths, from zero to infinity, one finds the total amount of energy of the blackbody radiation per unit volume. This amount turns out to be equal to:

$$u = 8\pi^{5}\left(kT\right)^{4} / 15\left(hc\right)^{3} = 7.56464 \times 10^{-16} T^{4} \left[\mathrm{J} / \mathrm{m}^{3}\right] \tag{D3}$$

where T is het temperature in Kelvins. This is *Stefan-Boltzmann's law*.

E. van den Heuvel, *The Amazing Unity of the Universe*,
Astronomers' Universe, DOI 10.1007/978-3-319-23543-1

Appendix E: The Equations for the Jeans Length and the Jeans Mass

The *Jeans length* R_J is the size of the smallest condensation which can form by gravitational instability in a gas of density ρ (in kg/m) and temperature T. The corresponding mass of such a condensation is the mass of a sphere with diameter R_J, which is called the *Jeans mass* M_J.

The equations for these quantities are:

$$R_J = \left(\gamma \bar{R} T / \mu \rho G\right)^{1/2} \tag{E1}$$

$$M_J = \left(\gamma \bar{R} T / \mu G\right)^{3/2} / \rho^{1/2} \tag{E2}$$

Here $\bar{R}$ is he universal gas constant from the ideal gas law $PV = \bar{R}T$ ($\bar{R}$ is the same for all gases), where P and V are pressure and volume of the gas (for an Avogradro number of gas molecules, for which the universal gas constant is defined), respectively. Further: γ is the so-called *adiabatic index*, which for an ideal gas is 5/3, and μ is the average mass of the molecules of the gas, counted in units in which the mass of a hydrogen atom is 1 (here electrons are also counted as molecules, so, for example, for ionized hydrogen $\mu = 0.5$), and G is the gravitational constant from Newton's law.

M_J is the mass of the smallest celestial object (star, planet, proto-galaxy) that can form by gravitational instability in a gas of density ρ and temperature T. The reason why there is such a lower limit is that, if by a random fluctuation a density enhancement forms in a part of the gas, the enhanced density will cause an *enhanced gas pressure* in this part of the gas. On the other hand, the density enhancement causes this part if he gas to exert a higher than average gravitational attraction on the surrounding gas, which causes the surrounding gas to start moving towards this region of enhanced density, which may cause the density enhancement to grow. However, the enhanced pressure in this density-enhanced region acts in a just opposite way: it causes the gas in this region to try to move outwards, and thus to wipe out the den-

E. van den Heuvel, *The Amazing Unity of the Universe*,
Astronomers' Universe, DOI 10.1007/978-3-319-23543-1

sity enhancement. In density-enhanced regions that are *smaller* than the Jeans length R_J the gas pressure wins from the tendency of this region to attract gas from its surroundings, and the density enhancement is wiped out. On the other hand, in density-enhanced regions larger than R_J the gravitational attraction wins from the pressure, and the region begins to contract and grow, such that a galaxy, a star or a planet can form.

Credits of the Figures

Figure above title of Chap. 1: STScI/NASA.

Figure above title of Chap. 2: Jet Propulsion Laboratory (JPL/NASA)

Fig. 2.1: California State Polytechnic Institute: www.calpoly.edu.

Fig. 2.2: Wikimedia Commons, uploaded by Orion 8; inset a: JPL/NASA; inset b: Japanese Space Agency JAXA and NASA.

Fig. 2.3: spaceguard.rm.iasf.cnr.it

Fig. 2.4: above: Sagor, Beek, Netherlands; middle and below: European Southern Observatory ESO

Fig. 2.5: Space Telescope Science Institute (STScI)/NASA

Fig. 2.6: NASA/spaceplace.nasa.gov

Fig. 2.7: NASA and ESA

Fig. 2.8a: NASA Mariner 10, 1974

Fig. 2.8b, c and d: NASA

Fig. 2.9 upper left (a): Wikimedia commons, uploaded from NASA

Fig. 2.9 upper right (b): NASA Earth Observatory

Fig. 2.9 below: A. C. Alekseev, A. V. Mikheeva, V. E. Petrenko and B. G. Mikhailenko, Institute of Computational Mathematics and Mathematical Geophysics, Siberian Branch Russian Academy of Sciences.

Fig. 2.10: NASA

Fig. 2.11a: NASA, Voyager 1 and 2

Fig. 2.11b: www.astrobio.net.

Fig. 2.12: NASA and ESA/Cassini-Huygens mission

Fig. 2.13: Jimmy Westlake December 1985/www.jpl.nasa.gov

E. van den Heuvel, *The Amazing Unity of the Universe*,
Astronomers' Universe, DOI 10.1007/978-3-319-23543-1

Fig. 2.14: Wikimedia Commons licensed under Creative Commons Attribution Share Alike 2.0 Generic license.

Fig. 2.15: Sagor, Beek, Netherlands

Fig. 2.16: *Top left figure*: thestonescryout.com; *right figure*: cambridgecarbonates.com; *lower left*: ucmp.berkeley.edu

Fig. 2.17: Lunar and Planetary Laboratories: lpl.arizona.edu

Figure above the title of Chap. 3: NASA, Space Telescope Science Institute (STScI)

Fig. 3.1: www.tychobrahesverden.dk

Fig. 3.2: Sagor, Beek, Netherlands

Fig. 3.3: Sagor, Beek, Netherlands

Fig. 3.4: University of Cambridge, Institute of Astronomy Library

Fig. 3.5: Sagor, Beek, Netherlands

Fig. 3.6a above: Astronomy Department of Ohio State University: astronomy.ohio-state.edu

Fig. 3.6a below: Sagor, Beek, Netherlands

Fig. 3.6b: Sagor, Beek, Netherlands, after: Tomas J. Stolte, Amsterdam (tomas@crispybison.nl)

Fig. 3.7: NASA astronomy picture of the day/apod.nasa.gov

Fig. 3.8: archive of Veen Magazines, Diemen, Netherlands

Fig. 3.9: Sagor, Beek, Netherlands, after: www.physics.csbsju.edu

Fig. 3.10: Sagor, Beek, Netherlands

Figure above title of Chap. 4: NASA, STScI

Fig. 4.1: European Southern Observatory ESO

Fig. 4.2: NASA-JPL, and: coolcosmos.ipac.caltech.edu

Fig. 4.3: W.F. Herschel and (e.g.): cfa.harvard.edu

Fig. 4.4: Kapteyn Astronomical Laboratory, Groningen University, Netherlands

Fig. 4.5: Sagor, Beek, Netherlands, after *Sky and Telescope*

Fig. 4.6: NASA/apod.nasa.edu

Fig. 4.7: Archive Professor Piet van de Kamp

Fig. 4.8: NASA/STScI

Fig. 4.9: Sagor, Beek, Netherlands

Fig. 4.10: H. Shapley and the Astrophysical Journal vol. 49, 1918

Fig. 4.11: Sterrewacht Leiden, the Netherlands

Fig. 4.12: Sagor, Beek, the Netherlands

Fig. 4.13: Jodrell Bank Radio Observatory/ jb.man.ac.uk

Fig. 4.14: Sterrewacht Leiden, the Netherlands

Fig. 4.15: Netherlands Foundation for Radio Astronomy ASTRON/NWO, Dwingeloo, the Netherlands

Fig. 4.16: *upper*: NASA/JPL (Robert Hurt); *lower*: Pennsylvania State University (www.astro.psu.edu)

Figure above title of Chap. 5: European Southern Observatory ESO

Fig. 5.1: Sagor, Beek, Netherlands

Fig. 5.2: Deutsches Museum, München, Germany

Fig. 5.3: Sagor, Beek, Netherlands

Fig. 5.4: Sagor, Beek, Netherlands

Fig. 5.5: hetutrechtsarchief.nl

Fig. 5.6: Sidney Observatory, Australia: Sidneyobservatory.com.au

Fig. 5.7: Sagor, Beek, Netherlands

Fig. 5.8: Sagor, Beek, Netherlands

Fig. 5.9a: Sagor, Beek, Netherlands; b. www.mit.images

Fig. 5.10: Ben McCall; bjmccall@uiuc.edu

Figure above the title of Chap. 6: STScI/NASA

Fig. 6.1: Bill Schoening, Vanessa Harvey/REU Program/NOAO/AURA/NSF

Fig. 6.2: right picture: www.astrosurf.com, after King, H. C. (1955) The History of the Telescope; left-hand picture: Veen Magazines, Diemen, Netherlands, from the book Astronomie.nl, Een Hollandse kijk op het heelal, by Govert Schilling 2008; also: nmspacemuseum.org.

Fig. 6.3a: STScI/NASA; b and c: European Southern Observatory/ESO.org

Fig. 6.4: STScI/NASA/apod.nasa.gov

Fig. 6.5: Kitt Peak/NOAO/AURA

Fig. 6.6: astronomy.com.au

Fig. 6.7: STScI/NASA/apod.nasa.gov

Fig. 6.8: Courtesy Alson Wong, California (alsonwongastro.com)

Fig. 6.9: STScI/NASA/ESA/ESO

Fig. 6.10: European Southern Observatory ESO

Fig. 6.11: Wikimedia Commons, uploaded by LOBS (File Nl-lobs.ogg) (CC BY-SA 4.0)

Fig. 6.12: E. Hubble, Proceedings National Acad. of Sciences USA pp. 168–173, 1929

Fig. 6.13: R. Kirshner/Proceedings National Academy of Sciences (PNAS.org)

Fig. 6.14: Sagor, Beek, Netherlands

Fig. 6.15: Sagor, Beek, Netherlands

Figure above the title of Chap. 7: www.wolverton-mountain.com and: www.wm.edu

Fig. 7.1: Archive Le Scienze, Milan, Italy

Fig. 7.2: Académie des Sciences, Paris

Fig. 7.3: Sagor, Beek, Netherlands

Fig. 7.4: Isaac Newton's Principia

Fig. 7.5: Sagor, Beek, Netherlands

Fig. 7.6: NASA 'Astronomy Picture of the Day' and Historical Museum of Berne; Swiss copyright expired 1992

Fig. 7.7: Sagor, Beek, Netherlands

Fig. 7.8a: Time Travel Research Center: http://www.zamandayolculuk.com/cetinbal/HTMLdosya1/RelativityFile.htm)

Fig. 7.8b: Sagor, Beek, Netherlands, after: physics.unc.edu

Fig. 7.9: Sagor, Beek, Netherlands

Fig. 7.10: Sterrewacht Leiden/Leiden University, the Netherlands

Figure above the title of Chap. 8: Museum Boerhave, Leiden/Sterrewacht Leiden/Leiden University, the Netherlands

Fig. 8.1: Sagor, Beek, Netherlands

Fig. 8.2: Museum Boerhave, Leiden/Sterrewacht Leiden/Leiden University, the Netherlands

Fig. 8.3: Wikimedia Commons, uploaded by Q Werk

Fig. 8.4: Sagor, Beek, Netherlands

Fig. 8.5: Sagor, Beek, Netherlands

Fig. 8.6: Archive of Catholic University Leuven, Belgium

Fig. 8.7: Sagor, Beek, Netherlands

Fig. 8.8: Sagor, Beek, Netherlands

Fig. 8.9: Astronomical Institute University of Amsterdam, Netherlands

Figure above the title of Chap. 9: STScI/NASA

Fig. 9.1: the biography by Rob van den Berg: George Gamow, *Van Atoomkern tot kosmos*, Veen Magazines, Diemen, Netherlands, 2011.

Fig. 9.2: University of Michigan/www.globalchange.umich.edu; also: Purdue University/www.physics.purdue.edu.

Fig. 9.3: the biography by Rob van den Berg: George Gamow, *Van Atoomkern tot kosmos*, Veen Magazines, Diemen, Netherlands, 2011.

Fig. 9.4: Sagor, Beek, Netherlands

Fig. 9.5: The Royal Swedish Academy of Science/www.nobelprize.org; also: NASA/cosmictimes.gsfc.nasa.gov

Fig. 9.6a and b: STScI/NASA

Fig. 9.7: STScI/NASA and NASA's Chandra X-ray observatory: Chandra.harvard.edu

Fig. 9.8: STScI/NASA

Fig. 9.9: Time magazine

Fig. 9.10: NASA/Chandra; Chandra.harvard.edu

Fig. 9.11: left: Kitt Peak National Observatory, NOAO, AURA, NSF; middle NASA/Chandra.harvard.edu; right: NRAO, NSF/nrao.edu

Fig. 9.12a: top: NASA/Chandra; Chandra.harvard.edu; lower left: NRAO, NSF/nrao.edu; lower right: STScI/NASA; b: Very Large Array, NRAO, NSF/nrao.edu; c: optical picture: STScI/NASA; radio overlay: Very Large Array, NRAO, NSF/nrao.edu.

Fig. 9.13: GSFC/NASA and the COBE Team; also: Wikimedia Commons.

Fig. 9.14: GSFC/COBE/NASA; also: Aether.lbl.gov.

Figure above the title of Chap. 10: Fermilab/www.fnal.gov

Fig. 10.1: Sagor, Beek, Netherlands, after: University of Oregon/Abyss.uoregon.edu.

Fig. 10.2: Sagor, Beek, Netherlands, after graphene.limited.

Fig. 10.3: Veen Media, Diemen, Netherlands: book "Tijd in machten van tien" by Gerard 't Hooft and Stefan Vandoren (p. 107), 2011; drawing by Sagor, Beek, Netherlands

Figure above the title of Chap. 11: STScI/NASA

Fig. 11.1: Institute of Astronomy, Cambridge University, UK/ast.cam.ac.uk

Fig. 11.2: European Southern Observatory/astroex.org

Fig. 11.3: Sagor, Beek, Netherlands, after: University of Oregon/abyss.uoregon.edu.

Fig. 11.4: Sagor, Beek, Netherlands

Fig. 11.5: Sagor, Beek, Netherlands, after: University of Oregon/zebu.uoregon.edu.

Fig. 11.6: Sagor, Beek, Netherlands

Fig. 11.7: NASA's Planetary Biology Program/ portaltotheuniverse.org/http://hubblesite.org/newscenter/archive

Fig. 11.8: Sagor, Beek, Netherlands

Fig. 11.9: Sagor, Beek, Netherlands

Figure above the title of Chap. 12: STScI/NASA

Fig. 12.1: NASA/GSFC

Fig. 12.2: Sagor Beek, Netherlands, after: University of California, Santa Cruz/scipp.ucsc.edu.

Fig. 12.3: from the book by Gerard 't Hooft and Stefan Vandoren "Tijd in Machten van Tien", Veen Magazines (Veen Media), Diemen, Netherlands, 2011. Drawing by Sagor, Beek, Netherlands

Fig. 12.4: Sagor, Beek, Netherlands, after: University of Arizona/icamera.as.arizona.edu.

Fig. 12.5: Sagor, Beek, Netherlands, after: University of Oregon/abyss.oregon.edu.

Figure above the title of Chap. 13: STScI/NASA

Fig. 13.1: www.zwicky-stiftung.ch

Fig. 13.2: STScI/NASA

Fig. 13.3: Sagor, Beek, Netherlands

Fig. 13.4: STScI/NASA, Chandra/NASA and Hubble/ESA (esa.int)

Fig. 13.5a (right): Chandra/NASA (chandra.harvard.edu); a (left): NRAO/NSF (nrao.edu and astro.psu.edu); b: Lawrence Berkeley Laboratory (lbl.gov).

Fig. 13.6: STScI/NASA

Fig. 13.7: Lawrence Berkeley Laboratory/lbl.gov and PNAS.org.

Fig. 13.8a: from left to right: panisse.lbl.gov; www.nobelprize.org; and Jewish Virtual Library/www.nobelprize.org, respectively.

Fig. 13.8b: STScI/NASA/spacetelescope.org.

Fig. 13.9: Wikimedia Commons, uploaded by Emok, CC BY-SA 3.0.

Fig. 13.10: Sagor, Beek, Netherlands, after: Wikimedia Commons, uploaded by Materialscientist, retrieved from NASA-online.

Figure above the title of Chap. 14: picture made for NASA by Tempshill; Wikimedia Commons.

Fig. 14.1: Map.gsfc.nasa.gov; NASA/WMAP Science Team.

Fig. 14.2: Map.gsfc.nasa.gov; NASA/WMAP Science Team.

Fig. 14.3a: ESA and the Planck Collaboration

Fig. 14.3b: ESA and the Planck Collaboration

Fig. 14.4: California Institute of Technology: ned.ipac.caltech.edu.

Fig. 14.5a and b: Volker Springel (mpa-garching.mpg.de).

Fig. 14.6: Sloan Digital Sky Survey: www.sdss.org.

Fig. 14.7: Royal Observatory Edinburgh: www.roe.ac.uk.

Figure above the title of Chap. 15: from the book *Tijd in machten van tien* by Gerard't Hooft and Stefan Vandoren, Veen Magazines, Diemen, Netherlands, 2011.

Fig. 15.1: Sagor, Beek, Netherlands after: J. Narlikar *The Structure of the Universe*, Oxford University Press, London, 1977.

Fig. 15.2: Sagor, Beek, Netherlands, after: Georgia State University/ hyperphysics. phys-astr.gsu.edu.

Figure above the title of Chap. 16: Victor Habbick; www.victorhabbick.com.

Fig. 16.1: Sagor, Beek, Netherlands

Fig. 16.2: Sagor, Beek, Netherlands

Fig. 16.3: Royal Swedish Academy of Sciences/Nobelprize.org.

Figure above the title of Chap. 17: NASA.org/picture by Jenny Mottar.

Fig. 17.1: Young Sun, Early Earth and the Origin of Life/www.springer.com.

Fig. 17.2: Atomic Heritage Foundation/www.atomicheritage.org.

Fig. 17.3: SETI Institute/www.seti.org.

Fig. 17.4: SETI Institute/www.seti.org.

Fig. 17.5: US Postage

Fig. 17.6: photo taken by David Morrison.

Fig. 17.7: NASA-Goddard Space Flight Centre/GSFC-BMeast.

Figure above title of Chap. 18: from Rob van den Berg: *George Gamow, van atoomkern tot kosmos*, Veen Magazines, Diemen, Netherlands 2011, p.24

Fig. 18.1: from: *Cosmology, fusion and other matters*, Gamow Memorial Volume, F. Reines, editor, Colorado Associated University Press, 1972. Also from: Rob van den Berg: *George Gamow, van atoomkern tot kosmos*, Veen Magazines, Diemen, Netherlands, 2011, p. 111.

Figures of the Appendices:

First figure of section B: Sagor, Beek, Netherlands

Second figure of section B: from: Gerard 't Hooft and Stefan Vandoren, *Tijd in machten van tien* (p. 114, 115), Veen Magazines, Diemen, Netherlands, 2011.

Fig. C1: Sagor, Beek, Netherlands

Fig. C2: Sagor, Beek, Netherlands

Author Index

E. van den Heuvel, *The Amazing Unity of the Universe*,
Astronomers' Universe, DOI 10.1007/978-3-319-23543-1

Subject Index

E. van den Heuvel, *The Amazing Unity of the Universe*,
Astronomers' Universe, DOI 10.1007/978-3-319-23543-1

C

Printed in the United States
By Bookmasters